The Design and Analysis of Parallel Algorithms

# The Design and Analysis
# of Parallel Algorithms

JUSTIN R. SMITH

New York        Oxford
OXFORD UNIVERSITY PRESS
1993

*To my dear wife, Brigitte*

Oxford University Press

Oxford   New York   Toronto
Delhi   Bombay   Calcutta   Madras   Karachi
Kuala Lumpur   Singapore   Hong Kong   Tokyo
Nairobi   Dar es Salaam   Cape Town
Melbourne   Auckland   Madrid

and associated companies in
Berlin   Ibadan

Published by Oxford University Press, Inc.,
200 Madison Avenue, New York, New York 10016

Oxford is a registered trademark of Oxford University Press

Library of Congress Cataloging-in-Publication Data
Smith, Justin R.
The design and analysis of parallel algorithms / Justin R. Smith.
p. cm.   Includes bibliographical references and index.
ISBN 0-19-507881-0
1. Parallel processing (Electronic computers)
2. Computer algorithms.
I. Title.
QA76.58.S62   1993   005.1'1—dc20
92-32052

1 3 5 7 9 8 6 4 2

Printed in the United States of America
on acid-free paper

# Preface

This book grew out of lecture notes for a course on parallel algorithms that I gave at Drexel University over a period of several years. I was frustrated by the lack of texts that had the focus that I wanted. Although the book also addresses some architectural issues, the main focus is on the development of parallel algorithms on "massively parallel" computers. This book could be used in several versions of a course on Parallel Algorithms. We tend to focus on SIMD parallel algorithms in several general areas of application:

- Numerical and scientific computing. We study matrix algorithms and numerical solutions to partial differential equations.
- "Symbolic" areas, including graph algorithms, symbolic computation, sorting, etc.

There is more material in this book than can be covered in any single course, but there are many ways this book can be used. I have taught a graduate course in parallel numerical algorithms by covering the material in:

1. The Introduction.
2. § 2 of chapter III (page 63).
3. Chapter IV, and;
4. Chapter V.

Another possible "track" through this book, that emphasizes symbolic algorithms, involves covering:

1. The Introduction;
2. Chapter II;
3. and Chapter VI.

A graduate course on parallel algorithms in general could follow the sequence:

1. The Introduction;
2. Chapter II — for a theoretical background;

3. §§ 1 and 1.2 of chapter V — for a taste of some numerical algorithms;

4. §§ 1, 2.1, 2.2, 2.3, 2.4.1 4, and if time permits, 4.2.1 of chapter VI.

I welcome readers' comments, and am particularly interested in reports of errors or suggestions for new topics or exercises. My address is:

Justin R. Smith
Department of Mathematics and Computer Science
Drexel University
Philadelphia, PA 19104
USA

and my electronic mail address is jsmith@mcs.drexel.edu. Although I will try to respond to readers' comments, I regret that it will probably not be possible to respond to all of them.

I generally used the C* language as a kind of "pseudocode" for describing parallel algorithms. Although Modula* might have been more suitable than C* as a "publication language" for parallel algorithms, I felt that C* was better known in the computing community[1]. In addition, my access to a C* compiler made it possible for me to debug the programs. I was able to address many issues that frequently do not arise until one actually attempts to program an algorithm. All of the source code in this book can actually be run on a computer, although I make no claim that the published source code is optimal.

The manuscript was typeset using $\mathcal{A}_{\mathcal{M}}S$-LaTeX — the extension of LaTeX developed by the American Mathematical Society. I used a variety of machines in the excellent facilities of the Department of Mathematics and Computer Science of Drexel University for processing the TeX source code, and used MacDraw on a Macintosh Plus computer to draw the figures. The main font in the manuscript is Adobe Palatino — I used the psfonts.sty package developed by David M. Jones at the MIT Laboratory for Computer Science to incorporate this font into TeX.

Many of the symbolic computations and graphs were made using the Maple symbolic computation package, running on my own Macintosh Plus computer. Although this package is not object-oriented, it is very powerful. I am particularly impressed with the fact that the versions of Maple on microcomputers are of "industrial strength".

---

[1] This will probably change as Modula* becomes more widely available.

# Acknowledgements

I am indebted to several people who read early versions of this manuscript and pointed out errors or made suggestions as to how to improve the exposition, including Jeffrey Popyack, Herman Gollwitzer, Bruce Char, David Saunders, and Jeremy Johnson. I am also indebted to the students of my M780 course in the Summer of 1992 who discovered some errors in the manuscript. This course was based upon a preliminary version of this book, and the students had to endure the use of a "dynamic textbook" that changed with time.

I would like to thank Donald Jackson and the proofreaders at Oxford University Press for an extremely careful reading of an early version of the manuscript, and many helpful suggestions.

I would also like to thank Christian Herter and Ernst Heinz for providing me with information on the Triton project at the University of Karlsruhe and the Modula* language, and Kenneth Traub for providing me with information on the Monsoon Project at M.I.T. I am indebted to Thinking Machines Corporation and the CMNS project for providing me with access to a CM-2 Connection Machine for development of the parallel programs that appear in this book.

Most of all, I am indebted to my wonderful wife Brigitte, for her constant encouragement and emotional support during this arduous project. Without her encouragement, I wouldn't have been able to complete it. This book is dedicated to her.

Philadelphia, PA
October 23, 1992

# Contents

# List of Figures

The Design and Analysis of Parallel Algorithms

# I
# Basic Concepts

## 1. Introduction

Parallel processing algorithms is a very broad field — in this introduction we will try to give some kind of overview.

Certain applications of computers require much more processing power than can be provided by today's machines. These applications include solving differential equations and some areas of artificial intelligence like image processing. Efforts to increase the power of sequential computers by making circuit elements smaller and faster have approached basic physical limits. Consequently, it appears that substantial increases in processing power can only come about by somehow breaking up a task and having processors work on the parts independently. The parallel approach to problem-solving is sometimes very different from the sequential one — we will occasionally give examples of parallel algorithms that convey the flavor of this approach.

A designer of custom VLSI circuits to solve some problem has a potentially large number of processors available that can be placed on the same chip. It is natural, in this context, to consider parallel algorithms for solving problems. This leads to an area of parallel processing known as the theory of *VLSI algorithms*. These are parallel algorithms that use a large number of processors that each have a small amount of local memory and that can communicate with neighboring processors only along certain predefined communication lines. The number of these communication lines is usually very limited.

The amount of information that can be transmitted through such a communication scheme is limited, and this provides a limit to the speed of computation. There is an extensive theory of the speed of VLSI computation based on *information-flow* arguments — see [153].

Certain tasks, like low level image processing, lend themselves to parallelization because they require that a large number of *independent* computations be carried out. In addition, certain aspects of computer design lead naturally to the question of whether

3

tasks can be done in parallel. For instance, in custom VLSI circuit design, one has a large number of simple processing elements available and it is natural to try to exploit this fact in developing a VLSI to solve a problem. We illustrate this point with an example of one of the first parallel algorithms to be developed and applied. It was developed to solve a problem in computer vision — the counting of distinct objects in a field of view. Although this algorithm has some applications, we present it here only to convey the *flavor* of many parallel algorithms. It is due to Levialdi (see [98]).

We are given a two-dimensional array whose entries are all 0 or 1. The array represents pixels of a black and white image and the 1's represent the darkened pixels. In one of the original applications, the array of pixels represented digitized images of red blood cells. Levialdi's algorithm solves the problem:

**How do we efficiently count the connected sets of darkened pixels?**

Note that this is a more subtle problem than simply counting the *number* of darkened pixels. Levialdi developed a parallel algorithm for processing the array that shrinks the objects in the image in each step — it performs transformations on the array in a manner reminiscent of Conway's well-known Game of Life:

Suppose $a_{i,j}$ denotes the $(i,j)^{\text{th}}$ pixel in the image array during some step of the algorithm. The $(i,j)^{\text{th}}$ entry of the array in the next step is calculated as

(1) $$h[h(a_{i,j-1} + a_{i,j} + a_{i+1,j} - 1) + h(a_{i,j} + a_{i+1,j-1} - 1)]$$

where $h$ is a function defined by:

$$h(x) = \begin{cases} 1, & \text{if } x \geq 1; \\ 0 & \text{otherwise.} \end{cases}$$

This algorithm has the effect of *shrinking* the connected groups of dark pixels until they finally contain only a *single pixel*. At this point the algorithm calls for removing the isolated pixels and incrementing a counter. We assume that each array element $a_{i,j}$ (or pixel) has a processor (or CPU) named $P_{i,j}$ associated with it, and that these CPU's can communicate with their neighbors. These can be very simple processors — they would have only limited memory and only be able to carry out simple computations — essentially the computations contained in the equations above, and simple logical operations. Each processor would have to be able to communicate with its neighbors in the array. This algorithm can be carried out in cycles, where each cycle might involve:

    1. Exchanging information with neighboring processors; or

    2. Doing simple computations on data that is stored in a processor's local memory.

In somewhat more detail, the algorithm is:

$C \leftarrow 0$

**for** $i \leftarrow 1$ **to** $n$ **do**

    **for** all processors **do in parallel**

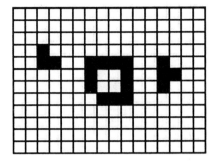

FIGURE I.1. Initial image

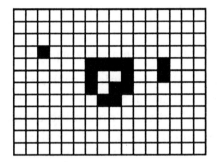

FIGURE I.2. Result of first iteration

$P_{i,j}$ receives the values of

$a_{i+1,j}, a_{i-1,j}, a_{i+1,j+1}, a_{i+1,j-1}$

$a_{i-1,j+1}, a_{i,j-1}, a_{i,j+1}$

from its neighbors (it already contains the value of $a_{i,j}$

**if** $a_{i,j} = 1$ and all neighboring elements are 0 **then**

$C \leftarrow C + 1$

**Perform** the computation in equation (1) above

**end** /* *do in parallel* */

**end** *for*

Here is an example — we are assuming that the $i$ axis is horizontal (increasing from left-to-right) and the $j$ axis is vertical. Here is an example of the Levialdi algorithm. Suppose the initial image is given by figure I.1.

The result of the first iteration of the algorithm is given by figure I.2.

In the *next* step the lone pixel in the upper right is removed and a counter incremented. The result is depicted in figure I.3. After a sufficient number of steps (in fact, $n$ steps, where $n$ is the size of the largest side of the rectangle) the screen will be blank, and all of the connected components will have been counted.

This implementation is an example of a particularly simple form of parallel algorithm

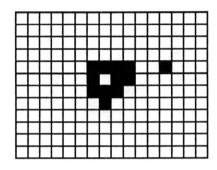

FIGURE I.3. Result of second iteration

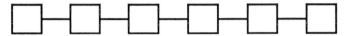

FIGURE I.4. A linear array of processors

called a *systolic algorithm*[1]. Another example of such an algorithm is the following:

Suppose we have a one-dimensional array (with $n$ elements) whose *entries* are processing elements. We assume that these processing elements can carry out the basic operations of a computer — in this case it must be able to store at least two numbers and compare two numbers. Each entry of this array is connected to its two neighboring entries by communication lines so that they can send a number to their neighbors — see figure I.4.

Now suppose each processor has a number stored in it and we want to sort these numbers. There exists a parallel algorithm for doing this in $n$ steps — note that it is well-known that a sequential sort using comparisons (other than a radix-sort) requires $\Omega(n \lg n)$ steps. Here lg denote the logarithm to the base 2. In odd-numbered steps odd-numbered processors compare their number with that of their next higher numbered even processors and exchange numbers if they are out of sequence. In even-numbered steps the even-numbered processors carry out a similar operation with their odd-numbered neighbors (this is a problem that first appeared in [89]). See figure I.5 for an example of this process.

Note that this algorithm for sorting corresponds to the *bubble sort* algorithm when regarded as a sequential algorithm.

At first glance it might seem that the way to develop the fastest possible *parallel* algorithm is to start with the fastest possible *sequential* algorithm. This is very definitely *not* true, in general. In fact, in many cases, the best parallel algorithm for a problem doesn't remotely resemble the best sequential algorithm. In order to understand this phenomena it is useful to think in terms of *computation networks* for computations. For

---

[1]See the discussion on page 18 for a (somewhat) general definition of systolic algorithms. Also see [94]

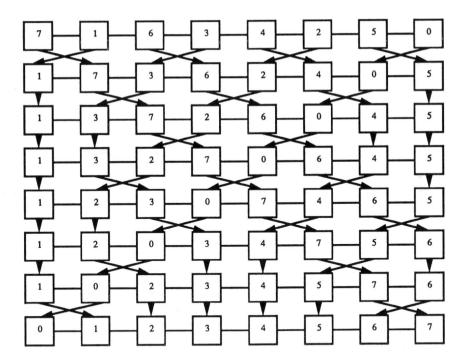

FIGURE I.5. The odd-even sorting algorithm

instance the expression $(a + b + c)/(e + f - g)$ can be represented by the directed graph in figure I.6.

The meaning of a computation-network is that we try to perform a computation at each vertex of the graph and transmit the result of the computation along the single exit edge. See § 7.1 on page 48 for a rigorous definition of a computation network. The computation cannot be performed until data arrives at the vertex along the *incoming* directed edges. It is not hard to see that the computations begin at the vertices that have no incoming edges and ends at the vertex (labeled the root in figure I.6) that has no outgoing edges. We will briefly analyze the possibilities for parallelization presented by this network. Suppose that each vertex of the network has a number associated with it, called its value. The numbers attached to the leaf vertices are just the numerical values of the variables $a$ through $g$ — we assume these are *given*. For a non-leaf vertex the number attached to it is equal to the result of performing the indicated operation on the values of the *children* of the vertex. It is not hard to see that the value of the root of the tree will equal the value of the whole expression. It is also not hard to see that to compute the value of a given non-leaf vertex, it is first necessary to compute the values of its children — so that the whole computation proceeds in a *bottom up* fashion. We claim that:

> If a computation is to be done *sequentially* the execution time is very roughly proportional to the *number of vertices* in the syntax tree. If the

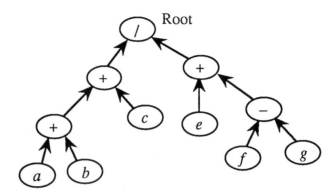

FIGURE I.6. Computation network

execution is to be *parallel*, though, computations in different branches
of the syntax tree are essentially independent so they can be done simultaneously. It follows that the parallel execution time is roughly
proportional to the *distance* from the root to the *leaves* of the syntax tree.

This idea is made precise by Brent's Theorem (7.2 on page 49). The task of finding a
good parallel algorithm for a problem can be regarded as a problem of *re-modeling* the
computation network in such a way as to make the distance from the root to the leaves
a *minimum*. This process of remodeling may result in an increase in the *total number* of
vertices in the network so that the efficient parallel algorithm would be very inefficient
if executed sequentially. In other words a relatively compact computation network (i.e.
small number of vertices) might be remodeled to a network with a large total number
of vertices that is relatively *balanced* or *flat* (i.e. has a shorter distance between the root
and the leaves).

For instance, if we want to add up 8 numbers $a_1, \ldots, a_8$, the basic sequential algorithm for this has the computation network given in figure I.7.

This represents the process of carrying out 8 additions sequentially. We can remodel
this computation network to get figure I.8. In this case the total execution-time is 3 units,
since the distance from the root of the tree to the leaves is 3. When we remodel these
computation-graphs and try to implement them on parallel computers, we encounter
issues of *interconnection topology* — this is a description of how the different processors
in a parallel computer communicate with each other. It is not hard to see that the linear
array of processors used in the odd-even sorting algorithm depicted in figure I.5 on
page 7 would have a hard time implementing the addition-algorithm shown in figure
I.8. The ideal situation would be for the communication-patterns of the processors to
be identical to the links of the computation graph. Although this ideal configuration
is not always possible, there are many interconnection topologies that are suitable for
a wide variety of problems — see chapter II, particularly § III on page 67 through § 6
on page 89.

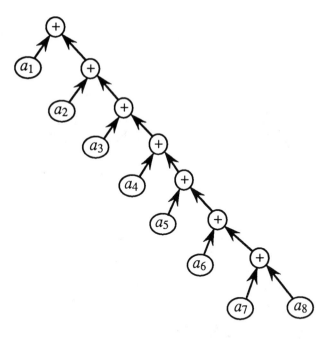

FIGURE I.7. Computation network for sequential addition

Assuming an ideal interconnection topology, we get an algorithm for adding $2^k$ numbers in $k$ steps, by putting the numbers at the leaves of a complete binary tree. Figure I.9 shows this algorithm adding $8 = 2^3$ numbers in 3 steps.

As simple as this algorithm for adding $2^k$ numbers is, it forms the basis of a *whole class* of parallel algorithms — see chapter VI.

Note that, since we can add $n$ numbers in $\lg n$ steps, with sufficiently many processors, we can do matrix multiplication of $n \times n$ matrices in $\lg n$ steps. See 1.1 on page 157 for the details. Consequently we can also perform other operations derived from matrix multiplication rapidly. For instance, we can find the distance between all pairs of vertices in a graph in $O(\lg^2 n)$-steps — see 2.5 on page 322 for the details.

While it is clear that certain tasks, like matrix addition, can be done rapidly in parallel

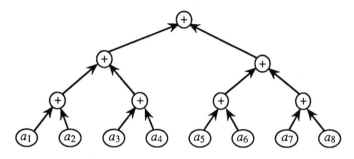

FIGURE I.8. Remodeled computation network

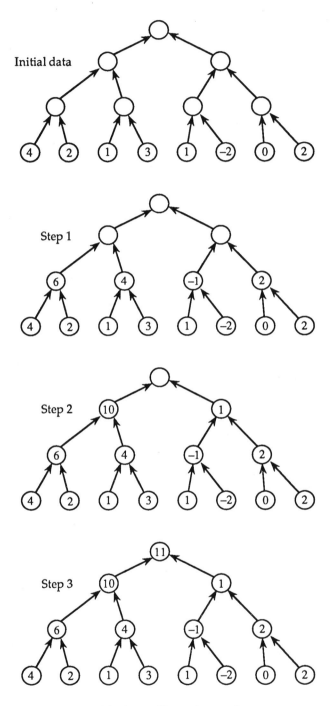

FIGURE I.9. Sum of 8 numbers in 3 steps

it doesn't follow that all tasks have this property. This leads to the question of whether there exist *inherently sequential problems* — problems that can't necessarily be done faster in parallel.

From the discussion connected with the example given above one might get the impression that many unbounded parallel algorithms have an execution time that is $O(\lg^k n)$ for some value of $k$. This turns out to be the case — for instance, many algorithms are loosely based upon the example given above.

This phenomena is so widespread that Nicholas Pippenger has defined a class of problems called **NC**. These are problems that can be solved on a parallel computer (like that in the preceding example) $O(\lg^k n)$-time using a polynomial number of processors (it seems reasonable to impose this restriction on the number of processors — in addition many problems can be rapidly solved in parallel if an unlimited number of processors are available). A rigorous definition is given in § 6.1 on page 39. In general, a problem is called *parallelizable* if it is in **NC**. Since any **NC** problem can be sequentially solved in polynomial time (by simulating a PRAM computer by a sequential one) it follows that **NC** ⊆ **P**. The natural question at this point is whether **NC** = **P** — do there exist any *inherently sequential problems?* This question remains open at present but there are a number of interesting things one can say about this problem. The general belief is that **NC** ≠ **P** but no one has come up with an example of an inherently sequential problem yet. A number of problems are known to be **P-complete**. Roughly speaking, a polynomial-time sequential problem is called **P-complete** if any other polynomial-time sequential problem can be transformed into an instance of it — see page 43. If a fast parallel algorithm can be found for any **P-complete** problem, then fast parallel versions can be found for *all* polynomial-time sequential algorithms. See § 6.2 in chapter II for examples and a discussion of the theoretical issues involved.

Chapter II discusses various unbounded parallel models of computation and Batcher's sorting algorithm. We discuss the distinction between "Procedure-level" and "Data-level" parallelism, and the question of whether "super-linear" speedup of a sequential algorithm is possible. We give comparisons of the so-called Concurrent-I/O and Exclusive-I/O models (where the "I/O" in question is between processors and the shared random access memory).

We also discuss some theoretical questions of when efficient parallel algorithms exist. These questions all involve the class **NC** mentioned above and its relations with the class of polynomial-time sequential problems **P**. We will discuss the relations between ·**NC** and **PLogspace**-sequential problems — known as the Parallel Processing Thesis, originally due to Fortune and Wyllie. We will also discuss potential candidates for inherently sequential problems here.

In addition, we discuss some of architectures of existing parallel computers and how algorithms are developed for these architectures. These architectures include:
- the Butterfly Network

- the Hypercube
- Shuffle-Exchange networks and
- Cube-Connected Cycles

We also briefly discuss dataflow computers, and issues connected with the question of how memory is accessed in network computers and the Granularity Problem.

Chapter IV discusses some concrete examples of parallel computers and programming languages. It shows how architectural considerations influence programming languages. It considers some of the languages for the Sequent Symmetry and the Connection Machine in some detail.

The programming constructs used on the Sequent provide examples of how one programs MIMD computers in general. The use of semaphores and synchronization primitives like **cobegin** and **coend** are discussed in this connection. We also discuss the LINDA system for asynchronous parallel programming. It has the advantage of simplicity, elegance, and wide availability.

The C* language for the Connection Machine is a model for programming SIMD machines in general. We use it as a kind of *pseudocode* for describing SIMD algorithms in the remainder of the text.

Chapter V considers *numeric* applications of parallel computing:

- Systems of linear equations. We discuss iterative techniques for solving systems of linear equations.
- Matrix operations. We discuss the Pan-Reif matrix-inversion algorithm as an example of an extraordinarily inefficient sequential algorithm that can be efficiently parallelized.
- The Fourier Transform. We develop the FFT algorithm and give a few applications.
- Wavelet transforms. This is a new area, related to the Fourier Transform, that involves expanding input-functions (which may represent time-series or other phenomena) in a series whose terms are *fractal functions*. Wavelet transforms lend themselves to parallel computations — in some respects to a greater extent than Fourier Transforms. We present parallels algorithms for wavelet transforms.
- Numerical integration. We give several parallel algorithms for approximately evaluating definite integrals.
- Numeric solutions to partial differential equations — including techniques peculiar to elliptic, parabolic, and hyperbolic differential equations.

Chapter VI considers several classes of non numeric parallel algorithms, including:

1. A general class of algorithms that may be called *doubling algorithms*. All of these roughly resemble the algorithms for adding 8 numbers given in this introduction. In general, the technique used for adding these 8 numbers can

be used in performing any *associative* operation. A large number of operations can be expressed in these terms:

- The solution of linear recurrences.
- Parsing of linear languages (with applications to pattern-recognition and the design of front-ends to compilers).

2. searching and various sorting algorithms including that of Cole and the Ajtai, Komlós, Szemerédi sorting algorithm.
3. Graph-theoretic algorithms like minimal spanning tree, connected components of a graph, cycles and transitive closure.
4. Fast Fourier Transforms (revisited) with applications to Computer Algebra.
5. Probabilistic algorithms including Monte Carlo Integration, and some symbolic algorithms. We introduce the class **RNC**, of problems that have parallel solutions that involve random choices, and whose *expected* execution time (this term is defined on page 446) is in $O(\lg^k n)$ for $k$ some integer, and for $n$ equal to the complexity-parameter of the problem.

**Exercises:**

1. Prove the correctness of the algorithm on page 69 for forming the cumulative sum of $2^k$ numbers in $k$ steps.
2. Find examples of PRAM and networked parallel computers that are commercially available. What programming languages are commonly used on these machines?

# II
# Models of parallel computation

## 1. Generalities

In this section we discuss a few basic facts about parallel processing in general. One very basic fact that applies to parallel computation, regardless of how it is implemented, is the following:

CLAIM 1.1. Suppose the fastest sequential algorithm for doing a computation with parameter $n$ has execution time of $T(n)$. Then the fastest parallel algorithm with $m$ processors (each comparable to that of the sequential computer) has execution time $\geq T(n)/m$.

The idea here is: If you could find a faster parallel algorithm, you could execute it sequentially by having a sequential computer *simulate* parallelism and get a faster sequential algorithm. This would contradict the fact that the given sequential algorithm is the fastest possible. We are making the assumption that the *cost* of simulating parallel algorithms by sequential ones is negligible.

This claim is called the "Principle of Unitary Speedup".

As usual, the parameter $n$ represents the relative size of the instance of the problem being considered. For instance, if the problem was that of sorting, $n$ might be the number of items to be sorted and $T(n)$ would be $O(n \lg n)$ for a sorting algorithm based upon comparisons.

As simple as this claim is, it is a bit controversial. It makes the tacit assumption that the algorithm in question is *deterministic*. In other words, the algorithm is like the usual idea of a computer program — it performs calculations and makes decisions based on the results of these calculations.

There is an interesting area of the theory of algorithms in which statement 1.1 is *not necessarily true* — this is the theory of *randomized* algorithms. Here, a solution to a problem may involve making random "guesses" at some stage of the calculation. In this case, the parallel algorithm using $m$ processors can run faster than $m \times$ the speed of the sequential algorithm ("Super-unitary speedup"). This phenomenon occurs in certain problems in which random search is used, and most guesses at a solution quickly lead to a valid solution, but there are a few guesses that execute for a long time without producing any concrete results.

We will give an example of this phenomenon. Although it is highly oversimplified, it does illustrate how super unitary speedup can occur.

Suppose there are a 100 possible approaches to an AI-type search problem and:

1. 99 out of the 100 possibilities arrive at a solution in 1 time unit.
2. 1 of the possibilities runs for 1000 time units, and then fails.

The expected execution-time of a *single (sequential) attempt* to find a solution is the *average* of all of these times, or 10.99 time-units.

If we attack this problem with a parallel computer that has 2 processors that try *distinct* possibilities, the expected time (and even the *worst-case* time) is 1 unit, since at least one of these two distinct possibilities will be a *fast solution*. We, consequently, see a *super-unitary* speedup in the parallel algorithm. In other words the *expected*[1] running-time of the algorithm is divided by $> 10$, which is much greater than the ratio of processors.

The opponents of the concept of super-unitary speedups (including the author) would argue that the original sequential algorithm was not *optimal* — and that the optimal sequential algorithm would have attempted two possible solutions with two distinct processes, run *concurrently*). A sequential algorithm that created two processes would have a running time of 2 units. The speedup that results by going to the sample parallel algorithm is 2, which is exactly equal to the ratio of processors. Thus, by modifying the sequential algorithm used, the validity of 1.1 is restored.

In [122], Parkinson argues that super-unitary speedup is possible, and in [48] Faber Lubeck White argue that it is *not*.

See [96], [99], and [110], for more information about this phenomenon.

We will will concentrate on deterministic algorithms in this text, so that we will assume that super-unitary speedup is essentially impossible.

The next question we consider is how the instructions to the processors are handled. There exist several different models of program control. In [50], Flynn listed several basic schemes:

**SIMD** Single Instruction Multiple Data. In this model the processors are controlled by a program whose instructions are applied to all of them simultaneously (with certain qualifications). We will assume that each of the processors has a unique number that is "known" to the processor in the sense that instructions to the parallel computer can refer to processor numbers. An example of

---

[1]Recall that the expected running-time of an algorithm like the one in the example is the average of actual running times, weighted by probabilities that these running times occur.

this type of machine is:

> The Connection Machine (models CM-1 and CM-2), from Thinking Machines Corporation. This is used as a generic example of a SIMD computer in this text.

A seemingly more powerful model is:

**MIMD** Multiple Instruction Multiple Data. In this model processors can each have independent programs that are read from the common RAM. This model is widely used in several settings:

1. Coarse-grained parallelism — this is a form of parallel computing that closely resembles *concurrent programming* in that it involves processes that run *asynchronously*. Several commercial parallel computers are available that support this model of computation, including the Sequent Balance and Symmetry and the Encore Multimax. Many interesting issues arise in this case. The data-movement and communications problems that occur in all parallel computation are more significant here because the instructions to the processors as well as the data must be passed between the common memory and the processors.

   Due to these data-movement problems, commercial MIMD computers tend to have a relatively small number of processors ($\approx 20$). In general, it is easier to program a MIMD machine if one is only interested in a very limited form of parallelism — namely the formation of processes. Conventional operating systems like UNIX form separate processes to carry out many functions, and these processes *really* execute in parallel on commercially-available MIMD machines. It follows that, with one of these MIMD machines, one can reap *some* of the benefits of parallelism without explicitly doing any parallel programming. For this reason, most of the parallel computers in commercial use today tend to be MIMD machines, run as general-purpose computers.

   On the surface, it would appear that MIMD machine are strictly more powerful than SIMD machines with the same number of processors. Interestingly enough, this is not the case — it turns out that SIMD machines are more suited to performing computations with a very *regular* structure. MIMD machines are not as suited to solving such problems because their processors must be *precisely synchronized* to implement certain algorithms — and this synchronization has a cost that, in some cases, can *dominate* the problem. See § 1.1, and particularly 7.8 for a discussion of these issues.

   Pure MIMD machines have no hardware features to guarantee synchronization of processors. In general, it is not enough to simply load multiple copies of a program into all of the processors and to start all of these copies at the same time. In fact many such computers have hardware features that tend to *destroy* synchronization, once it has been achieved. For instance, the manner in which memory is accessed in the Sequent Symmetry series, generally causes processes to run at different rates even if they are synchronized at some time. Many Sequents even have processors that run at different clock-rates.

2. The BBN Butterfly Computer — this is a computer whose architecture is based upon that of the shuffle-exchange network, described in section 5, on page 83.

Three other terms that fill out this list are:

**SISD** Single Instruction, Single Data. This is nothing but conventional *sequential computing*.

**MISD** This case is often compared to computation that uses *Systolic Arrays*. These are arrays of processors that are developed to solve specific problems — usually on a single VLSI chip. A clock coordinates the data-movement operations of all of the processors, and output from some processors are pipelined into other processors. The term "Systolic" comes from an analogy with an animal's circulatory system — the data in the systolic array playing the part of the blood in the circulatory system. In a manner of speaking, one can think of the different processors in a systolic array as constituting "multiple processors" that work on one set of (pipelined) data. We will not deal with the MISD case (or systolic arrays) very much in this text. See [94] for a discussion of systolic computers.

**SIMD-MIMD Hybrids** This is a new category of parallel computer that is becoming very significant. These machines are also called SAMD machines (Synchronous-Asynchronous Multiple Data). The first announced commercial SAMD computer is the new Connection Machine, the CM-5. This is essentially a MIMD computer with hardware features to allow:

- Precise synchronization of processes to be easily achieved.
- Synchronization of processors to be maintained with little or no overhead, once it has been achieved (assuming that the processors are all executing the same instructions in corresponding program steps). It differs from pure MIMD machines in that the hardware maintains a uniform "heartbeat" throughout the machine, so that when the same program is run on all processors, and all copies of this program are started at the same time, it is possible to the execution of all copies to be kept in lock-step with essentially no overhead. Such computers allow efficient execution of MIMD and SIMD programs.

Many new systems under development will be of this type, including:

1. The Paragon system from Intel.

2. The Triton Project, at the University of Karlsruhe (Germany). The Triton Project is currently developing a machine called the Triton/1 that can have up to 4096 processors.

Flynn's scheme for classifying parallel computers was somewhat refined by Händler in 1977 in [62]. Händler's system for classifying parallel computers involves three pairs of integers:

$$T(C) = < K \times K', D \times D', W \times W' >$$

where:

1. $K$: the number of *processor control units* (PCU's). These are portions of CPU's that interpret instructions and can alter flow of instructions. The number $K$ corresponds, in some sense, to the number of instruction-streams that can execute on a computer.

2. $K'$: The number of processor control units that can be pipelined. Here, pipelining represents the process of sending the output of one PCU to the input of another without making a reference to main memory. A MISD computer (as described above) would represent one in which $K' > K$.

3. $D$: The number of arithmetic-logical units (ALU's) controlled by each PCU. An ALU is a computational-element — it can carry out arithmetic or logical calculations. A SIMD computer would have a $K$-number of 1 and a $D$ number that is $> 1$.

4. $D'$: The number of ALU's that can be *pipelined*.

5. $W$: The number of bits in an ALU word.

6. $W'$: The number of pipeline segments in all ALU's controlled by a PCU or in a single processing element.

Although Händler's scheme is much more detailed than Flynn's, it still leaves much to be desired — many modern parallel computers have important features that do not enter into the Händler scheme[2] at *all*. In most case, we will use the much simpler Flynn scheme, and we will give additional details when necessary.

With the development of commercially-available parallel computers, some new terms have come into common use:

*Procedure-level parallelism:* This represents parallel programming on a computer that has a relatively small number of processors, and is usually a MIMD-machine. Since the number of processors is small, each processor usually does a large chunk of the computation. In addition, since the machine is usually MIMD each processor must be programmed with a separate program. This is usually done in analogy with *concurrent programming* practices — a kind of *fork-statement* is issued by the main program and a procedure is executed on a separate processor. The different processors wind

---

[2]For instance, the CM-5 computer mentioned above.

up executing *procedures* in parallel, so this style of programming is called *procedure-level parallelism*. It has the flavor of concurrent programming (where there is real concurrency) and many standard concurrent programming constructs are used, like semaphores, monitors, and message-passing.

*Data-level parallelism:* This represents the style of parallel programming that is emphasized in this book. It is used when there is a large number of processors on a machine that may be SIMD or MIMD. The name 'data-level parallelism' is derived from the idea that the number of processors is so great that the data can be broken up and sent to different processors for computations — originally this referred to a **do**-loop in FORTRAN. With a large number of processors you could send each iteration of the loop to a separate processor. In contrast to procedure-level parallelism, where you broke the code up into procedures to be fed to different processors, here you break up the data and give it to different processors.

This author feels that these terms are not particularly meaningful — they are only valid if one does parallel programming in a certain way that is closely related to ordinary sequential programming (i.e. the terms arose when people tried to parallelize sequential programs in a fairly straightforward way). The terms are widely used, however.

## 2. The PRAM model and a sorting algorithm.

In this section we will consider some simple algorithms that can be implemented when we have a SIMD computer in which every processor can access common RAM. In general, a computer in which many processors can access common RAM in a single program-step is called the **PRAM** model of computation. There are, of course, many forms of the PRAM-model:

- The PRAM-SIMD model of computation, in which all processors execute the same instructions in each program-step. To *some extent*, we can view the Connection Machine as falling into this category. See section 2 for a discussion of the differences between the operation of the Connection Machine and the PRAM-SIMD model of computation.
- The PRAM-MIMD model of computation. Several commercially-available parallel computers fit this model very nicely — for instance the Sequent Symmetry and the Encore Multimax.

  Even within the PRAM model there are several distinct possibilities. For instance, we have:

  *EREW — Exclusive Read, Exclusive Write.* In this case at most one processor can read from or write to a memory location in one step.

The next section will compare the EREW case with two others.

Consider the sorting algorithm discussed in the introduction. It isn't hard to see that it is optimal in the sense that it will always take at least $n$ steps to sort $n$ numbers on that computer. For instance, some numbers might start out $n - 1$ positions away from their final destination in the sorted sequence and they can only move one position per program step. On the other hand it turns out that the EREW-SIMD computer described above can sort $n$ numbers in $O(\lg^2 n)$ program steps using an old algorithm due to Batcher. This difference in execution time throws some light on the fundamental property of the PRAM model of computation: there is an unbounded flow of information. In other words even if only one processor can access one memory location at a time it is very significant that all processors can access all of the available memory in a single program step.

Batcher's sorting algorithm involves recursively using the Batcher Merge algorithm, which merges two sequences of length $n$ in time $O(\lg n)$.

In order to discuss this algorithm, we must first look at a general result that is indispensable for proving the validity of sorting algorithms. This is the 0-1 Principle — already alluded to in the exercise at the end of the introduction. See § 2 in chapter 28 of [34] for more information. This result applies to sorting and merging *networks*, which we now define.

DEFINITION 2.1. A *comparator* is a type of device (a computer-circuit, for instance) with two inputs and two outputs:

such that:

- $OUT_1 = \min(IN_1, IN_2)$
- $OUT_2 = \max(IN_1, IN_2)$

The standard notation for a comparator (when it is part of a larger network) is the more compact diagram:

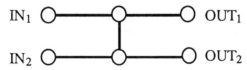

A *comparator network* is a directed graph with several special properties:

1. Data (numbers) can flow along the edges (in their natural direction). You can think of these edges as "pipes" carrying data, or "wires".
2. One set of vertices constitute the *inputs* for the graph, and another are called its *outputs*. These are the exterior vertices of the graph.
3. The interior vertices of the graph are comparators.

A *sorting network* is a comparator network that has the additional property:

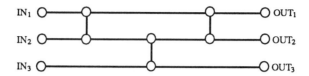

FIGURE II.1. A sorting network on three inputs

The data that appears at the output vertices is the result of sorting the data that was at the input vertices.

A *merging network* is defined to be a comparator network with the property that, if we subdivide the inputs into two subsets of equal sizes, and insert sorted data into each of these subsets, the output is the result of merging the input-sequences together.

PROPOSITION 2.2. *If a comparator network correctly sorts all input-sequences drawn from the set* $\{0,1\}$*, then it correctly sorts any input-sequence of numbers, so that it constitutes a sorting network.*

Similarly, if a comparator-network whose inputs are subdivided into two equal sets correctly merges all pairs of 0-1-sequences, then it correctly merges all pairs of number-sequences.

The proof is given in the appendix to this section on page 27. Figure II.1 shows a sorting network that correctly sorts all possible sequences of three numbers.

Although the 0-1 Principle is stated for comparator networks it also applies to many algorithms. If a sort or merge algorithm performs fixed sequences of comparisons between elements, it clearly defines a network in the sense defined above.

### 3. Bitonic Sorting Algorithm

In this section we will discuss one of the first parallel sorting algorithms to be developed.

DEFINITION 3.1. A sequence of numbers will be called *bitonic* if either of the following two conditions is satisfied:

- It starts out being monotonically increasing up to some point and then becomes monotonically decreasing.
- It starts out being monotonically decreasing up to some point and then becomes monotonically increasing.

A sequence of 0's and 1's will be called *clean* if it consists entirely of 0's or entirely of 1's.

For instance the sequence $\{4, 3, 2, 1, 3, 5, 7\}$ is bitonic. We will present an algorithm that correctly sorts all bitonic sequences. This will turn out to imply an efficient

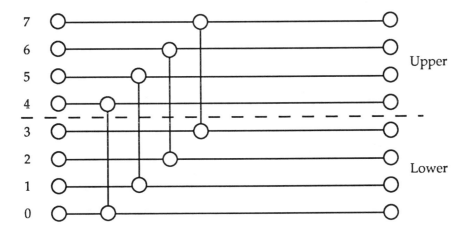

FIGURE II.2. A bitonic halver of size 8

algorithm for merging all pairs of sorted sequences, and then, an associated algorithm for sorting all sequences.

DEFINITION 3.2. Given a bitonic sequence of size $\{a_0, \ldots, a_{n-1}\}$, where $n = 2m$, a *bitonic halver* is a comparator network that performs the following sequence of compare-exchange operations:

**for** $i \leftarrow 0$ **to** $m - 1$ **do in parallel**
   **if**$(a_i < a_{i+m})$ **then swap**$(a_i, a_{i+m})$
**endfor**

Figure II.2 shows a bitonic halver of size 8.

Note that a bitonic halver performs some limited sorting of its input (into ascending order). The following proposition describes the precise sense in which sorting has been performed:

PROPOSITION 3.3. Suppose $\{a_0, \ldots, a_{n-1}\}$, where $n = 2m$ is a bitonic sequence of 0's and 1's, that is input to a bitonic halving network, and suppose the output is $\{b_0, \ldots, b_{n-1}\} = \{r_0, \ldots, r_{m-1}, s_0, \ldots, s_{m-1}\}$. Then one of the two following statements applies:

- The sequence $\{r_0, \ldots, r_{m-1}\}$ consists entirely of 0's and the sequence $\{s_0, \ldots, s_{m-1}\}$ is bitonic, or
- The sequence $\{r_0, \ldots, r_{m-1}\}$ is bitonic, the sequence $\{s_0, \ldots, s_{m-1}\}$ consists entirely of 1's.

Consequently, the smallest element of the lower half of the output is $\geq$ the largest element of the upper half.

The two cases are distinguished by the number of 1's that were in the original input.

PROOF. We have four cases to contend with:

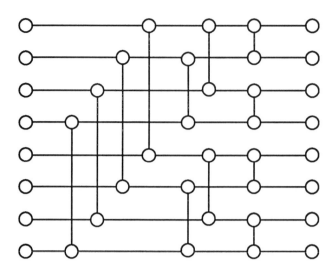

FIGURE II.3. Bitonic sorting network for 8 inputs

1. The input first increases and later decreases, and has a preponderance of 1's:

$$\{0_0, \ldots, 0_{\alpha-1}, 1_\alpha, \ldots, 1_{m-1}, \ldots, 1_\beta, 0_{\beta+1} \ldots, 0_{n-1}\}$$

where $\beta - \alpha \geq m - 1$ or $\beta \geq \alpha + m - 1$, or $n - 1 - \beta \leq m - 1 - \alpha$. This inequality implies that when we compare 0's in the upper half of the input with 1's in the lower half, each 0 will be compared with a 1 (i.e., there will be enough 1's) and, therefore, will be swapped with it. A straightforward computation shows that the output will be

$$\{0_0, \ldots, 0_{\alpha-1}, 1_\alpha, \ldots, 1_{\beta-m+1}, 0_{\beta-m+2}, \ldots, 0_{m-1}, 1_m, \ldots, 1_{n-1}\}$$

so the conclusion is true.

2. The input first decreases and later increases, and has a preponderance of 1's:

$$\{1_0, \ldots, 1_{\alpha-1}, 0_\alpha, \ldots, 0_{m-1}, \ldots, 0_\beta, 1_{\beta+1} \ldots, 1_{n-1}\}$$

where $\beta - \alpha < m$. In this case each 0 in the lower half of the input will also be compared with a 1 in the upper half, since $\beta \leq \alpha + m - 1$. The output is

$$\{0_0, \ldots, 0_{\beta-1}, 1_\beta, \ldots, 1_{\alpha-m+1}, 0_{\alpha-m+2}, \ldots, 0_{m-1}, 1_m, \ldots, 1_{n-1}\}$$

The two cases with a preponderance of 0's follow by symmetry.  □

It is not hard to see how to completely sort a bitonic sequence of 0's and 1's:

ALGORITHM 3.4. **Bitonic Sorting Algorithm.** Let $n = 2^k$ and let $\{a_0, \ldots, a_{n-1}\}$ be a bitonic sequence of 0's and 1's. The following algorithm sorts it completely:

**for** $i \leftarrow k - 1$ **downto** 1 **do in parallel**
    Subdivide the data into disjoint sublists of size $2^i$
    Perform a Bitonic Halving operation on each sublist
**endfor**

Since this correctly sorts all bitonic sequences of 0's and 1's, and since it is a sorting network, the 0-1 Principle implies that it correctly sorts all bitonic sequences of numbers.

Figure II.3 shows a bitonic sorting network. Since the bitonic halving operations can be carried out in a single parallel step (on an EREW computer, for instance), the running time of the algorithm is $O(\lg n)$, using $O(n)$ processors.

PROOF. Each Bitonic Halving operation leaves one half of its data correctly sorted, the two halves are in the correct relationship with each other, and the other half is bitonic. It follows that in phase $i$, each sublist is either sorted, or bitonic (but in the proper sorted relationship with the other sublists). In the end the intervals will be of size 1, and the whole list will be properly sorted. $\square$

Here is an example of the bitonic sorting algorithm:

EXAMPLE 3.5. We set $n$ to 8. The input data is:

$$\{1, 2, 3, 6, 7, 4, 2, 1\}$$

and we will sort in ascending order. After the first bitonic halving operation we get

$$\{1, 2, 2, 1, 7, 4, 3, 6\}$$

Now we apply independent bitonic halving operations to the upper and lower halves of this sequence to get

$$\{1, 1, 2, 2, 3, 4, 7, 6\}$$

In the last step we apply bitonic halving operations to the four sublists of size 2 to get the sorted output (this step only interchanges the 6 and the 7)

$$\{1, 1, 2, 2, 3, 4, 6, 7\}$$

Now we have a fast algorithm for sorting bitonic sequences of numbers. Unfortunately, such sequences are very rare in the applications for sorting. It turns out, however, that this bitonic sorting algorithm gives rise to a fast algorithm for *merging* two arbitrary sorted sequences. Suppose we are given two sorted sequences $\{a_0, \ldots, a_{m-1}\}$ and $\{b_0, \ldots, b_{m-1}\}$. If we reverse the second sequence and concatenate it with the first sequence the result is

$$\{a_0, \ldots, a_{m-1}, b_{m-1} \ldots, b_0\}$$

which is bitonic. This give rise to the following Batcher Merge algorithm:

ALGORITHM 3.6. **Batcher Merge.** Suppose $\{a_0, \ldots, a_{m-1}\}$ and $\{b_0, \ldots, b_{m-1}\}$ are two sorted sequences, where $m = 2^{k-1}$. Then the following algorithms merges them together:

1. Preparation step. Reverse the second input-sequence and concatenate it with the first, forming
$$\{a_0, \ldots, a_{m-1}, b_{m-1} \ldots, b_0\}$$

2. Bitonic sort step.

   **for** $i \leftarrow k - 1$ **downto** 1 **do in parallel**
   Subdivide the data into disjoint sublists of size $2^i$
   Perform a Bitonic Halving operation on each sublist
   **endfor**

The result is a correctly-sorted sequence.

This merge algorithm executes in $O(\lg n)$ steps, using $O(n)$ processors.

It is now straightforward to come up with an algorithm for sorting a sequence of $n$ numbers on an EREW-PRAM computer with $n$ processors. We will also analyze its running time. Suppose $T(n)$ represents the time required to sort $n = 2^k$ data-items.

ALGORITHM 3.7. **Batcher Sort**

1. *Sort* right and left halves of the sequence (recursively). This runs in $T(2^{k-1})$-time (assuming that the right and left halves are sorted in parallel.
2. Perform a Batcher Merge (3.6) of the two sorted halves.

The overall running time satisfies the relation $T(2^k) = T(2^{k-1}) + c_1(k-1) + c_2$. Here $c_1$ is equal to the constant factors involved in the Merge algorithm and $c_2$ is the total contribution of the constant factors in the unshuffle and shuffle-steps, and the last step. We get

$$T(2^k) = \sum_{j=1}^{k} (c_1(j-1) + c_2) = c_1 \frac{(k-1)(k-2)}{2} + c_2 k = O(k^2)$$

Since $n = 2^k$, $k = \lg n$ and we get $T(n) = O(\lg^2 n)$.

**Exercises:**

1. Is it possible for a sorting algorithm to *not* be equivalent to a sorting network? What operations would the algorithm have to perform in order to destroy this equivalence?
2. Prove that the odd-even sorting algorithm on page 7 works. Hint: Use the 0-1 principle.

3. Consider the example of super-unitary speedup on page 16. Is such super-unitary speedup *always* possible? If not, what conditions must be satisfied for it to happen?

4. At about the time that he developed the Bitonic Sorting Algorithm, and the associated merging algorithm, Batcher also developed the *Odd-Even Merge Algorithm* .

   a. Assume that we have two sorted sequences of length $n$: $\{A_i\}$ and $\{B_i\}$. Unshuffle these sequences forming four sequences of length $n/2$: $\{A_{2j-1}\}, \{A_{2j}\}, \{B_{2j-1}\}, \{B_{2j}\}$;

   b. (Recursively) merge $\{A_{2j-1}\}$ with $\{B_{2j-1}\}$ forming $\{C_i\}$ and $\{A_{2j}\}$ with $\{B_{2j}\}$ forming $\{D_i\}$;

   c. Shuffle $\{C_i\}$ with $\{D_i\}$ forming $\{C_1, D_1, C_2, D_2, \ldots C_n, D_n\}$;

   d. Sorting this result requires at most 1 parallel step, interchanging $C_i$ with $D_{i-1}$ for some values of $i$.

The correctness of this algorithm depends upon the lemma:

LEMMA 3.8. Let $n$ be a power of 2 and let $A_1, \ldots, A_n$ and $B_1, \ldots, B_n$ be two sorted sequences such that $A_1 \leq B_1$. Merge the sequences $A_1, A_3, A_5, \ldots$ and $B_1, B_3, B_5, \ldots$ to form $C_1, \ldots, C_n$ and the sequences $A_2, A_4, A_6, \ldots$ and $B_2, B_4, B_6, \ldots$ to form $D_1, \ldots, D_n$. Now shuffle the $C$-sequence with the $D$- sequence to get $C_1, D_1, C_2, D_2, \ldots C_n, D_n$. Then sorting this last sequence requires, at most interchanging $C_i$ with $D_{i-1}$ for some values of $i$.

Here is an example: Suppose the original $A$-sequence is $1, 5, 6, 9$ and the original $B$-sequence is $2, 3, 8, 10$. The sequence of odd $A$'s is $1, 6$ and the sequence of even $A$'s is $5, 9$. The sequence of odd $B$'s is $2, 8$ and the sequence of even $B$'s is $3, 10$. The result of merging the odd $A$'s with the odd $B$'s is the sequence of $C$'s — this is $1, 2, 6, 8$. Similarly the sequence of $D$'s is $3, 5, 9, 10$. The result of shuffling the $C$'s with the $D$'s is $1, 3, 2, 5, 6, 9, 8, 10$. Sorting this sequence only involves interchanging $C_2 = 2$ and $D_1 = 3$ and $D_3 = 9$ and $C_4 = 8$.

Prove this lemma, using the 0-1 Principal.

# 4. Appendix: Proof of the 0-1 Principal

We begin by recalling the definition of *monotonically increasing functions*:

DEFINITION 4.1. A real-valued function $f$ is called *monotonically increasing* if, whenever $x \leq y$, then $f(x) \leq f(y)$.

For instance $f(x) = x$ or $x^2$ are monotonically increasing functions. These functions (and 0-1 comparator networks) have the following interesting property:

PROPOSITION 4.2. Let $\{a_0, \ldots, a_{k-1}\}$ be a set of numbers that are input to a network, $N$, of comparators, and let $\{b_0, \ldots, b_{k-1}\}$ be the output. If $f(x)$ is any monotonically increasing function, the result of inputting $\{f(a_0), \ldots, f(a_{k_1})\}$ to $N$ will be $\{f(b_0), \ldots, f(b_{k-1})\}$.

In other words, the way a comparator network permutes a set of numbers is not affected by applying a monotonically increasing function to them. This is intuitively clear because:

- the decisions that a comparator network makes as to whether to permute two data items are based solely upon the relative values of the data items;
- and monotonically increasing functions preserve these relative value relationships.

PROOF. We use induction on the number of comparators in the comparator network. If there is only one comparator in the network, then each data item must traverse at most one comparator. In this case, the proof is clear:

$$\min(f(x), f(y)) = f(\min(x, y))$$
$$\max(f(x), f(y)) = f(\max(x, y))$$

Now suppose the conclusion is true for all comparator networks with $n$ comparators. If we are given a comparator network with $n + 1$ comparators, we can place one comparator, $C$, adjacent to the input-lines and regard the original network as a composite of this comparator, and the subnetwork, $N \setminus C$, that remained after we removed it. The notation $A \setminus B$ represents the *set-difference* of the sets $A$ and $B$.

We have just shown that comparator $C$ (i.e., the one that was closest to the input lines in the original network) satisfies the conclusion of this result. It follows that the input-data to $N \setminus C$ will be $f$(output of $C$). The inductive hypothesis implies that the rest of the original network (which has $n$ comparators) will produce output whose relative order will not be modified by applying the function $f$.   $\square$

COROLLARY 4.3. If a comparator network correctly sorts all input-sequences drawn from the set $\{0, 1\}$, then it correctly sorts any input-sequence of numbers, so that it constitutes a sorting network.

Similarly, if a comparator-network whose inputs are subdivided into two equal sets correctly merges all pairs of 0-1-sequences, then it correctly merges all pairs of number-sequences.

PROOF. If we have $k$ inputs $\{a_0, \ldots, a_{k-1}\}$, define the $k$ monotonically increasing functions:

$$f_i(x) = \begin{cases} 0 & \text{if } x < a_i \\ 1 & \text{if } x \geq a_i \end{cases}$$

The conclusion follows immediately from applying 4.2, above, to these monotonically increasing functions.  □

## 5. Relations between PRAM models

In this section we will use the sorting algorithm developed in the last section to compare several variations on the PRAM models of computation. We begin by describing two models that appear to be substantially stronger than the EREW model:

**CREW** — Concurrent Read, Exclusive Write. In this case any number of processors can read from a memory location in one program step, but at most one processor can write to a location at a time. In some sense this model is the one that is most commonly used in the development of algorithms.

**CRCW** — Concurrent Read, Concurrent Write. In this case any number of processors can read from or write to a common memory location in one program step. The outcome of a concurrent write operation depends on the particular model of computation being used (i.e. this case breaks up into a number of sub-cases). For instance, the result of a concurrent write might be the boolean **OR** of the operands; or it might be the value stored by the lowest numbered processor attempting the write operation, etc.

This model of computation is more powerful than the CREW model — see § 2.4 in chapter VI (on page 325) for an example of a problem (connected components of a graph) for which there exists an improved algorithm using the CRCW model. The Hirschberg-Chandra-Sarawate algorithm runs on a CREW computer in $O(\lg^2 n)$ time and the Shiloach-Vishkin algorithm runs on a CRCW computer in $O(\lg n)$ time. There is no known algorithm for this problem that runs on a CREW computer in $O(\lg n)$ time.

It is a somewhat surprising result, due to Vishkin (see [160]) that these models can be effectively simulated by the EREW model. The original statement is as follows:

THEOREM 5.1. If an algorithm on the CRCW model of memory access executes in $\alpha$ time units using $\beta$ processors then it can be simulated on the EREW model using $O(\alpha \lg^2 n)$ -time and $\beta$ processors. The RAM must be increased by a factor of $O(\beta)$.

This theorem uses the Batcher sorting algorithm in an essential way. If we substitute the (equally usable) EREW version of the Cole sorting algorithm, described in § 4.2.3 in chapter VI (see page 396) we get the following theorem:

THEOREM 5.2. **Improved Vishkin Simulation Theorem** If an algorithm on the CRCW model of memory access executes in $\alpha$ time units using $\beta$ processors then it can be simulated on the EREW model using $O(\alpha \lg n)$ -time and $\beta$ processors. The RAM must be increased by a factor of $O(\beta)$.

Incidentally, we are assuming the *SIMD* model of program control.

The algorithm works by *simulating* the read and write operations in a single program step of the CRCW machine.

**CRCW Write Operation.** This involves sorting all of the requests to write to a single memory location and picking only one of them per location. The request that is picked per memory location is the one coming from the processor with the *lowest number* — and that request is actually performed.

Suppose that processor $i$ wants to write to address $a(i)(0 \le i \le \beta - 1)$.

1. Sort the pairs $\{(a(i), i), 0 \le i \le \beta - 1\}$ in lexicographic order using the Batcher Algorithm (or the Cole sorting algorithm on page 396, for the improved version of the theorem) presented in the last section. Call the resulting list of pairs $\{(a(j_i), j_i)\}$.

2. Processor 0 writes in $a(j_0)$ the value that processor $j_0$ originally intended to write there. Processor $k$ $(k > 0)$ tests whether $a(j_{k-1}) = a(j_k)$. If not it writes in $a(j_k)$ the value that processor $j_k$ originally intended to write there.
   Here is an example of the simulation of a CRCW-write operation:

| Processor | 0 | 1 | 2 | 3 | 4 | 5 | 6 | 7 |
|---|---|---|---|---|---|---|---|---|
| Target | 2 | 6 | 3 | 1 | 5 | 7 | 1 | 0 |
| $D(i)$ | 3 | 4 | 5 | 6 | 7 | 8 | 9 | 0 |

Here $D(i)$ is the data that processor $i$ wants to write to location $a(i)$. This is converted into a list of pairs:

$$(0,2), (6,1), (3,2), (1,3), (5,4), (7,5), (7,6), (0,7)$$

This list is sorted by the second element in each pair:

$$(0,7), (1,3), (2,0), (3,2), (5,4), (6,1), (7,5), (7,6)$$

Suppose the $i^{\text{th}}$ pair in the sorted list is called $(a(j_i), j_i)$, and the memory in the $i^{\text{th}}$ processor is called $M_i$. These pairs are processed via the following sequence of operations:

| Processor | Action |
|-----------|--------|
| 0 | $M_0 \leftarrow D(7) = 0$ |
| 1 | Test $a(j_1) = 1 \neq a(j_0) = 0$ and do $M_1 \leftarrow D(3) = 6$ |
| 2 | Test $a(j_2) = 2 \neq a(j_1) = 1$ and do $M_2 \leftarrow D(0) = 3$ |
| 3 | Test $a(j_3) = 3 \neq a(j_2) = 2$ and do $M_3 \leftarrow D(2) = 5$ |
| 4 | Test $a(j_4) = 5 \neq a(j_3) = 3$ and do $M_5 \leftarrow D(4) = 7$ |
| 5 | Test $a(j_5) = 6 \neq a(j_4) = 5$ and do $M_6 \leftarrow D(1) = 4$ |
| 6 | Test $a(j_6) = 7 \neq a(j_5) = 6$ and do $M_5 \leftarrow D(6) = 9$ |
| 7 | Test $a(j_7) = a(j_6) = 7$ and do nothing |

**CRCW Read Operation.** Here $a(i)(0 \leq i \leq \beta - 1)$ denotes the address from which processor $i$ wants to read in the CRCW machine.

1. Identical to step 1 in the Write Operation. In addition introduce an auxiliary $\beta \times 3$ array denoted $Z$.

   For $i$, $0 \leq i \leq \beta - 1$: $Z(i, 0)$ contains the content of memory address $a(j_i)$ at the end of the read-operation.

   $Z(i, 1)$ contains **YES** if the content of $a(j_i)$ is already written in $Z(i, 1)$, and **NO** otherwise. It is set to **NO** before each simulated CRCW read-operation.

   $Z(i, 2)$ contains the content of address $a(i)$ at the end of the read-operation.

2. Processor 0 copies the content of $a(j_0)$ into $Z(0, 0)$; $Z(0, 1) \leftarrow$ **YES**. If $a(j_i) \neq a(j_{i-1})$ then processor $j_i$ copies the content of $a(j_i)$ into $Z(j_i, 0)$; $Z(j_i, 1) \leftarrow$ **YES**. The array now has the unique values needed by the processors. The next step consists of *propagating* these values throughout the portions of $Z(0, *)$ that correspond to processors reading from the same location. This is accomplished in $\lg n$ iterations of the following steps (for processors $0 \leq i \leq \beta - 1$):

   $k(i) \leftarrow 0$ (once, in the first iteration);
   **Wait** until $Z(i, 1)$ is turned to **YES**;
       **while** $(i + 2^{k(i)} \leq \beta - 1$ **and**
           $Z(i + 2^{k(i)}, 1) = $ NO) **do**
           $Z(i + 2^{k(i)}, 1) \leftarrow$**YES**;
           $Z(i + 2^{k(i)}, 0) \leftarrow Z(i, 0)$;
           $k(i + 2^{k(i)}) \leftarrow k(i) + 1$;
           $k(i) \leftarrow k(i) + 1$; $Z(j_i, 2) \leftarrow Z(i, 0)$;
   **endwhile**

Note that the EREW design of the computer makes it necessary for us to have *separate counters* — the $k(i)$ — for each of the processors.

This operation copies the values that have been read from memory as many times as are necessary to satisfy the original read-requests. The final step consists in having the processors read $Z(*, 2)$.

Here is an example of the simulation of a CRCW-read operation:

| Processor | 0 | 1 | 2 | 3 | 4 | 5 | 6 | 7 |
|---|---|---|---|---|---|---|---|---|
| Reads from | 2 | 6 | 7 | 1 | 5 | 7 | 1 | 0 |
| $D(i)$ | 3 | 4 | 5 | 6 | 7 | 8 | 9 | 0 |

In this example $D_i$ is the data that processor $i$ initially contains. The first step is the same as in the simulation of the CRCW-write operation. The set of desired read-operations is converted into a list of pairs:

$$(2,0), (6,1), (7,2), (1,3), (5,4), (7,5), (1,6), (0,7)$$

This list is sorted by the second element in each pair:

$$(0,7), (1,3), (1,6), (2,0), (5,4), (6,1), (7,2), (7,5)$$

Now we set up the $Z$ array. Initially it looks like the following:

| $i$ | 0 | 1 | 2 | 3 | 4 | 5 | 6 | 7 |
|---|---|---|---|---|---|---|---|---|
| $Z(i,0)$ | | | | | | | | |
| $Z(i,1)$ | NO | NO | NO | NO | NO | NO | NO | NO |
| $Z(i,2)$ | | | | | | | | |

The first processor copies $D(a(j_0))$ into position $Z(0,0)$. Every other processor tests whether $a(j_i) \neq a(j_{i-1})$ and, if the values are not equal, copies its value of $D(a(j_i))$ into $Z(i,0)$. Each position of the $Z$-array that receives one of the $a(i)$ is marked by having its value of $Z(i,1)$ set to **YES**. We also set up the variables $k(i)$. We get the following array:

| $i$ | 0 | 1 | 2 | 3 | 4 | 5 | 6 | 7 |
|---|---|---|---|---|---|---|---|---|
| $Z(i,0)$ | 0 | 6 | | 3 | 7 | 4 | 5 | |
| $Z(i,1)$ | YES | YES | NO | YES | YES | YES | YES | NO |
| $Z(i,2)$ | | | | | | | | |
| $k(i)$ | 0 | 0 | 0 | 0 | 0 | 0 | 0 | 0 |

Now we begin the iterations of the algorithm. After the first iteration we get

| $i$ | 0 | 1 | 2 | 3 | 4 | 5 | 6 | 7 |
|---|---|---|---|---|---|---|---|---|
| $Z(i,0)$ | 0 | 6 | 6 | 3 | 7 | 4 | 5 | 5 |
| $Z(i,1)$ | YES | YES | YES | YES | YES | YES | YES | YES |
| $Z(i,2)$ | | | | | | | | |
| $k(i)$ | 1 | 1 | 1 | 1 | 1 | 1 | 1 | 1 |

In this particular example the iterations are completed in the first step. No computations occur in the remaining (2) iterations. In the last step of the algorithm the data is copied into $Z(*,2)$.

| $i$ | 0 | 1 | 2 | 3 | 4 | 5 | 6 | 7 |
|---|---|---|---|---|---|---|---|---|
| $Z(i,0)$ | 0 | 6 | 6 | 3 | 7 | 4 | 5 | 5 |
| $Z(i,1)$ | YES | YES | YES | YES | YES | YES | YES | YES |
| $Z(i,2)$ | 3 | 4 | 5 | 6 | 7 | 5 | 6 | 0 |
| $k(i)$ | 3 | 3 | 3 | 3 | 3 | 3 | 3 | 3 |

**Exercises:**

1. Modify the CRCW Write phase of the simulation algorithm described above (page 30) so that, whenever multiple processors attempt to write numbers to the same simulated location, their *sum* is actually written.[3].

## 6. Theoretical Issues

**6.1. Complexity Classes and the Parallel Processing Thesis.** In this chapter we will be concerned with various *theoretical* issues connected with parallel processing. We will study the question of what calculations can be efficiently done in parallel and in what sense. We present the so-called *Parallel Processing Thesis* of Fortune and Wyllie — see [51]. It essentially shows that *execution-time* on a *parallel computer* corresponds in some sense to *space* (i.e., memory) on a sequential computer. The *arguments* used by Fortune and Wyllie *also* give some insight into why the execution time of *many* parallel algorithms is a power of a *logarithm* of the complexity of the problem.

One of the most interesting theoretical questions that arise in this field is whether there exist *inherently sequential problems*. These are essentially computations for which it is *impossible* to find parallel algorithms that are substantially faster than the fastest sequential algorithms. This is a subtle question, because there are many problems that appear to be inherently sequential at first glance but have fast parallel algorithms. In many cases the fast parallel algorithms approach the problem from a completely different angle than the preferred sequential algorithms. One of the most *glaring examples* of this is the problem of *matrix inversion*, where:

---

[3]This situation actually arises in existing parallel computers — see the description of census-operations on the CM-2 on page 117.

1. the fastest sequential algorithm (i.e., a form of Gaussian Elimination) only lends itself to a limited amount of parallelization (see the discussion below, on page 47);
2. the (asymptotically) fastest parallel algorithm would be extremely bad from a sequential point of view.

This should not be too surprising — in many cases the fastest sequential algorithms are the ones that reduce the amount of parallelism in the computations to a minimum.

First it is necessary to make precise what we mean by a parallel algorithm being substantially faster than the corresponding sequential algorithm. Here are some of the algorithms that have been considered so far:

1. Forming cumulative sums of $n$ numbers. The sequential algorithm has an execution time of $O(n)$. The parallel algorithm has an execution time of $O(\lg n)$ using $O(n)$ processors;
2. Sorting $n$ numbers by performing comparisons. The best sequential algorithms have an asymptotic execution time of $O(n \lg n)$. The best parallel algorithms have asymptotic execution times of $O(\lg n)$ using $O(n)$ processors — see Chapter V, § 1.3 (page 182);
3. Inversion of an $n \times n$ non-sparse matrix. The best sequential algorithms use Gaussian Elimination and have an execution time of $O(n^3)$. The asymptotically fastest known parallel algorithms have an execution time of $O(\lg^2 n)$ using $n^{2.376}$ processors.

The general pattern that emerges is:

- we have a sequential algorithm that executes in an amount of time that is bounded by a *polynomial* function of the input-size. The *class* of such problems is denoted **P**;
- we have parallel algorithms that execute in an amount of time that is bounded by a polynomial of the *logarithm* of the input-size, and use a number of *processors* bounded by a polynomial of the input size. The class of these problems is denoted **NC**;

As has been remarked before, **NC** ⊆ **P** — any algorithm for a problem in **NC** can be sequentially *simulated* in an amount of time that is bounded by a polynomial function of the original input.

Our question of whether inherently sequential problems exist boils down to the question of whether there exist any problems in **P** \ **NC** — or the question of whether **NC** = **P**.

As of this writing (1991) this question is still open. We will discuss some *partial results* in this direction. They give a natural relationship between parallel execution *time* and the amount of *RAM* required by sequential algorithms. From this we can deduce some rather weak results regarding sequential execution time.

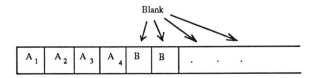

FIGURE II.4. Tape in a Turing Machine

It is first necessary to define the complexity of computations in a fairly rigorous sense. We will consider general problems equipped with

- An *encoding scheme* for the input-data. This is some procedure, chosen in advance, for representing the input-data as a string in some language associated with the problem. For instance, the general sorting problem might get its input-data in a string of the form $\{a_1, a_2, \ldots\}$, where the $a_i$ are bit-strings representing the numbers to be sorted.

- A *complexity parameter* that is proportional to the size of the input-string. For instance, depending on how one defines the sorting problem, the complexity parameter might be equal to the number of items to be sorted, or the total number of symbols required to represent these data-items.

  These two definitions of the sorting problem differ in significant ways. For instance if we assume that inputs are all *distinct*, then it requires $O(n \lg n)$ symbols to represent $n$ numbers. This is due to the fact that $\lg n$ bits are needed to count from 0 to $n - 1$ so (at least) this many bits are needed to represent each number in a set of $n$ distinct numbers. In this case, it makes a big difference whether one defines the complexity-parameter to be the number of data-items or the size (in bits) of the input. If (as is usual) we assume the all inputs to a sorting-algorithm can be represented by a bounded number of bits, then the number of input items is proportional to the *actual size* of the input.

Incidentally, in defining the complexity-parameter of a problem to be the size of the string containing the input-data, we make *no* assumptions about *how* this input string is processed — in particular, we do not assume that it is processed sequentially.

Having defined what we mean by the size of the input, we must give a rigorous definition of the various parameters associated with the execution of an algorithm for solving the problem like running time, memory, and number of processors. This is usually done with a *Turing Machine*.

A Turing Machine is a kind of generalization of a finite state automaton — it is a kind of fictitious computer for which it is easy to define the number of steps required to perform a computation. It consists of the following elements:

1. A Tape, shown in figure II.4.
   This tape is infinite in length and bounded at one end. It is divided into cells that may have a symbol marked on them or be blank. Initially at most a finite

number of cells are nonblank.
2. A Control Mechanism, which consists of:
   - $Q$ = a finite set of states;
   - $\Gamma$= set of tape symbols, called the *alphabet* of the Turing machine;
   - $B$=blank symbol;
   - $\Sigma$= input symbols;
   - $\delta$=next move function:$Q \times \Gamma \rightarrow Q \times \Gamma \times \{L, R\}$;
   - $q_0$=start state;
   - $F \subseteq Q$ = final states.

In each step the machine reads a symbol from the tape, changes state, optionally writes a symbol, and moves the tape head left (L) or right (R) one position — all of these actions are encoded into the *next move function*. There are a number of auxiliary definitions we make at this point:

DEFINITION 6.1. Suppose we have a Turing machine $T$. Then:
   1. An input string, $S$, is *accepted* by $T$ if $T$ ends up in one of its *stop-states* after executing with the string as its original tape symbols.
   2. The set, $L(T)$, of input strings accepted by $T$ is called the *language recognized* by $T$. The reader might wonder what all of this has to do with performing computations.
   3. If a string, $S$, is recognized by $T$, the symbols *left* on the tape after $T$ has stopped will be called the *result of $T$ performing a computation* on $S$. Note that this is only well-defined if $T$ recognizes $S$.

It can be shown that any computation that can be done on a *conventional* sequential computer can also be done on a suitable Turing machine (although in much more time). It is also well-known that any computation that can be done in *polynomial time* on a conventional computer can be done in polynomial time on a *Turing machine*. Thus a Turing machine presents a simple model of computation that can be analyzed theoretically.

EXAMPLE 6.2. Here is a Turing machine whose language consists of arbitrary finite strings on two symbols $\{0, 1\}$ (in other words, it *accepts* all such strings). It has five states and its actions are described by table II.1.

Careful examination of the action of this Turing Machine shows that it *sorts* its input string.

We will also need a variation on this form of Turing machine. An *offline* Turing machine is defined to be like the one above except that there are three tapes. The first one is a *read-only* tape that contains the input string. The second is like the tape in the definition above but it is required to be initially *blank*. The third is a *write-only* tape — it receives the output of the computation. Note that this form of Turing machine clearly

| States | Input Symbols | | |
|---|---|---|---|
|  | space | 0 | 1 |
| 1 (Start state) | Move left Go to state 4 | Write 1, move left, go to state 2 | Move right, go to state 1 |
| 2 | Move right, go to state 3 | Move left, go to state 1 | Move left, go to state 1 |
| 3 |  | Move right | Write 0, move right, go to state 1 |
| 4 | Move right, go to state 5 | Move left | Move left |
| 5 | Stop | | |

TABLE II.1. Actions of a Turing machine for sorting a string of 0's and 1's

distinguishes input from output, unlike the general Turing machines. It is possible to show that offline Turing machines are essentially equivalent to general Turing machines in the sense that any language recognized by a general Turing machine can also be recognized by a suitable offline Turing machine. Furthermore, the computations of the two types of Turing machines agree under this equivalence. Clearly, if we are only interested in what computations can be computed by a Turing machine, there is no need to define offline Turing machines — we only define them so we can rigorously define the *memory* used by an algorithm. See [71] for more information on Turing machines.

DEFINITION 6.3. The *space* required to solve a problem on an offline Turing machine is defined to be the number of cells of the second tape that was used in the course of computation when the problem was solved.

DEFINITION 6.4. 1. A problem with complexity parameter $n$ will be said to be in the class $T(n)$-space if there exists an offline Turing machine that uses space $T(n)$.
2. A problem will be said to be in **Plogspace($k$)** if there exists an offline Turing machine that solves the problem using space that is $O(\lg^k n)$.

It is well-known that **Plogspace(1)** $\subset$ **P**, i.e. any problem that can be solved on an offline Turing machine in *space* proportional to the logarithm of the complexity-parameter can be solved in polynomial *time* on a conventional Turing machine. This is not hard to see — if the total amount of RAM used by a sequential algorithm is $c \lg n$ then the total number of possible *states* (or sets of data stored in memory) is a power of $n$.

The converse question is *still open* — and probably very difficult. In fact it is not known whether every problem in **P** is in **Plogspace**($k$) for any value of $k$.

One question that might arise is: "How can problems with an input-size of $n$ use an amount of RAM that is *less* than $n$?" The answer is that we can use an inefficient algorithm that doesn't store much of the input data in RAM at any one time. For instance it turns out that *sorting* $n$ quantities by comparisons is in **Plogspace** — we simply use a kind of *bubblesort* that requires a great deal of *time* to execute but very little RAM. It turns out that essentially all of the typical algorithms discussed (for instance) in a course on algorithms are in **Plogspace**.

This will be our model of *sequential* computation. Our model for *parallel* computation will be somewhat different from those described in chapter IV. It is a *MIMD* form of parallel computation:

We will have a large (actually *infinite*) number of independent processors that can access common RAM and can execute the following instructions:

> **LOAD** op
> **STORE** op
> **ADD** op
> **SUB** op
> **JUMP** label
> **JZERO** label
> **READ** reg#
> **FORK** label
> **HALT**

Here an operand can be either: an *address*; an *indirect address*; or a *literal*. Initially input is placed in *input registers* that can be accessed via the **READ** instruction. Each processor has one register called its *accumulator*. Binary arithmetic instructions use the accumulator as the *first operand* and store the *result* in the accumulator. New processors are introduced into a computation via the **FORK** instruction which, when executed by processor $i$:

1. activates the next available processor — say this is processor $j$;
2. copies the accumulator of processor $i$ to processor $j$;
3. and makes processor $j$ take its first instruction from label.

When a processor executes a **HALT** instruction it stops running (and re-enters the pool of processors available for execution by **FORK** instructions). Execution of a program continues until processor 0 executes a **HALT** instruction, or two or more processors try to write to the same location in the same program step.

One processor can initiate two other processors in constant time via the **FORK** command. It can pass (via its accumulator) the address of a block of memory containing *parameters* for a subtask. It follows that a processor can initiate a tree of $n$ other processors in $O(\lg n)$-time.

Processors have *no* local memory — they use the global memory available to all processors. It is not hard to "localize" some of this memory via a suitable allocation scheme. For instance, suppose a given processor has every $k^{\text{th}}$ memory location allocated to it (as local memory). When this processor initiates a new processor, it can allocate local memory to the *new* processors from its own local memory. It can allocate every memory location (of its local memory) to the $i^{\text{th}}$ processor it directly initiates, where $p_i$ is the $i^{\text{th}}$ prime number.

Note that this is a very *generous* model of parallel computation — it is much more powerful than the *Connection Machine*, for instance.

We are in a position to give a rigorous definition of the term **NC**:

DEFINITION 6.5. A problem with complexity parameter $n$ is in the class **NC** if there exists a parallel algorithm on the computer described above, that executes in time $O(\lg^k n)$ and uses $O(n^{k'})$ processors, where $k$ and $k'$ are two integers $\geq 0$.

Our first result is:

THEOREM 6.6. For $\mathrm{T}(n) \geq \lg n$:

$$\bigcup_{k=1}^{\infty} \mathrm{T}(n)^k\text{-time-P-RAM} = \bigcup_{k=1}^{\infty} \mathrm{T}(n)^k\text{-space}$$

In particular,

$$\bigcup_{k=1}^{\infty} \lg^k n\text{-time-P-RAM} = \bigcup_{k=1}^{\infty} \lg^k n\text{-space}$$

The proof consists in the following two lemmas:

LEMMA 6.7. Let $L$ be a language accepted by a deterministic $\mathrm{T}(n)$-space bounded Turing machine $M$, for $\mathrm{T}(n) \geq \lg n$. Then $L$ is accepted by a deterministic $c\mathrm{T}(n)$-time bounded **P-RAM** $P$, for some constant $c$.

In the theory of Turing machines, acceptance of a language can be regarded as performance of a computation.

PROOF. We will *simulate* the behavior of $M$ by $P$. Initially our simulation will assume that the value of $\mathrm{T}(n)$ is available at the beginning of the computation (possibly an unrealistic assumption) and then we will show how to conduct the simulation in such a way that this assumption is removed.

Given $\mathrm{T}(n)$, $P$ constructs a directed graph representing all possible configurations of $M$ during the computation. We will regard a configuration of $M$ as a state of $M$ together with the data required to compute the next state. This consists of:

    1. the data in the *memory* tape;

2. a *pointer* to a position of the *input* tape;

Statement 1 above results in $2^{T(n)}$ possible configurations (since the memory tape has $T(n)$ slots for data) and since the parameter defining the original problem's size is $n$ we assume $O(n)$ possible positions on the *input tape*. The total number of *possible configurations* is thus $wn2^{T(n)}$, where $w$ is some constant. This can be regarded as $2^{T(n)+\lg n+\lg w} \leq 2^{dT(n)}$, for some constant $d$ depending on $M$. Leaving each node of the graph will be a single edge to the node of its successor configuration. *Accepting* configuration-nodes of $M$ are their *own* successors. Thus there exists a path from the initial configuration node to an *accepting* node *if and only if* $M$ accepts its input within $T(n)$-space.

To build the graph, $P$ first initiates $2^{dT(n)}$ processors in $O(T(n))$-steps, each holding a different integer, representing a different configuration of $M$. Each processor then, in $O(T(n))$-time:

1. unpacks its configuration integer (we assume this integer contains an encoded representation of a state of the Turing machine);
2. computes the successor configuration (simulating the behavior of the Turing machine when it is in this state);
3. packs this successor configuration into integer form.

The graph is stored in global memory and the parallel computer then determines whether there exists a path connecting the initial node to some accepting node. This is done as *follows:* Each processor computes the successor of its successor node and stores this as its immediate successor. In $k$ steps each processor will point to the $2^k$-th successor node and in $O(T(n))$-steps the question of whether the language will be accepted will be decided because the successor nodes are all a distance of at most $2^{dT(n)}$ from the start node.

As pointed out above this algorithm has the unfortunate drawback that it requires knowledge of $T(n)$ before it can be carried out. This can be corrected as follows: At the beginning of the simulation processor 0 starts other processors out that assume that the value of $T(n)$ is $1, 2, \ldots$, respectively. These processors carry out the simulation above, using these assumptions and if any of them accepts the language, then processor 0 accepts also. We assume that memory is allocated so that the areas of memory used by the different simulations is disjoint — this can be accomplished by adding $c2^{dT(n)}$ to the addresses of nodes for the simulation of a given value of $T(n)$.   □

LEMMA 6.8. Let $L$ be accepted by a deterministic $T(n)$-time bounded P-RAM. Then $L$ is accepted by a $T(n)^2$-space bounded Turing machine.

PROOF. We will first construct a *nondeterministic* $T(n)^2$-space bounded Turing Machine accepting $L$. We will then show how to make it deterministic. Recall that a *nondeterministic Turing machine* is one that may take *many different* actions in a given step — imagine that the machine "splits" into *many different machines*, each taking a different action and attempting to complete the computation using that alternate. The language is ultimately accepted if *any one* of these split machines accepts it.

In order to determine whether the **P-RAM** accepts its input, the Turing Machine needs to:

1. know the contents of processor 0's accumulator when it halts; and
2. to verify that no two writes occur simultaneously into the same global memory location.

The simulation is based on a recursive procedure **ACC** which checks the contents of a processor's accumulator at a particular time. By applying the procedure to processor 0 we can determine if the **P-RAM** accepts the language. **ACC** will check that at time $t$, processor $j$ executed the $i^{\text{th}}$ instruction of its program leaving $c$ in its accumulator. In order to check this, **ACC** needs to know

1. the instruction executed by processor $j$ at time $t - 1$ and the ensuing contents of its accumulator, and
2. the contents of the memory locations referenced by instruction $i$.

**ACC** can nondeterministically *guess* 1 and recursively *verify* it. To determine 2, for each memory location, $m$, referenced **ACC** guesses that $m$ was last written by some processor $k$ at time $t' < t$. **ACC** can recursively verify that processor $k$ did a **STORE** of the proper contents into $m$ at time $t'$. **ACC** must also check that no other processor writes into $m$ at any time between $t'$ and $t$. It can do this by guessing the instructions executed by *each* processor at *each* such time, recursively verifying them, and verifying that none of the instructions changes $m$. Checking that two writes do not occur into the same memory location at the same time can be done in a similar fashion. For each time step and each pair of processors, **ACC** nondeterministically guesses the instructions executed, recursively verifies them, and checks that the two instructions were not write-operations into the same memory location. The correctness of the simulation follows from the determinism of the **P-RAM**. In general, each instruction executed by the **P-RAM** will be guessed and verified many times by **ACC**. Since the **P-RAM** is deterministic, however, there can be only one possible instruction that can be executed by a processor in a program step — each *verified* guess must be the same. Now we analyze the *space* requirements:

Note that there can be at most $2^{T(n)}$ processors running on the **P-RAM** after $T(n)$ program steps so writing down a processor number requires $T(n)$ space. Since addition and subtraction are the only arithmetic operators, numbers can increase by at most one bit each step. Thus, writing down the contents of the accumulator takes at most $T(n) + \lg n = O(T(n))$ space. Writing down a time step takes $\lg T(n)$ space and the program counter requires only constant space. Hence the arguments to a recursive call require $O(T(n))$ space. Cycling through time steps and processor numbers to verify that a memory location was not overwritten also only takes $T(n)$ space, so that the total space requirement at each level of the recursion is $O(\mathrm{T}(n))$. Since there can be at most $\mathrm{T}(n)$ levels of recursion (one for each time step of the program on the parallel machine) the total space requirement is $O(\mathrm{T}(n)^2)$. Note that *this* simulation can be performed by a *deterministic* Turing Machine — in each step, simply loop through all of the possibilities when performing the recursive calls to **ACC** — i.e., all instructions in the instruction set; all prior times, etc.

This requires that we modify the **ACC** procedure slightly so that it returns information on whether a given guess was correct. This could be accomplished by having it returns either the value of the accumulator, if a given processor really did execute a given instruction at a given time or return **NULL**, indicating that the guess was wrong. This increases the *execution time* of the algorithm *tremendously* but has no effect on the *space* requirement.  □

**Exercises:**

1. Show that the example of a Turing machine given in 6.2 on page 36 is in *Plogspace*. Do this by explicitly transforming it into an offline Turing machine.

**6.2. P-Completeness and Inherently Sequential Problems.** The previous section provided some important definitions and showed that problems were solvable in polylogarithmic time on a parallel computer if and only if they were solvable in polylogarithmic space on an offline Turing machine. In the present section we will apply these results to study the question of whether there exist inherently sequential problems. Although this question is still (as of 1991) open, we can study it in a manner reminiscent of the theory of NP-complete problems. As is done there, we restrict our attention to a particular class of problems known as *decision problems*. These are computations that produce a boolean 0 or 1 as their result. It will turn out that there are reasonable candidates for inherently sequential problems even among this restricted subset.

DEFINITION 6.9. A *decision problem* is a pair $\{T, L\}$ where:

1. $T$ recognizes *all* strings made up of the basic alphabet over which it is defined.
2. $L$ is the set of strings over this alphabet for which the computation that $T$ performs results in a 1.

For instance, a *decision-problem* version of the problem of sorting $n$ numbers would be the problem of whether the $k^{\text{th}}$ element of the result of sorting the $n$ numbers had some given value.

We will be able to give likely candidates for inherently sequential problems by showing that there exist problems in **P** with the property that, if there exists an **NC** algorithm for solving them, then **P** = **NC**. We need the very important concept of *reducibility* of decision-problems:

DEFINITION 6.10. Let $P_1 = \{T_1, L_1\}$ and $P_2 = \{T_2, L_2\}$ be two decision-problems and let $\Sigma_1$ and $\Sigma_2$ be the respective alphabets for the Turing machines that resolve these decision-problems. Then $P_1$ will be said to be *reducible* to $P_2$, denoted $P_1 \propto P_2$, if there exists a function $f : \Sigma_1^* \to \Sigma_2^*$ such that:

- a string $s \in L_1$ if and only if $f(s) \in L_2$;
- the function $f$ is computable in polynomial time.

$P_1$ will be said to be *logspace reducible* to $P_2$, denoted $P_1 \propto_{\text{logspace}} P_2$, if the conditions above are satisfied, and in addition, the algorithm computing the function $f$ is in **logspace**. In other words, there must exist an offline Turing machine using logspace, that computes $f$.

Note that $P_1 \propto P_2$ implies that a polynomial-time solution to $P_2$ gives rise to a similar solution to $P_1$: if we want to decide whether a string $s \in \Sigma_1^*$ results in a 1 when executed on $T_1$, we just apply the function $f$ (which can be done in polynomial time) and execute $T_2$ on the result. If there exists an **NC** algorithm for $P_2$, $P_1 \propto P_2$ implies that, in some sense, the decision problem $P_1$ *is solvable* in polylogarithmic time with a polynomial number of processors, but doesn't give any clue as to *how* this can be accomplished. If there exists an **NC** algorithm for $P_2$ and $P_1 \propto_{\text{logspace}} P_2$, then there is also an **NC** algorithm for $P_1$ — it consists of the following steps

1. Convert an input-string of $P_1$ into a corresponding input-string of $P_2$ via the transformation-function $f$, in $P_1 \propto_{\text{logspace}} P_2$. The *computation* of this function can be done via an **NC** algorithm due to the results of the previous section — see 6.6 on page 39.
2. Execute the **NC** algorithm for $P_2$.

As in the theory of NP completeness, we distinguish certain problems in **P** that are the "hardest":

DEFINITION 6.11. A problem $Z$ in **P** will be called *P-complete* if it has the property that:

For every problem $A \in$ **P**, $A \propto Z$.

The problem $Z$ will be called *logspace-complete for P* or *strongly P-complete* if:

For every problem $A \in$ **P**, $A \propto_{\text{logspace}} Z$.

The first problem *proved* to be P-complete was the problem of "directed forest accessibility" — see [31].

We will conclude this section with an example of a problem that is known to be strongly P-complete — i.e., logspace complete for **P**, as defined above. It is fairly simple to state:

EXAMPLE 6.12. **Circuit Value Problem.**
- **Input:** A list $L = \{L_1, \ldots, L_n\}$ of $n$-terms (where $n$ is some number $> 0$, where each term is either:
  1. A 0 or a 1, or;
  2. A boolean expression involving strictly lower-numbered terms — for instance, we might have $L_5 = (L_2 \vee L_4) \wedge (L_1 \vee L_3)$.
- **Output:** The boolean value of $L_n$. Note that this is a *decision problem*.

Note that the requirement that boolean expressions involve *strictly* lower-numbered terms means that the *first* term must be a 0 or a 1.

This problem is *trivial* to solve in $n$ steps sequentially: just scan the list from left to right and evaluate each term as you encounter it. It is interesting that there is no known **NC** algorithm for solving this problem. Ladner proved that this problem is logspace complete for **P**, or *strongly P-complete* — see [95].

LEMMA 6.13. The circuit value problem, as defined above, is logspace-complete for **P**.

PROOF. Let CVP denote the Circuit Value Problem. We must show, that if $Z$ is any problem in **P**, then $Z \propto_{\text{logspace}}$ CVP. We will assume that $Z$ is computed by a Turing machine $T$ that always halts in a number of steps that is bounded by a polynomial of the complexity parameter of the input. We will construct a (fairly large) circuit whose final value is always the same as that computed by $T$. We will also show that the construction can be done in logspace. We will assume:
- The number of characters in the input-string to $T$ is $n$;
- The execution-time of $T$ is $E = O(n^k)$, where $k$ is some integer $\geq 1$.
- During the course of the execution of $T$, the number of states that it ever passes through is $O(n^k)$, since it stops after this number of steps. The binary representation of a state-number of $T$ clearly requires $O(\lg n)$ bits.
- The number of tape-symbols that are nonblank at any step of the algorithm is $O(n^k)$. This is because the tape started out with $n$ nonblank (input) symbols, and at most $O(n^k)$ write-operations were executed while $T$ was active.
- The transitions of $T$ are encoded as a large set of boolean formulas, computing:
  1. the bits of the next state-number from the bits of the current state-number, the bits of the current input-symbol.
  2. whether the tape is moved to the left or the right.

3. whether anything is written to the tape in this step (and what is written). Although it might seem that this assumption is an attempt to make life easy for ourselves, it is nevertheless a reasonable one. The transition function may well be encoded as such a sequence of boolean formulas. If it is simply represented as a table like that on page 36, it is fairly trivial to convert this table into a series of boolean formulas representing it.

We will build a large circuit that is the concatenation of sublists $\{L_1, \ldots, L_E\}$, one for each program-step of the execution of $T$. The list $L_1$ consists of bits representing the input-data of $T$, and a sequence of bits representing its *start-state*. In general $L_i$ will consist of a concatenation of two sublists $S_i$ and $Q_i$.

- the sublist $S_i$ consists of a sequence of boolean formulas computing the bits of the new *state-number* of $T$ in program-step $i$. These formulas use the bits of $S_{i-1}$ and $Q_{i-1}$ as inputs.
- $Q_i$ lists the bits of the nonblank *tape symbols* in program-step $i$ (of $T$). For tape-positions that are not written to in step $i$, the corresponding entries in $Q_i$ simply *copy* data from corresponding positions of $Q_{i-1}$. In other words, these formulas are *trivial* formulas that are just equal to certain previous terms in the giant list. The entries of $Q_i$ that represent tape-symbols that are *written to* in program-step $i$ have boolean formulas that compute this data from:
  1. The bits of the current state-number, computed by the boolean formulas in $S_i$.
  2. Bits of $Q_{i-1}$, representing the tape in the previous program-step.

Now we consider how this conversion of the input data and the transition-function of $T$ into the $L_i$ can be accomplished:

1. We must be able to count the number of program-steps of $T$ so that we will know when to stop generating the $L_i$. Maintaining such a counter requires $O(\lg n)$ memory-locations in the offline Turing machine that converts $T$ into the circuit.
2. We must be able to copy formulas from rows of the table representing the transition-function of $T$ into entries of the circuit we are generating. This requires looping on the number of bits in these boolean formulas. Again, we will need $O(\lg n)$ memory-locations.
3. We must keep track of the current location of the read-write head of $T$. This tells us how to build the formulas in $Q_i$ that represent tape-symbols that are *written to* in program-step $i$. Since the total number of such tape-positions is $O(n^k)$, we will need $O(\lg n)$ *bits* to represent this number. We increment this number every time the transition-function calls for movement to the *right* (for instance) and decrement it every time we are to move to the left.

The offline Turing machine will contain this memory, and its input program. The total memory used will be $O(\lg n)$, so the transformation from $T$ to the Circuit Value Problem will be logspace. $\square$

Many problems are known to be P-complete. We will give a list of a few of the more interesting ones.

- **The Monotone Circuit Problem** This is a circuit whose only operations are $\vee$ and $\wedge$. Goldschlager gave a logspace reduction of the general circuit-value problem to this — see [57]. The Planar Circuit Value Problem is also P-complete. A planar circuit is a circuit that can be drawn on a plane without any "wires" crossing. It is interesting that the Monotone, Planar Circuit Value problem is in **NC** — see [55].

- **Linear Inequalities** The input to this problem is an $n \times d$ integer-valued matrix $A$, and an integer-valued $n \times 1$ vector $c$. The problem is to answer the question of whether there *exists* a rational-valued vector $x$ such that

$$Ax \leq c$$

In 1982, Cook found the following reduction of the Circuit Value Problem to the Linear Inequality Problem:

1. If input $x_i$ (of the Circuit Value Problem) is $\left\{ \begin{matrix} \textbf{True} \\ \textbf{False} \end{matrix} \right\}$, it is represented by the equation $\left\{ \begin{matrix} x_i = 1 \\ x_i = 0 \end{matrix} \right\}$.

2. A **NOT** gate with input $u$ and output $w$, computing $w = \neg u$, is represented by the inequalities $\left\{ \begin{matrix} w = 1 - u \\ 0 \leq w \leq 1 \end{matrix} \right\}$

3. An **OR** gate with inputs $u$ and $v$, and output $w$ is represented by the inequalities $\left\{ \begin{matrix} 0 \leq w \leq 1 \\ u \leq w \\ v \leq w \\ w \leq u + v \end{matrix} \right\}$

4. An **AND** gate with gate with inputs $u$ and $v$, and output $w$ is represented by the inequalities $\left\{ \begin{matrix} 0 \leq w \leq 1 \\ w \leq u \\ w \leq v \\ u + v - 1 \leq w \end{matrix} \right\}$

This is interesting because this decision problem is a crucial step in performing Linear Programming. It follows that Linear Programming is P-complete. It is interesting that Xiaotie Deng has found an **NC** algorithm for *planar* Linear Programming. This is linear programming with only two variables (but, perhaps, many inequalities) — see [45] for the algorithm.

- **Gaussian elimination** This is the standard sequential algorithm for solving a system of simultaneous linear equations or inverting a matrix. Among other things, it involves clearing out rows of a matrix at certain *pivot points*, which are computed as the algorithm executes. The paper [158], by Stephen Vavasis proves that the problem of deciding whether a given element will be a pivot-element in the Gaussian elimination algorithm is P-complete. This implies that fast parallel algorithms (i.e., **NC** algorithms) *cannot* be based upon Gaussian elimination. This conclusion is interesting because many people have worked on the problem of parallelizing Gaussian elimination by trying to exploit certain parallelisms in the problem. This paper implies that *none* of this work will ever result in an **NC** algorithm for solving linear equations. See § 1 in chapter V for an example of an **NC** algorithm for matrix inversion. This algorithm is not based upon Gaussian elimination.

- **Maximum flows** This is an important problem in the theory of networks. In order to understand it, imagine a network of pipes with the property that each pipe in the network has a certain *carrying capacity* — a maximum rate at which fluid can flow through it. Suppose that one point of the network is a source of fluid and some other point is where fluid leaves the network. The maximum flow problem asks the question "What is the maximum rate fluid can flow from the source to the exit?". This must be computed from characteristics of the network, including the carrying capacities of the pipes and how they are arranged. The paper [58], by L. Goldschlager, L. Shaw and J. Staples proves that this problem is **Plogspace**-complete.

- **Inference problem for multivalued dependencies** This is a problem from the theory of databases. Essentially, a large part of the problem of designing a database consists in identifying *dependencies* in the data: aspects of the data that determine other aspects. A multivalued dependency represents a situation where one aspect of the data *influences* but doesn't completely determine another[4]. These multivalued dependencies imply the existence of *other* multi-valued dependencies among aspects of data. The problem of *determining* these *other* multivalued dependencies from the *given* ones turns out to be P-complete — see [44].

**6.3. Further reading.** Greenlaw, Hoover, and Ruzzo have compiled a large list of P-complete problems — see [59]. This list will be periodically updated to incorporate newly discovered P-complete problems.

In [162] Wilson shows that the class **NC** has a "fine-structure" — it can be decomposed into subclasses that have natural descriptions.

---

[4]There is more to it than this, however. See a good book on database design for more information

In [30] Cook gives a detailed description of parallel complexity classes and how they relate to each other. Also see [29].

## 7. General Principles of Parallel Algorithm Design

In this section we discuss some general principles of algorithm-design. Later chapters in this book will explore these principles in more detail. We have already seen some of these concepts in the Introduction.

**7.1. Brent's Theorem.** Brent's Theorem makes precise some of the heuristic arguments in the introduction relating computation networks with time required to compute something in parallel.

We will need a rigorous definition of a combinational network, or computational network. As with comparator networks defined in § 2 we consider a kind of "circuit" whose "wires" can *transmit* numerical data and nodes that can *modify* the data in prescribed ways[5].

DEFINITION 7.1. A *computational network* is a directed acyclic[6] graph whose vertices are subdivided into three sets:

> **Input vertices** These vertices have no incoming edges.
> **Output vertices** These vertices have no outgoing edges.
> **Interior vertices** These vertices have incoming edges and a single outgoing edge.

Each interior vertex is labeled with an elementary operation (to be discussed below). The number of incoming edges to an interior vertex is called its *fan-in*. The number of outgoing edges is called the *fan-out*. The maxima of these two quantities over the entire graph is called, respectively, the fan-in and the fan-out of the graph.

The length of the longest path from any input vertex to any output vertex is called the *depth* of the computational network. The *computation* performed by a computation network, on a given set of inputs is defined to be the data that appears on the output vertices as a result of the following procedure:

1. Apply the input data to the input vertices.
2. Transmit data along directed edges. Whenever an interior vertex is encountered, wait until data arrives along all of its incoming edges, and then perform the indicated elementary computation. Transmit the result of the computation along all of the outgoing edges.
3. The procedure terminates when there is no data at interior vertices.

---

[5]These are, of course, precisely the properties of the electronic circuits used in a computer. We do not want to become involved with issues like the representation of arithmetic operations in terms of boolean logic elements.

[6]"Acyclic" just means that the graph has no directed loops.

Each edge of the computational network is assumed to be able to transmit a number of a certain size that is fixed for the rest of this discussion (a more detailed consideration of these networks would regard the carrying capacity of each wire to be a *single bit* — in this case the magnitudes of the numbers in question would enter into complexity-estimates). The elementary operations are operations that can be carried out on the numbers transmitted by the edges of the network within a fixed, bounded, amount of time on a RAM computer. As before, a more *detailed* discussion would define an elementary operation to be an elementary boolean operation like AND, OR, or NOT. In *our* (higher level) discussion, elementary operations include $+$, $-$, min, max, $*$, $/$, and many more.

Now we are in a position to state Brent's Theorem:

THEOREM 7.2. Let $N$ be a computational network with $n$ interior nodes and depth $d$, and bounded fan-in. Then the computations performed by $N$ can be carried out by a CREW-PRAM computer with $p$ processors in time $O(\frac{n}{p} + d)$.

The total time depends upon the fan-in of $N$ — we have absorbed this into the constant of proportionality.

PROOF. We simulate the computations of $N$ in a fairly straightforward way. We assume that we have a data-structure in the memory of the PRAM for encoding a vertex of $N$, and that it has a field with a pointer to the vertex that receives the output of its computation if it is an interior vertex. This field is nil if it is an output vertex. We define the depth of a vertex, $v$, in $N$ to be the maximum length of any path from an input vertex to $v$ — clearly this is the greatest distance any amount of input-data has to travel to reach $v$. It is also clear that the depth of $N$ is the maximum of the depths of any of its vertices. We perform the simulation inductively — we simulate all of the computations that take place at vertices of depth $\leq k - 1$ before simulating computations at depth $k$. Suppose that there are $n_i$ interior vertices of $N$ whose depth is precisely $i$. Then $\sum_{i=1}^{d} n_i = n$. After simulating the computations on vertices of depth $k - 1$, we will be in a position to simulate the computations on nodes of depth $k$, since the inputs to these nodes will now be available.

CLAIM 7.3. When performing the computations on nodes of depth $k$, the order of the computations is irrelevant.

This is due to the definition of depth — it implies that the output of any vertex of depth $k$ is input to a vertex of *strictly higher* depth (since depth is the length of the *longest path* from an input vertex to the vertex in question).

The simulation of computations at depth $k$ proceeds as follows:

1. Processors read the data from the output areas of the data-structures for vertices at depth $k - 1$.
2. Processors perform the required computations.

Since there are $n_k$ vertices of depth $k$, and the computations can be performed in any order, the execution-time of this phase is

$$\left\lceil \frac{n_k}{p} \right\rceil \le \frac{n_k}{p} + 1$$

The total execution-time is thus

$$\sum_{i=1}^{d} \left\lceil \frac{n_k}{p} \right\rceil \le \sum_{i=1}^{d} \frac{n_k}{p} + 1 = \frac{n}{p} + d$$

□

We normally apply this to modifying parallel algorithms to use fewer processors —see § 7.2.2 below (page 51). If a computational network has bounded fan-*out* as well as bounded fan-in we get:

COROLLARY 7.4. Let $N$ be a computational network with $n$ interior nodes and depth $d$, and bounded fan-in and fan-out. Then the computations performed by $N$ can be carried out by a CREW-EREW computer with $p$ processors in time $O(\frac{n}{p} + d)$.

Brent's theorem has interesting implications for the question of *work efficiency* of an algorithm.

DEFINITION 7.5. The amount of *work* performed by a parallel algorithm is defined to be the product of the execution time by the number of processors.

This measures the number of distinct computations a parallel algorithm performs.

We can think of a computation network with $n$ vertices as requiring $n$ units of work to perform its associated computation — since there are $n$ distinct computations to be performed. The work required by a simulation of a computation network is (by Brent's Theorem) $O(n + dp)$. This is proportional to the number of vertices in the original computation network if $p$ is proportional to $n/d$.

DEFINITION 7.6. Let $A$ be a parallel algorithm that performs a computation that can be represented by a computation network. Then $A$ will be said to be a *work-efficient* parallel algorithm if it executes in time $O(d)$ using $p$ processors, where $p = n/d$, and $n$ is the number of vertices in a computation network for performing the computations of algorithm $A$ with the smallest possible number of vertices.

Work-efficient parallel algorithms are *optimal*, in some sense.

**Exercises:**

1. Find a computation network for evaluating the expression $(x^2 + 2)(x^3 - 3) - 2x^3$.

2. Show that sorting networks can be expressed in terms of computation networks. This implies that they are a special case of computation networks.

## 7.2. SIMD Algorithms.

7.2.1. *Doubling Algorithms.* The program for adding up $n$ numbers in $O(\lg n)$ time is an example of a general class of parallel algorithms known by several different names:

- Parallel-prefix Operations.
- Doubling Algorithms.

In each case a single operation is applied to a large amount of data in such a way that the amount of relevant data is *halved* in each step. The term "Doubling Algorithms" is somewhat more general than "Parallel Prefix Operations". The latter term is most often used to refer to generalizations of our algorithm for adding $n$ numbers — in which the operation of addition is replaced by an arbitrary associative operation. See §1 in chapter VI for several examples of how this can be done.

Another variation on this theme is a class of techniques used in Graph Algorithms called "Pointer Jumping". This involves assigning a processor to each node of a *directed graph* and:

1. Having each processor determine the successor of its successor;
2. Causing the processors to effectively regard the nodes computed in step 1 as their *new* successors;
3. Go to step 1.

In this way, the *end* of a directed path of length $n$ can be found in $O(\lg n)$ steps. This is a special case of parallel prefix operations, since we can regard the edges of the graph as the objects under study, and regard the operation of combining two edges into "one long edge" (this is basically what the pointer jumping operation amounts to) as an *associative operation*. See § 2 for examples of algorithms of this type (and definitions of the terminology used in graph theory).

7.2.2. *The Brent Scheduling Principle.* One other general principle in the design of parallel algorithm is the *Brent Scheduling Principle*. It is a very simple and ingenious idea first described by R. Brent in [20], that often makes it possible to reduce the number of processors used in parallel algorithms, without increasing the asymptotic execution time. In general, the execution time increases somewhat when the number of processors is reduced, but not by an amount that increases the *asymptotic* time. In other words, if an algorithm has an execution time of $O(\lg^k n)$, then the execution-time might increase by a constant factor. In order to state this result we must recall the concept of computation network, defined in 7.1 on page 48.

COROLLARY 7.7. Suppose algorithm $A$ has the property that its computations can be expressed in terms of a computation network with $x$ vertices and depth $t$ that has bounded fan-in. Then algorithm $A$ can be executed on a CREW-PRAM computer with $p$ processors in time $O(\frac{x}{p} + t)$.

The proof is a direct application of Brent's Theorem (7.2 on page 49).

See page 307 for some applications of this principle.

This result has some interesting consequences regarding the relationship between data-representation and execution-time of an algorithm. Consider the algorithm for adding up numbers presented on page 9. Since the data is given in an array, we can put it into any computation network we want — for instance, the one in figure I.8 on page 9. Consequently, the Brent Scheduling Principle states that the algorithm on page 9 can be executed with $\lceil n/\lg(n) \rceil$ processors with no asymptotic degradation in execution time (i.e., the execution time is still $O(\lg n)$).

If the input-data was presented to us as elements of a linked list, however, it is not clear how we could apply the Brent Scheduling Principle to this problem. The linked list can be regarded as a computation network of depth $n$, so Brent's Theorem would imply an execution time of $O(n)$. We can actually get an execution time of $O(\lg n)$ by using the technique of pointer jumping in § 7.2.1 above, but this actually requires $n$ processors. The parallel algorithms for list-ranking in this case are more complicated than straightforward pointer-jumping — see [7]. Also see chapter VII for a simple *probabilistic* algorithm for this problem.

7.2.3. *Pipelining.* This is another technique used in parallel algorithm design. Pipelining can be used in situations where we want to perform several operations in sequence $\{P_1, \ldots, P_n\}$, where these operations have the property that some steps of $P_{i+1}$ can be carried out before operation $P_i$ is finished. In a parallel algorithm, it is often possible to overlap these steps and decrease total execution-time. Although this technique is most often used in MIMD algorithms, many SIMD algorithms are also able to take advantage of it. Several algorithms in this book illustrate this principle:

- The Shiloach-Vishkin algorithm for connected components of a graph (see page 333). Pipelining in this case is partly responsible for reducing the execution-time of this algorithm from $O(\lg^2 n)$ to $O(\lg n)$.
- The Cole sorting algorithm (see page 385). This is, perhaps, the most striking example of pipelining in this book. This sorting algorithm is based upon ideas like those in 3.7 on page 26, but ingeniously "choreographs" the steps in such a way that the execution-time is reduced from $O(\lg^2 n)$ to $O(\lg n)$.

**7.2.4.** *Divide and Conquer.* This is the technique of splitting a problem into small independent components and solving them in parallel. There are many examples of this technique in this text:

- The FFT algorithm (at least if we consider its recursive definition) in § 2.3 of chapter V (page 205);
- The parallel prefix, or doubling algorithms of § 1 in chapter VI (page 305);
- All of the algorithms for connected components and minimal spanning trees in § 2 of chapter VI(page 313);
- The Ajtai, Komlós, Szemerédi sorting algorithm in § 4.3 of chapter VI (page 402);

The reader will doubtless be able to find many other examples in this book.

**7.3. MIMD Algorithms.**

**7.3.1.** *Generalities.* this section is longer than the corresponding section on SIMD algorithm design because many of the important issues in SIMD algorithm design are dealt with throughout this book, and frequently depend upon the problem being studied. The issues discussed in this section are fairly constant throughout all MIMD algorithms. As is shown in §7.4 on page 62, design of a MIMD algorithm can sometimes entail

- the design of a good SIMD algorithm,
- conversion to a MIMD algorithm

The second step requires the computations to be *synchronized*, and this involves using the material in this section.

The issues that arise in the design of MIMD algorithm are essentially identical to those that occur in *concurrent programming*. The problems that occur and their solutions are basically the same in both cases. The main difference between MIMD algorithm design and concurrent algorithm design involves questions of *when* and *why* one creates multiple processes:

1. When designing concurrent algorithms to run on a single-processor computer, we usually create processes in order to handle *asynchronous events* in situations in which we expect little *real* concurrency to occur. The generic example of this is waiting for I/O operations to complete. Here we expect one process to be dormant most of the time, but are unable to accurately predict when it will be dormant. We usually avoid creating processes to handle computations or other operations that will be truly concurrent, because there is no *real* concurrency on a single-processor machine — it is only simulated, and this simulation has an associated cost.

2. When designing MIMD algorithms to run on a parallel computer, we try to maximize the amount of concurrency. We look for operations that can be

carried out simultaneously, and try to create multiple processes to handle them. Some of the considerations in writing concurrent programs still apply here. Generally, it is not advisable to create many more processes than there are *processors* to execute them. The overhead involved in creating processes may be fairly large. This is the problem of *grain-size*. This is in contrast to SIMD algorithms, in which there is little overhead in creating parallel threads of data and computation.

We will discuss a few of the very basic issues involved in concurrent and MIMD algorithm design. The most basic issue is that we should analyze the computations to be performed and locate parallelism. This can be done with a *dependency graph*. Page 9 gives two examples of syntax trees. These are like dependency graphs for arithmetic operations. The cost of creating processes on most MIMD computers almost always makes it necessary to work on a much coarser scale.

We generally take a high-level description of the computations to be performed, and make a directed graph whose nodes represent discrete blocks of independent operations. The edges represent situations in which one block of operations depends upon the outcome of performing other blocks. After this has been done we can design the MIMD algorithm by:

1. Creating one process for each *node* of the graph, and make the processes wait for the completion of other processes upon which they are dependent.
2. Creating one process for each directed path through the dependency graph, from its starting point to its end.

7.3.2. *Race-conditions.* However we do this, we encounter a number of important issues at this point. If two processes try to access the same shared data, they may interfere with each other:

> Suppose two processes update a shared linked list *simultaneously* — the head of the list is pointed to by a pointer-variable named **head**, and each entry has a **next** pointer to the next entry.

Process A wants to *add* a record to the beginning of the list by:

**A.1** making the new record's **next** pointer equal to the head **pointer** (so it points to the same target);

**A.2** making the **head** pointer point to the new record;

Process B wants to *delete* the first record by:

**B.1** Making the **head** pointer equal to the **next** pointer of its target;

**B.2** Deleting the record that was originally the target of the head pointer;

Through *unlucky timing*, these operations could be carried out in the sequence A.1, B.1, A.2, B.2. The result would be that the head would point to the new record added by process A, but the next pointer of that record would point to the record deleted by process B. The rest of the list would be *completely inaccessible*. This is an example of a *race-condition* — the two processes are in a "race" with each other and the outcome of the computations depend crucially on which process reaches certain points in the code first.

Here is a program in C for the Sequent Symmetry computer that illustrates race-conditions:

```c
#include <stdio.h>

/* The next two include files refer to system libraries for
 * sharing memory and creating processes. */

#include <parallel/microtask.h>
#include <parallel/parallel.h>
shared int  data;
void dispatch ();
void child1 ();
void child2 ();
void main ()
{
   m_set_procs (2);  /* Causes future calls to 'm_fork' to
                      * spawn two processes */
   m_fork (dispatch);/* This call to 'm_fork' spawn two processes
              * each of which, is a contains a copy of the
              *routine 'dispatch' */
   exit (0);
}
void dispatch ()     /* This routine is executed in two
              * concurrent copies. */
{
   int    i,
        j;
   int    p = m_get_myid ();  /* Each of these copies determines
                      * its identity (i.e., whether it is
                      * process number 0 or 1) */
   if (p == 0)
     child1 ();
   else
     child2 ();
}
```

```
void child1 ()        /* 'child1' contains the actual
                      * code to be executed by process 0. */
{
   int    i,
          j;
   for (i = 0; i < 10; i++)
      {
        data = 0;
        for (j = 0; j < 500; j++);
           printf ("Child 1, data=%d\n", data);
      }
}
void child2 ()        /* 'child2' contains the actual
                      * code to be executed by process 1. */
{
   int    i,
          j;
   for (i = 0; i < 10; i++)
      {
        data++;
        for (j = 0; j < 500; j++);
           printf ("Child 2, data=%d\n", data);
      }
}
```

Here two processes are generated, called child1 and child2. Since **mfork** normally generates many processes at once, we have to make it spawn both child-processes in a single statement. This is done by making a routine named dispatch be the child-process. Each copy of this routine calls m_get_myid to determine its identity and call child1 or child2 depending on the result. Note that child1 *zeroes* a shared data item named data, and child2 *increments* it. They both then wait a short time and print the value out. Since data is shared though, it is possible that while one process is waiting to print out the data, the other process can *sneak* in and change it's value. This actually happens, as you can see if you run this program:

```
cc name.c -lpps
a.out
```

The results are unpredictable (this is usually true with race-conditions) — you will probably never get the same set of results twice. Most of the time, however, child1 will occasionally print out values other than 0, and child2 will sometimes print 0.

This type of problem is solved in several ways:

a. One involves *locking operations* that prevent more than one process from accessing shared data (access to data that is exclusive to a single process is called atomic). Essentially, the first process that calls the **m_lock** system-function continues to execute and any other process that calls this function is suspended until the first process calls **m_unlock**. If two processes call **m_lock** simultaneously, the system makes a decision as to which gets priority.

Here is how the program above looks when semaphores are used to prevent race-condition:

```
#include <stdio.h>
#include <parallel/microtask.h>
#include <parallel/parallel.h>
shared int  data;
void child ();
void child1 ();
void child2 ();
void main ()
{
   m_set_procs (2);
   m_fork (child);
   exit (0);
}
void child ()
{
   int    i,
          j;
   int    p = m_get_myid ();
   if (p == 0)
      child1 ();
   else
      child2 ();
}
void child1 ()
{
   int    i,
          j;
   for (i = 0; i < 10; i++)
      {
      m_lock ();
      data = 0;
```

```
    for (j = 0; j < 500; j++);
        printf ("Child 1, data=%d\n", data);
    m_unlock ();
    }
}
void child2 ()
{
  int    i,
       j;
  for (i = 0; i < 10; i++)
    {
      m_lock ();
      data++;
      for (j = 0; j < 500; j++);
          printf ("Child 2, data=%d\n", data);
      m_unlock ();
    }
}
```

The functions **m_lock** and **m_unlock()** are system calls available on the Sequent line of parallel computers.

The standard term (i.e. the term you will see most often in the literature) for a locking operation (in the theory of concurrent programming) is a *semaphore*. The lock and unlock-operations are called *semaphore down* and *semaphore up* operations.

One characteristic of processes that are under the control of a lock (or semaphore) is that the amount of speedup that is possible due to parallel processing is *limited*. This is due to the fact that the semaphore *forces* certain sections of code to be executed *sequentially*. In fact:

LEMMA 7.8. Suppose the optimal sequential algorithm for performing a computation requires time $T$, and accessing a semaphore requires time $k$. Then an optimal parallel version of this computation using processes under the control of a single semaphore requires an execution time of at least $O(\sqrt{T/k})$.

PROOF. If we use $m$ processors the execution time must be at least $T/m$ (see 1.1 in chapter II). On the other hand, since the semaphore-operations are executed sequentially, they will require an execution time of $km$ — i.e. the time required to carry out the semaphore-operations increases with the number of processors. The total execution time will be $\geq (T/m + km)$. The value of $m$ that *minimizes* this occurs when the derivative of this expression with respect to $m$ vanishes. This means that

$$-\frac{T}{m^2} + k = 0$$

This occurs when $m = \sqrt{T/k}$. $\square$

It is interesting to note that this result makes essential use of the fact that there is a *single* semaphore involved — and access of this semaphore by $n$ processes requires a time of $kn$. Recent unpublished results of David Saunders shows that it is possible to set up a kind of *tree* of semaphores that will permit the synchronization of $n$ processes in time that is $O(\lg n)$.

b. Another solution to this problem involves synchronizing the parallel processes. One common notation for this construct is:

cobegin

coend;

The idea here is that all processes execute the **cobegin** statement simultaneously and remain synchronized until the coend statement is reached. This solves the problem of processes interfering with each other when they access shared data by allowing the programmer to "choreograph" this common access. For instance, in the sorting algorithm on the Butterfly Computer, no semaphores were used, but processes never interfered with each other. The DYNIX operating system provides the **m_sync()** system call to implement cobegin. When a process calls **m_sync()** it *spins* (i.e. loops) until all processes call **m_sync()** — then all processes execute. The operating system uses a process scheduling algorithm that causes child processes to execute to completion without interruption (except by higher-priority processes). Consequently, once processes have been synchronized, they remain in sync if they execute the same instructions.

See § 1.1 (page 109) for information on another interesting paradigm for synchronizing parallel processes.

7.3.3. *Optimization of loops.* The theory of semaphores give rise to a number of issues connected with parallelizing *loops* in an algorithm. Suppose we have an algorithm that requires a number of computations for be performed in a loop, with very little dependency between different iterations of the loop. We assume that the loop has been divided up between several parallel processes — each process will execute a few iterations of the loop. Data that the loop references may be divided up into several categories:

1. *Local data.* This is data that is not shared — it is declared locally in each process. There is no need to use a semaphore in accessing this data.

2. *Read-only shared data.* This is data that is only read by different processes. There is no need to use a semaphore or lock to control access to it.

3. *Reduction data*. This is shared data that read and written by each process, but in a limited way. The data is used in a single associative commutative operation by each iteration of the loop and always, read and then written. Although it is shared, we do not need to use a semaphore every time we access such data. Since it is used in an associative commutative operation, the order in which the operations are applied is not significant. We can replace this reference variable in the loop by a local variable in each process. Only when the data from different processes is being combined need we use a semaphore to prevent simultaneous access. This saves a little execution time if each process is performing many iterations of the loop, because a semaphore inhibits parallelism. Here is an example in C — again this program is for the Sequent Symmetry computer:

```
for (i=0; i < 1000;i++)
for (j=0;j < 1000;j++) sum=sum+a[i][j];
```

Here 'sum' is a *reduction variable*. Suppose each processor performs all iterations of the loop on 'j'. Then we could replace this nested loop by:

```
for (i=0; i<1000;i++)
{
    int local_sum=0;
    for(j=0;j<1000;j++)
    {
      local_sum=local_sum+a[i][j];
    }
    m_lock();  /* Set semaphore. */
    sum=sum+local_sum; /* Now accumulate values
                        computed by different processors.*/
    m_unlock();      /* Release semaphore. */
}
```

4. *Ordered data*. This is shared data whose final numerical value depends upon the iterations of the loop being carried out in *precisely the same order* as in a *sequential execution* of the loop. A loop containing such data is *not suitable* for parallelization (at least not in an asynchronous program). There are situations in which such a loop might be contained in a much larger block of code that *does* lend itself to parallelization. In this case we must guarantee that the loop is question is executed *sequentially* (even if execution of different parts of the loop is done on different processors). There are several ways this can be done: a. we can isolate this code in a procedure and allow only *one processor* to *call* the procedure. b. We can *share* the variable that describes the index of the loop (i.e. iteration-count) and make each processor *wait* for that to reach

an appropriate value.

Alternative a is probably the more *structured* solution to this problem.

5. Shared variables that are read and written, but for which the order of execution of the iterations of the loop is not significant. Such variables must be locked via a semaphore before they can be accessed.

**7.3.4. Deadlocks.** The last general issue we will discuss is that of *deadlock conditions*. Suppose we have two processes that each try to access two data-items. Since we are aware of the problems that can arise involving race-conditions, we use semaphores to control access to the data-items. Now suppose, for some reason (unlucky choices or timing), the first process locks up the first data-item at the same time that the second process locks up the second. Now both processes try to lock the other data-item. Since they can't complete their computations (and release the data they have locked) until they get *both* data-items, they both wait *forever*. This is known as a deadlock condition.

> Five philosopher's sit at a round table with a huge plate of spaghetti in the center. There are five forks on the table — they lie between the philosophers. Each philosopher alternates between meditating and eating, and a philosopher needs two forks in order to eat. The philosophers are very egotistical, so no philosopher will relinquish a fork once they have picked it up until they finish eating[7].

The classic problem that illustrates the issue of deadlocks is the Dining Philosopher's Problem, described by Dijkstra in [46].

Deadlocks can only occur if the following conditions are satisfied:

1. Processes can request (and lock) only part of the resources they need.
2. Processes can never relinquish resources they have requested until their computations are completed.
3. Processes cannot take resources away from other processes.
4. A circular chain of requests for resources can exist. Each process in the chain requests two or more resources and at least one of these is also requested by the next process in the chain.

We prevent deadlock by eliminating at least one of these conditions. It is generally impractical to try to eliminate conditions 2 and 3 , but the other two conditions can be eliminated.

- We can prevent condition 1 from being satisfied by implementing semaphore *sets*. These are sets of semaphores with semaphore-down operations that apply atomically to the entire set. When a process performs a semaphore-down operation on the set, it is suspended if any of the semaphores in the set is 0.

---

[7]The reader may have noticed a few puns in this example!

In this case none of the semaphores is lowered. In the case where all of the semaphores in the set are 1, they are all lowered simultaneously. In the context of the Dining Philosopher's Problem, this is as if the philosophers could grab both of the forks at the same instant, so they either get both forks, or they get nothing. The ATT System V implementation of UNIX has such semaphore set operations.

- We can prevent condition 4 in several ways:
    - Careful algorithm design.
    - Use of resources in a fixed order.

**7.4. Comparison of the SIMD and MIMD models of computation.** As the title of this section suggests, we will prove that these two very general models of computation are essentially equivalent. They are equivalent in the sense that, with some restrictions, any algorithm that executes in $T$ time units on one type of computer can be made to execute in $kT$ time units on the other, where $k$ is some constant. Before the reader concludes that this means the type of parallel computer one uses is unimportant, it is necessary to point out that the constant $k$ may turn out to be very large. Many problems will naturally lend themselves to a SIMD or MIMD implementation, and any other implementation may turn out to be substantially slower. First we need the following definition:

DEFINITION 7.9. An algorithm for a SIMD parallel computer will be called *calibrated*, if whenever processors access memory, they also compute the program-step in which the memory was last written. This means that there is a function $f: P \times T \times M \to T$, where

1. $P$ is the set of processors in the SIMD computer.
2. $T$ is the set of possible time-steps — a range of integers.
3. $M$ is the set of memory-locations.

In addition, it means that the algorithm effectively computes this function $f$ in the course of its execution.

Many *highly regular* algorithms for SIMD computers have this property. For instance, an algorithm that accesses *all* of memory in each program step can be easily converted into a calibrated algorithm.

ALGORITHM 7.10. Suppose $A$ is a calibrated algorithm that runs in $T$ time units on a SIMD-EREW parallel computer with $n$ processors and uses $m$ memory locations. Then it is possible to execute this algorithm on a MIMD-EREW computer with $n$ processors in $kT$ time units, using $mT$ distinct semaphores, where $k$ is the number of instruction-steps required to:

1. Check a semaphore;
2. Suspend a process;

3. Awaken a suspended process;

This result suggests that a MIMD computer is strictly better than a SIMD computer, in the sense that

- It looks as though MIMD computers can execute any program that runs on a SIMD computer.
- A MIMD computer can also run programs that require asynchronous processes.

The "catch" here is that:

1. Many SIMD algorithms are not calibrated, and there is a very significant cost associated with converting them into calibrated algorithms.
2. Most MIMD computers today (1992) have far fewer processors than SIMD computers;
3. The constant $k$ may turn out to be very large.

PROOF. The basic idea here is that race-conditions will not occur, due to the fact that the SIMD algorithm was designed to execute on a EREW computer. Race conditions only occur when multiple processors try to *read and write* to the same location in a given program step. The only problem with carrying out the simulation in an entirely straightforward way is that of synchronizing the processors. This is easily handled by using the fact that $A$ is calibrated. Simply associate a $time - stamp$ with each data-item being computed. Each processor of the MIMD computer executes instructions of the SIMD program and maintains a program-counter. We attach a single semaphore to each simulated SIMD-memory location, and for each simulated time-step. This gives a total of $mT$ semaphores, and they are all initially down except the ones for all of the processors at time 0. When a processor is about to read the data that it needs for a given program-step, it checks the semaphore for that data-item at the required time. When a processor completes a computation in a given simulated time-step, it executes an *up* operation on the corresponding semaphore.

We must prove that the execution-time of the simulated algorithm is as stated. We use induction on the number of program-steps in the SIMD algorithm. Certainly the conclusion is true in the ground-case of the induction. Suppose that it is true after $t$ simulated program-steps. This means that all processors of the MIMD machine have simulated $t$ program steps of the SIMD algorithm after $kt$ time-units have elapsed. All processors are ready to begin simulating at least the $t + 1^{st}$ program-step at this point. If any processors require data *from* the $t^{th}$ program step, they must access a semaphore that is attached to that data. Consequently, the elapsed time may be $k$ before the algorithm can simulate the next SIMD program-step. □

The results of § 5, particularly 5.1 on page 29 imply:

COROLLARY 7.11. Suppose $A$ is a calibrated algorithm that runs in $T$ time units on a SIMD-CRCW parallel computer with $n$ processors. Then it is possible to execute this algorithm on a MIMD-EREW computer with $n$ processors in $kT \lg^2 n$ time units, where $k$ is the number of instruction-steps required to:

1. Check a semaphore;
2. Suspend a process;
3. Awaken a suspended process;

In more generality, we have:

ALGORITHM 7.12. Suppose $A$ is an algorithm that runs in $T$ time units on a SIMD-EREW parallel computer with $n$ processors. Then it is possible to execute this algorithm on a MIMD-EREW computer with $3n$ processors in $kT \lg n$ time units, where $k$ is a constant.

This algorithm eliminates the requirement that the SIMD algorithm be calibrated. It is based upon the tree-of-semaphores construction of David Saunders (see page 59). Essentially, we insert a *barrier* construct after each simulated SIMD instruction. We set up a tree of semaphores and, when processors finish a given simulated SIMD instruction-step (say instruction $j$) they *wait* until *all* processors complete this instruction-step. Then they begin their simulation of the next SIMD instruction-step.

In like fashion, it is possible to simulate MIMD algorithms on a SIMD machine. Suppose we have a MIMD algorithm that makes use of an instruction-set with $I$ instructions $\{a_1, \ldots, a_I\}$. Suppose that a SIMD computer can simulate $a_j$ in time $t_j$. Then we get:

ALGORITHM 7.13. Let $A$ be an algorithm that runs in $T$ time units on the MIMD computer described above. Then $A$ can be run in $T \sum_{j=1}^{I} t_j$ time units on the SIMD computer described above.

The idea of this simulation is extremely simple. Each program-step of the MIMD computer is simulated on the SIMD computer by a loop with $I$ iterations. In iteration $j$ all processors that should be executing instruction $a_j$ run a simulation of this instruction — this requires time $t_j$.

Note that simulations of this kind are extremely slow unless the simulated instruction-set is made as small as possible. Mr. Jeffrey Salvage is writing such a simulator to run on a Connection Machine (CM-2) computer[8].

The question of how one can simulate a MIMD machine by a SIMD machine has also been considered by M. Wloka in his Doctoral Dissertation ([164]) and by Michael Littman and Christopher Metcalf in [100].

**Exercises:**

---

[8]This is his Master's Thesis at Drexel University

1. Why doesn't the simulation algorithm 7.10 run up against the limitations implied by 7.8 on page 58?
2. Is it possible for the MIMD simulation in 7.10 to run *faster* than the original SIMD algorithm being simulated? Assume that the processors of the MIMD computer run at the same rate as those of the SIMD computer, and that the operations of checking semaphores take negligible time (so the constant $k$ is 0).

# III
# Distributed-Memory Models

## 1. Introduction.

The PRAM models of computation requires that many processors access the same memory locations in the same program-steps. This creates engineering problems that have only been solved in a few cases. Most practical parallel computers are built along a *Distributed Memory Model* of some kind. In a distributed-memory architecture, the processors of the computer are interconnected in a *communication-network* and the RAM of the computer is *local* to the processors. Each processor can access its own RAM easily and quickly. If it needs access to a larger address-space than is contained in its local RAM, it must communicate with *other processors* over the network

This leads to several interesting questions:

1. How do the various interconnection-schemes compare with each other in regard to the kinds of algorithms that can be developed for them?
2. How easy is it for processors to "share" memory over the communication network? This is related to the question of how easy it might be to port PRAM-algorithms to a network-computer.

It turns out that the answer to question 1 is that "almost any network will do". Work of Vishkin and others shows that algorithms that are fast most of the time (i.e. probabilistic algorithms) can be developed for any network in which it is possible to reach $2^k$ other processors from a given processor in $k$ steps. Question 2 has a similar answer — there exist efficient algorithms for *simulating* a PRAM-computer via a network-computer, if the network satisfies the condition mentioned above.

## 2. Generic Parallel Algorithms.

We will consider how several different networks handle many common algorithms. In order to do this, we follow Preparata and Vuillemin in [125] in defining a pair of *generic* parallel algorithms that can be easily implemented on the common network-computers. Many interesting parallel algorithms can be expressed in terms of these two.

We will assume that we have $n = 2^k$ data items stored in storage locations $T[0]$, $T[1],\ldots,T[n-1]$.

The notation $\text{OPER}(m, j; U, V)$ will denote some operation that modifies the data present in locations $U$ and $V$, and depends upon the parameters $m$ and $j$, where $0 \leq m < n$ and $0 \leq j < k$. We will also define the function $\text{bit}_j(m)$ to be the $j^{\text{th}}$ bit in the binary representation of the number $m$. Given these definitions, we will say that an algorithm is in the *DESCEND* class if it is of the form:

ALGORITHM 2.1. **DESCEND Algorithm.**

**for** $j \leftarrow k - 1$ **downto** 0 **do**
    **for** each $m$ such that $0 \leq m < n$
        **do in parallel**
            **if** $\text{bit}_j(m) = 0$ **then**
                $\text{OPER}(m, j; T[m], T[m + 2^j])$
    **endfor**
**endfor**

and we will say that an algorithm is of *ASCEND* class if it is a special case of

ALGORITHM 2.2. **ASCEND Algorithm.**

**for** $j \leftarrow 0$ **to** $k - 1$ **do**
    **for** each $m$ such that $0 \leq m < n$
        **do in parallel**
            **if** $\text{bit}_j(m) = 0$ **then**
                $\text{OPER}(m, j; T[m], T[m + 2^j])$
    **endfor**
**endfor**

In many cases, algorithms that do not entirely fall into these two categories can be decomposed into a sequence of algorithms that do — we will call these algorithms *composite*. We will often want to regard $\text{OPER}(m, j; T[m], T[m + 2^j])$ as a *pair* pf functions $(f_1, f_2)$:

$$T[m] \leftarrow f_1(m, j, T[m], T[m + 2^j])$$
$$T[m + 2^j] \leftarrow f_2(m, j, T[m], T[m + 2^j])$$

We will consider some common parallel algorithms that can be re-stated in the context of ASCEND and DESCEND algorithms. Throughout the remainder of the text, we will occasionally encounter more of these.

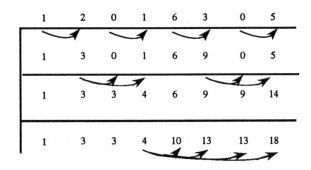

FIGURE III.1. Computing the cumulative sums of 8 integers, in parallel

EXAMPLE 2.3. The Bitonic Sorting algorithm (3.4 on page 24) is a DESCEND algorithm — here the operation $\mathrm{OPER}(m, j; U, V)$ is just compare-and-swap (if the elements are out of sequence).

Here is a refinement of the algorithm for forming the sum of $n = 2^k$ numbers given on page 9 on the Introduction: We will not only want the sum of the $n = 2^k$ numbers — we will also want all of the *cumulative partial sums*. Suppose we have $n$ processors and they are numbered from 0 to $n - 1$ and $n$ numbers in an array, $A$. In step $i$ (where steps are also numbered from 0 on) let processor number $u$, where $j2^{i+1} + 2^i \leq u \leq (j+1)2^{i+1} - 1$ (here $j$ runs through all possible values giving a result between 1 and $n$) add the value of $A(j2^{i+1} + 2^i - 1)$ to $A(u)$. Figure III.1 illustrates this process.

It is easy for a processor to determine if its number is of the correct form — if its identifying number is encoded in binary it can compute its value of $j$ by right-shifting that number $i + 1$ positions, adding 1 and right-shifting 1 additional position. Then it can compute the corresponding value of $j2^{i+1} + 2^i - 1$ and carry out the addition. It isn't difficult to see that this procedure will terminate in $\lg n$ steps and at the end $A$ will contain the cumulative sums of the original numbers. A single extra step will compute the absolute values of the differences of all of the partial sums from half of the sum of all of the numbers and the minimum of the differences is easily determined.

In step $i$ processors numbered $j$ compare the numbers in $A(j2^i - 2^{i-1})$ with those in $A(j2^i)$ and exchange them if the former is smaller than the latter. In addition subscript numbers (which are stored in another array of $n$ elements) are exchanged. Clearly, this determination of the minimum requires $\lg n$ steps also. Later, we will examine an improvement of this algorithm that only requires $n / \lg n$ processors — see 1.6 on page 308.

EXAMPLE 2.4. The algorithm for forming a cumulative sum of $n$ numbers is an AS-

CEND algorithm where $\mathrm{OPER}(m, j; U, V)$ has the following effect:

$$U \leftarrow U$$
$$V \leftarrow U + V$$

These two examples are fairly clear — the original statements of the algorithms in question were almost exactly the same the the statements in terms of the ASCEND and DESCEND algorithms.

Now we will consider a few less-obvious examples:

PROPOSITION 2.5. Suppose we have $n = 2^k$ input data items. Then the permutation that exactly reverses the order of input data can be expressed as a DESCEND algorithm — here $\mathrm{OPER}(m, j; U, V)$ simply *interchanges* $U$ and $V$ in all cases (i.e., independently of the values of $U$ and $V$).

PROOF. We prove this by induction. It is clear in the case where $k = 1$. Now suppose the conclusion is known for all values of $k \leq t$. We will prove it for $k = t + 1$. The key is to assume that the original input-data was $T[i] = i$, $0 \leq i \leq n - 1$. If we prove the conclusion in this case, we will have proved it in all cases, since the numerical values of the data-items never enter into the kinds of permutations that are performed by $\mathrm{OPER}(m, j; U, V)$. The next important step is to consider the binary representations of all of the numbers involved. It is not hard to see that the first step of the DESCEND algorithm will alter the high-order bit of the input-data so that the high-order bit of $T[i]$ is the same as that of $n - 1 - i$. The remaining bits will not be effected since:

Bits 0 through $t$ of the numbers $\{0, \ldots, n/2 - 1\}$ are the same as bits 0 through $t$ of the numbers $\{n/2, \ldots, n - 1\}$, respectively.

The remaining iterations of the DESCEND algorithm correspond exactly to the algorithm in the case where $k = t$, applied independently to the lower and upper halves of the input-sequence. By the inductive hypothesis, these iterations of the algorithm have the correct effect on bits 0 through $t$ of the data. It follows that, after the algorithm executes, we will have $T[i] = n - 1 - i$. $\square$

For instance:

EXAMPLE 2.6. Suppose $n = 2^3$ and our input-data is:

$$\{3, 7, 2, 6, 1, 8, 0, 5\}$$

After the first iteration of the DESCEND algorithm (with $\mathrm{OPER}(m, j; U, V)$ defined to always interchange $U$ and $V$), we have:

$$\{1, 8, 0, 5, 3, 7, 2, 6\}$$

After the second iteration, we get:

$$\{0, 5, 1, 8, 2, 6, 3, 7\}$$

And after the final iterations we get:

$$\{5, 0, 8, 1, 6, 2, 7, 3\}$$

This is the reversal of the input sequence.

The reason we took the trouble to prove this is that it immediately implies that:

PROPOSITION 2.7. The Batcher merge algorithm (3.6 on page 26) can be expressed as a composite of two DESCEND algorithms. The first phase reverses the upper half of the input-data via 2.5 above, and the second performs a Bitonic sort using 2.3.

This implies that the general Batcher sorting algorithm can be implemented with $O(\lg n)$ DESCEND algorithms via the reasoning of 3.7 on page 26.

We will conclude this section by showing that the operation of shifting data can be implemented as an ASCEND algorithm:

PROPOSITION 2.8. Suppose we have $n = 2^k$ input data-items. Then the cyclic shift operation

$$T[i] \leftarrow T[i-1]$$
$$T[0] \leftarrow T[n-1]$$

for all $i$ such that $0 \leq i \leq n-1$, occurs as the result of an ASCEND algorithm with $OPER(m, j; U, V)$ defined to

- Interchange $U$ and $V$, if $m$ is a multiple of $2^{j+1}$.
- Leave $U$ and $V$ unchanged otherwise.

PROOF. We use induction on $k$. Clearly, the conclusion is true for $k = 1$ — the algorithm just interchanges the two input data items. As with the previous algorithm, we will assume the input is given by $T[i] = i$, and will assume that the algorithm works for $k \leq t$. Now suppose we are given a sequence of $2^{t+1}$ data items:

$$\{0, 1, \ldots, 2^{t+1}\}$$

The inductive hypothesis implies that the first $t$ iterations of the algorithm will produce:

$$\{T[0] = 2^t - 1, T[1] = 0, \ldots, T[2^t - 1] = 2^t - 2,$$
$$T[2^t] = 2^{t+1} - 1, T[2^t + 1] = 2^t, \ldots, T[2^{t+1} - 1] = 2^{t+1} - 2\}$$

The last iteration interchanges $T[0]$ and $T[2^t]$ so we get the sequence

$$\{T[0] = 2^{t+1} - 1, T[1] = 0, \ldots, T[2^{t+1} - 1] = 2^{t+1} - 2\}$$

□

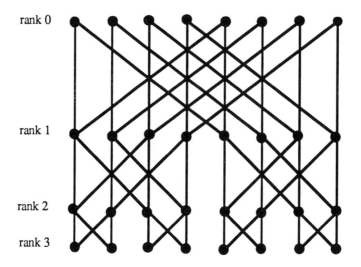

FIGURE III.2. The butterfly architecture

### 3. The Butterfly Network.

Naturally, some networks are better than others for developing parallel algorithms. Essentially, they have structural properties that make it easier to describe the data-movement operations necessary for parallel computations. We will consider several such architectures in this chapter. In every case, physical computers have been constructed that implement the given architecture.

We first consider the *butterfly layout* — see figure III.2 for a diagram that illustrates the basic idea. The nodes represent processors and the lines represent communication links between the processors.

Here the nodes of this graph represent processors (each of which has some local memory) and the edges are communication lines. This particular layout is of rank 3. The general rank $k$ butterfly layout has $(k + 1)2^k$ processors arranged in $2^k$ columns with $k + 1$ processors in each column. The processors in a column are numbered from 0 to $k$ (this number is called the *rank* of the processor) — these processors are denoted $d_{r,j}$, where $r$ is the rank and $j$ is the column number. Processor $d_{r,j}$ is connected to processors $d_{r-1,j}$ and $d_{r-1,j'}$, where $j'$ is the number whose $k$-bit binary representation is the same as that of $j$ except for the $r - 1^{\text{st}}$ bit from the *left*.

In some cases the $0^{\text{th}}$ and $k^{\text{th}}$ (last) ranks are *identified* so that every processor has exactly 4 communication lines going out of it.

The fundamental properties of butterfly networks that interest us are the following:

3.1.     1. If the rank-0 processors (with all of their communication arcs) of a butterfly network of rank $k$ are deleted we get two butterfly networks of rank $k - 1$.

rank 0

rank 1

rank 2

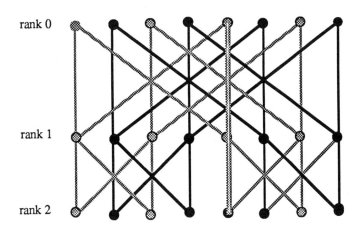

FIGURE III.3. The Even Sub-Butterfly

2. If the rank-$k$ processors (with all of their communication arcs) of a butterfly network of rank $k$ are deleted we get two interwoven butterfly networks of rank $k - 1$.

Statement 1 is immediately clear from figure III.2 — in this case the ranks of the remaining processors have to be changed. Statement 2 is not hard to see from figure III.3.

Here the even columns, lightly shaded, form one butterfly network of rank 2, the the odd columns form another.

The organization of the diagonal connections on the butterfly networks makes it easy to implement the generic ASCEND and DESCEND algorithms on it:

ALGORITHM 3.2. **DESCEND Algorithm on the Butterfly computer.** Suppose the input data is $T[0], \ldots, T[n-1]$, with $n = 2^k$ and we have a rank-$k$ butterfly network. Start with $T[i]$ in processor $d_{0,i}$, for $0 \le i \le n-1$ — the top of the butterfly diagram. In each phase of the Butterfly version of the DESCEND algorithm (2.1 on page 68) perform the following operations:

**for** $j \leftarrow k-1$ **downto** 0 **do**
    Transmit data from rank $k-1-j$ to rank $k-j$
    along both vertical and diagonal lines
        The processors in rank $r+1$ and columns
        $m$ and $m + 2^j$ now contain the old values
            of $T[m], T[m+2^j]$.
        **Compute** $\mathrm{OPER}(m, j; T[m], T[m+2^j])$
**endfor**

The ASCEND algorithm is very similar:

ALGORITHM 3.3. **ASCEND Algorithm on the Butterfly computer.** Suppose the input data is $T[0], \ldots, T[n-1]$, with $n = 2^k$ and we have a rank-$k$ butterfly network. Start with $T[i]$ in processor $d_{k,i}$, for $0 \le i \le n-1$ — the bottom of the butterfly diagram. In each phase of the Butterfly version of the ASCEND algorithm (2.1 on page 68) perform the following operations:

**for** $j \leftarrow 0$ to $k-1$ **do**
    Transmit data from rank $k-j$ to rank $k-j-1$
    along both vertical and diagonal lines
        The processors in rank $r+1$ and columns
        $m$ and $m + 2^j$ now contain the old values
            of $T[m], T[m + 2^j]$.
        **Compute** $\text{OPER}(m, j; T[m], T[m + 2^j])$
**endfor**

Note that the execution-times of the original ASCEND and DESCEND algorithms have at mostly been multiplied by a constant factor.

The results of the previous section immediately imply that that we can efficiently[1] implement many common parallel algorithms on butterfly computers:

- Bitonic sort (see 2.3 on page 69).
- Computation of cumulative sums of $n$ numbers (see 2.4 on page 69).
- the generalized Batcher sorting algorithm (see 2.7 on page 71).

In the future we will need the following:

DEFINITION 3.4. An algorithm for the butterfly computer will be called *normal* if in each step *either* (but not both) of the following operations is performed:

1. data is copied from one rank to a neighboring rank;
2. calculations are performed within processors and no data movement occurs.

Normality is a fairly restrictive condition for an algorithm to satisfy — for instance it implies that the significant data is contained in a single rank in the computer in each step (which may vary from one step to the next). Most of the processors are *inactive* in each step. The DESCEND and ASCEND algorithms described above are clearly normal. Normal algorithms are important because it will turn out that they can be simulated on computers that have only $O(n)$ processors (rather than $O(n \lg n)$ processors, like the butterfly). In a sense normal algorithms are wasteful — they only utilize a single row of the butterfly in each time-step[2].

---

[1]In this context "efficiently" means that the program on the butterfly-computer has the same asymptotic execution-time as the PRAM implementation.

[2]Such algorithms, however, can be used to create *online* algorithms that continually input new data and process it.

The fact that the Batcher sorting algorithm sort can be implemented on the butterfly computer implies that:

LEMMA 3.5. An algorithm for an EREW-unbounded computer that executes in $\alpha$-steps using $n$ processors and $\beta$ memory locations can be simulated on a rank $\lg n$ butterfly computer executing in $O(\alpha \lceil \beta/n \rceil \lg^2 n)$-time using $n(1 + \lg n)$ processors. Here each processor is assumed to have at least $\beta/n$ memory locations of its own.

Here is the algorithm:

1. At the beginning of each simulated EREW machine cycle each processor makes up memory requests: we get an array of lists like (op(i), (a(i) **mod** n), $\lfloor a(i)/n \rfloor$,d(i), dummyflag), where op(i)=**READ** or **WRITE**, a(i) is the address on the EREW computer, and d(i) is the data to be written, or empty, and dummyflag=**FALSE**. We assume that memory in the EREW computer is simulated by the memory of the processors of the butterfly computer: location $k$ in the EREW computer is represented by location $\lfloor k/n \rfloor$ in processor $k \pmod n$ — it follows that the two middle entries in the list above can be interpreted as:

(processor containing simulated location $a(i)$,
    memory location within that processor of simulated location $a(i)$);

2. We now essentially *sort* all of these memory requests by their second entry to route them to the correct processors. This sorting operation is basically what was outlined above but we must be a little careful — more than one memory request may be destined to go to the same processor. In fact we must carry out $\lceil \beta/n \rceil$ complete sorting operations. In the first step we sort all memory requests whose third list element is 0, next all memory requests whose third element is 1, and so on. We must modify the sort algorithm given above slightly, to take into account incomplete lists of items to be sorted. In the beginning of the algorithm, if a processor doesn't have any list item with a given value of the third element is makes up a dummy item with the dummyflag equal to **TRUE**. When the a given sorting operation is complete the dummy items are discarded.

3. When the $\lceil \beta/n \rceil$ sorting operations are completed all processors will contain the (maximum of $\lceil \beta/n \rceil$) memory requests for accesses to memory locations that they contain. These requests are then *processed sequentially* in $O(\lceil \beta/n \rceil)$-time.

4. The results of the memory accesses are then sorted again (in one sorting operation) to send them back to the processors that originated the requests.

Note that the algorithm *would* be poly-logarithmic if these was some small *upper-bound* on $\lceil \beta/n \rceil$. If $\beta$ is very *large* (i.e. $O(n^2)$ the algorithm *seriously* falls short of being poly-logarithmic. The problem here is clearly not the *amount* of data to be sorted — that is always $O(n)$ and it should be possible to sort this much data in $O(\lg^2 n)$-time. The problem is that a large amount of data might have to go to the *same processor* (in which case many other processors would receive no data). There is also a problem with the processing of the memory requests once they have been routed to the correct processors. In the algorithm above this is done sequentially within processors, but this is clearly wasteful since many other processors will be idle (because there are at most $n$ data items to be processed overall). This situation is known as the *Granularity Problem* — it has been studied in connection with the theory of *distributed databases* as well as parallel processing algorithms. See §8 for a solution to this problem (and, consequently, an improved algorithm for simulating an EREW computer with large memory on a butterfly).

A similar result is possible for the CRCW computer. In order to prove this it is first necessary to develop an algorithm for moving data through a butterfly efficiently.

We immediately conclude:

LEMMA 3.6. An algorithm for an CRCW-unbounded computer that executes in $\alpha$-steps using $n$ processors and $\beta$ memory locations can be simulated on a rank $\lg n$ butterfly computer executing in $O(\alpha \lceil \beta/n \rceil \lg^2 n)$-time using $n(1 + \lg n)$ processors. Here each processor is assumed to have $\beta/n + 3$ memory locations of its own.

This algorithm is clearly normal since each of its steps are. That will turn out to imply that a CRCW computer can be simulated via a network that has $O(n)$ processors, and with no degradation in the time estimates.

PROOF. We copy the proof of 5.1, substituting the sorting algorithm for the butterfly computer for that of §5 and the data movement algorithm above for the simple shift that takes place in the CRCW-read simulation of 5.1.

In addition, in the step of 5.1 in which processors compare data values with those of their neighbors to determine whether they contain the lowest indexed reference to a memory location, they can use the butterfly implementation of the ASCEND algorithm and 2.8 to first move the data to be compared to the same processors. Note that, due to the Granularity Problem mentioned above we will have to actually carry out the routing operation $\lceil \beta/n \rceil$ times. This accounts for the factor of $\lceil \beta/n \rceil$ in the time estimate. □

Here is a sample programming language (Butterfly Pascal) that might be used on a butterfly-SIMD computer:

DEFINITION 3.7. The language is like Pascal except that there exist:

1. pre-defined (and reserved) variables: **ROW, COLUMN, COPY_EXCHANGE, COPY** — **COPY_EXCHANGE** and **COPY** are areas of memory of size 1K — that represents the maximum amount of data that may be copied or copy-exchanged at any given time. **ROW** and **COLUMN** are integers which may never be the target of an assignment (and may never be used as a var-type parameter in a procedure or function call). They are equal, respectively, to the row- and column- numbers of a processor (so their values vary from one processor to another).

2. Procedures **copy_exch()** and copy_up(), **copy_down()**. In order to copy-exchange some data it must be plugged into **COPY_EXCHANGE** and the procedure **copy_exch()** must be called. Assume that the pascal assignment operators to and from **COPY** and **COPY_EXCHANGE** are size-sensitive — i.e. **COPY:=x;** copies a number of bytes equal to the size of the variable x and the corresponding statement is true for x:=**COPY**.

3. Assume an additional block structure:

   **if** <condition> **paralleldo** <stmt>; This statement evaluates the condition (which generally tests COLUMN number in some way) and executes the statement if the condition is true. This differs from the usual if-statement only in the sense that a subset of the processors may execute the statement in <stmt> and the remainder of the processors will attempt to "mark time" — they will not execute <stmt> but will attempt to wait the appropriate amount of time for the active processors to finish. This is accomplished as follows: in the machine language for the butterfly each processor has a flag that determines whether it "really" executes the current instruction or *merely waits*. This flag is normally true (for active execution) but when the pascal compiler translates the parallel-if statement above it sets this flag in each processor according to whether the condition is true or not (for that processor). At the end of the parallel-if block the flag is again set to true.

Note: this statement is not necessary in order to execute a parallel program — execution of a program is normally done in parallel by all processors. This construct merely facilitates the *synchronization* of all processors across a row of the butterfly.

Assume that a given program executes simultaneously on all processors in the computer.

**Exercises:**

1. Suppose we had a computer that used the SIMD model of computation and the Butterfly architecture. It is, consequently, necessary for each processor to have a copy of the program to be run on it. Devise an algorithm to transmit a program from one processor to all of the others. It should execute in $O(l \lg n)$-time, where $l$ is the length of the program.

2. Is it possible for the processors to get "out of synchronization" in butterfly pascal even though they use the parallel-if statement?

3. Why would it be difficult to synchronize processors in butterfly pascal without a parallel-if statement? (I.e. why couldn't we force some processors to wait via conventional if statements and loops, for instance?)

4. Suppose we the Butterfly Pascal language available on a Butterfly computer of
rank 5: Program (in butterfly pascal):
   a. the butterfly sorting algorithm algorithm (Hints:
      i. In each case, pass a parameter to subroutines telling which column
         the processor is supposed to regard itself as being in — within the
         appropriate sub-butterfly;
      ii. Confine activity to a single row of the butterfly at a time;
      iii. use the parallel-if statement described above whenever you want to
         make a subset of the processors in a row execute some statements);
   b. the simulation algorithm for a CRCW computer.

**3.1. Discussion and further reading.** The BBN Butterfly computer indirectly uti-
lizes a butterfly network. It has a number of processors that communicate via a system
called an *Omega switch*. This is a butterfly network whose vertices are not completely-
functional processors — they are *gates* or *data-switches*. See [97] for a discussion of the
issues involved in programming this machine.

We will discuss some of the literature on algorithms for the butterfly network. In
[16], Bhatt, Chung, Hong, Leighton, and Rosenberg develop algorithms for simulations
that run on a butterfly computer. network.

## 4. The Hypercube Architecture

**4.1. Description.** An $n$-dimensional hypercube is a graph that looks, in the 3-
dimensional case, like a wire frame that models a *cube*. The rigorous definition of
an $n$-dimensional hypercube is a graph $H_n$, where

1. The vertices of $H_n$ are in a 1-1 correspondence with the $n$-bit binary sequences
   $a_0 \cdots a_{n-1}$ (so there are $2^n$ such vertices). Each vertex has an identifying number.
2. Two vertices $a_0 \cdots a_{n-1}$ and $a'_0 \cdots a'_{n-1}$ are connected by an edge if and only if
   these sequences differ in *exactly one bit* — i.e., $a_i = a'_i$ for $0 \leq i \leq n - 1, i \neq k$
   for some value of $k$ and $a_k \neq a'_k$.

An $n$-dimensional hypercube computer has a *processing-element* at each vertex of $H_n$
and connecting *communication lines* along the edges.

It is not hard to see that each vertex has exactly $n$ edges incident upon it. Its connectivity is, consequently, higher than that of the butterfly or perfect-shuffle architectures.

One might think that such a hypercube-computer is harder to implement than a butterfly or perfect-shuffle computer.

The generic ASCEND and DESCEND algorithms (2.1 and 2.2 on page 68) are easy to implement on the hypercube architecture:

ALGORITHM 4.1. **DESCEND Algorithm on the Hypercube computer.** Suppose the input data is $T[0], \ldots, T[n-1]$, with $n = 2^k$ and we have an $n$-dimensional hypercube. Start with $T[m]$ in vertex $m$, for $0 \leq m \leq n-1$ — where vertex-numbers are as defined above. In iteration $j$ of the Hypercube version of the DESCEND algorithm perform the following operations:

**for** $j \leftarrow k - 1$ **downto** 0 **do**
    **for** each $m$ such that $0 \leq m < n$
        **do in parallel**
            **if** $\mathrm{bit}_j(m) = 0$ **then**
                vertices $m$ and $m + 2^j$ exchange copies
                of their data via the unique
                common communication line
            Each processor **computes** $\mathrm{OPER}(m, j; T[m], T[m + 2^j])$
                (Now having the necessary input-data:
                the old values of $T[m]$ and $T[m + 2^j]$)
    **endfor**
**endfor**

ALGORITHM 4.2. **ASCEND Algorithm on the Hypercube computer.** Suppose the input data is $T[0], \ldots, T[n-1]$, with $n = 2^k$ and we have an $n$-dimensional hypercube. Start with $T[i]$ in vertex $i$, for $0 \leq i \leq n - 1$ — where vertex-numbers are as defined above. In iteration $j$ of the Hypercube version of the ASCEND algorithm perform the following operations:

**for** $j \leftarrow 0$ **to** $k - 1$ **do**
    **for** each $m$ such that $0 \leq m < n$
        **do in parallel**
            **if** $\mathrm{bit}_j(m) = 0$ **then**
                vertices $m$ and $m + 2^j$ exchange copies
                of their data via the unique
                common communication line
             Each processor **computes** $\mathrm{OPER}(m, j; T[m], T[m + 2^j])$
                (Now having the necessary input-data:
                the old values of $T[m]$ and $T[m + 2^j]$)
    **endfor**
**endfor**

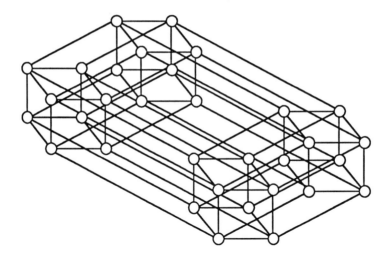

FIGURE III.4. A 5-dimensional hypercube

Note that the implementations of ASCEND and DESCEND on the hypercube are more straightforward than on the butterfly network. These implementations immediately imply that we have efficient implementations of all of the ASCEND and DESCEND algorithms in § 2 on page 67. The hypercube architecture is interesting for many reasons not related to having good implementations of these generic algorithms. It turns out to be very easy to map certain other interesting architectures into the hypercube. We will spend the rest of this section looking at some of these.

DEFINITION 4.3. Let $a$ and $b$ be sequences of bits of the same length. The *Hamming distance* between these sequences is the number of bit-positions in the two sequences, that have different values.

It is not hard to see that the distance between two different vertices in a hypercube is equal to the Hamming distance between the binary representations of their vertex-numbers.

Figure III.4 shows a 5 dimensional hypercube.

A *six* dimensional hypercube is the result of taking two copies of this graph and attaching each vertex of one copy to a corresponding vertex of the other — and each time the dimension is raised by 1 the complexity of the graphs doubles *again*. This is meant to convey the idea that high-dimensional hypercubes might be difficult to implement. Nevertheless, such computers are commercially available. The Connection Machine from Thinking Machines (CM-2 model) is a 12-dimensional hypercube computer with 64000 processors (it actually has 16 processors at each vertex of a 12-dimensional hypercube).

It turns out that an $n$-dimensional hypercube is equivalent to an order-$n$ butterfly network with all of the columns collapsed to *single vertices*, and half as many edges.

Basically, the result of collapsing the columns of a butterfly to vertices is a hypercube with all of its edges *doubled*.

It is not hard to see that any *normal* algorithm for a degree-$n$ butterfly network can easily be ported to an $n$-dimensional hypercube computer with *no* time degradation.

LEMMA 4.4. Every normal algorithm that runs on a degree-$k$ butterfly network (that has $k2^k$ processors) in $t$ time units can run on a $k$-dimensional hypercube computer in $O(t)$ time.

Hypercubes are interesting as models for parallel communication because of the fact that many other communication-schemes can be mapped into hypercubes. In order to see how this is done, we discuss the subject of *Gray codes*. We will be particularly interested in how one can map *lattice organizations* into hypercubes.

DEFINITION 4.5. The $k$-bit reflected Gray code sequence is defined recursively via:

- The 1-bit sequence is $\{0, 1\}$;
- If $\{s_1, \ldots, s_m\}$ is the $k - 1$-bit reflected Gray code sequence, then the $k$-bit sequence is $\{0s_1, \ldots, 0s_m, 1s_m, \ldots, 1s_1\}$.

The $k$-bit reflected Gray code sequence has $2^k$ elements.

Here are the first few reflected Gray code sequences:

1. $\{0, 1\}$
2. $\{00, 01, 10, 11\}$
3. $\{000, 001, 010, 011, 111, 110, 101, 100\}$

The important property of Gray codes that will interest us is:

PROPOSITION 4.6. In a $k$-bit Gray code sequence, the Hamming distance between any two successive terms, and the Hamming distance between the first term and the last term is 1.

PROOF. This follows by a simple induction. It is clearly true for the 1 bit Gray codes. If it is true for $k - 1$ bit Gray codes, then the inductive definition implies that it is true for the $k$ bit Gray codes, because each half of this sequence is just the concatenation of the $k - 1$ bit Gray codes with a fixed bit (0 of the first half, and 1 for the second). This leaves us with the question of comparing:

- the two middle terms — but these are just $0s_m, 1s_m$, and they differ in only 1 bit.
- the first and last elements — but these are $0s_1, 1s_1$, and they just differ in 1 element.

□

Since vertices whose numbers differ in only one bit are adjacent in a hypercube, the $k$ bit Gray code sequences provides us with a way to map a loop of size $2^k$ into a hypercube:

PROPOSITION 4.7. Suppose $S = \{s_1, \ldots, s_n\}$ be the $k$ bit Gray code sequence. In addition, suppose we have a $d$ dimensional hypercube, where $d \geq k$ — its vertices are encodes by $d$ bit binary numbers. If $z$ is any $d - k$ bit binary number, then the sequences of vertices whose encoding is $\{zs_1, \ldots, zs_n\}$ is an embedding of a loop of size $n$ in the hypercube.

We can use multiple Gray code sequences to map a multi-dimensional lattice of vertices into a hypercube of sufficiently high dimension. This lattice should actually be regarded as a *torus*. Here is an example:

EXAMPLE 4.8. Suppose we have a two dimensional lattice of processors that we want to simulate on a hypercube computer whose dimension is at least 7. Each processor in this lattice is denoted by a pair of numbers $(u, v)$, where (we suppose) $u$ runs from 0 to 3 (so it takes on 4 values) and $v$ runs from 0 to 15 (so it takes on $2^4 = 16$ values). We use the 3 and 4 bit Gray codes:

- 3 bit Gray code, $S_1$:
  000,001,010,011,111,110,101,100
- 4 bit Gray code, $S_2$:
  0000,0001,0010,0011,
  0111,0110,0101,0100,
  1100,1101,1110,1111,
  1011,1010,1001,1000

and we will assume that both of these sequences are numbered from 0 to 7 and 0 to 15, respectively. Now we map the processor numbered $(u, v)$ into the element of the hypercube location whose binary representation has low-order 7 bits of $\{S_1(u), S_2(v)\}$. Processor $(2, 3)$ is sent to position $0100011 = 67$ in the hypercube.

Note that size of each dimension of the lattice must be an exact power of 2. The general statement of how we can embed lattices in a hypercube is:

PROPOSITION 4.9. Let $L$ be a $k$-dimensional lattice of vertices such that the $i^{\text{th}}$ subscript can take on $2^{r_i}$ possible values. Then this lattice can be embedded in an $r$-dimensional hypercube, where $r = \sum_{i=1}^{k} r_i$. An element of $L$ can be represented by a sequence of $k$ numbers $\{i_1, \ldots, i_k\}$, and the embedding maps this element to the element of the hypercube whose address has the binary representation $\{S(i_1, r_1), \ldots, S(i_k, r_k)\}$, where $S(q, r)$ denotes the $q^{\text{th}}$ element of the $r$-bit reflected Gray code.

This mapping has been implemented in the hardware of the Connection Machine (CM-2 model), so that it is easy to define and operate upon arrays of processors that have the property that the range of each subscript is exactly a power of 2.

Many parallel computers have been built using the hypercube network architecture including:

- the nCUBE computers, including the nCUBE/7 and nCUBE 2;
- The Cosmic Cube.
- the Connection Machines (CM-1 and CM-2) from Thinking Machines Corporation. These are SIMD computers discussed in more detail later in this book (see § 2 on page 114).
- the Intel iPSC series, including the iPSC/1, iPSC/2, iPSC/860, and the Touchstone Gamma.
- the Caltech Mark II.
- the MasPar MP-1 computer.

Most of these are MIMD computers

**Exercises:**

1. Prove the statement that an $n$-dimensional hypercube is equivalent to an order-$n$ butterfly network with all of the columns collapsed to *single vertices*, and half as many edges.
2. Compute an explicit embedding of a 3-dimensional lattice of size $8 \times 4 \times 2$ into a 6-dimensional hypercube.
3. Suppose an $n \times n$ matrix is embedded in a hypercube (as an $n \times n$ lattice — we assume that $n$ is a power of 2). Find an algorithm for transposing this matrix in $O(\lg n)$ time. Hint: Consider algorithm 2.5 on page 70, and the implementation of DESCEND on the hypercube.

## 5. The Shuffle-Exchange Network

The last network-computer we will consider is the shuffle-exchange network. Like the others it is physically realizable and has the ability to efficiently simulate unbounded parallel computers. It has the added advantage that this simulation can be done without any increase in the number of processors used.

A degree-$n$ shuffle exchange network is constructed as follows: Start with $n$ processors, numbered 0 to $n-1$, arranged in a linear array except that processor $i$ is connected to:

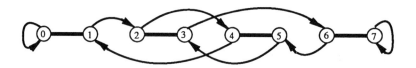

FIGURE III.5. The shuffle exchange network

1. processor $i + 1$ if $i$ is even;
2. processor $j$, where $j \equiv 2i \pmod{n - 1}$;
3. itself, if $i = n - 1$ (rather than processor 0, as rule b would imply).

Figure III.5 shows an 8-node shuffle-exchange network.

Here the shaded lines represent exchange lines: data can be swapped between processors connected by such lines — this movement will be denoted $EX(i)$. The dark curved lines represent the shuffle lines — they connect processor $i$ with $2i \bmod n - 1$ (in the pascal sense). Although data can move in both directions along these lines one direction is regarded as forward and the other is regarded as reverse.

There are two main data-movement operations that can be performed on the shuffle-exchange network:

1. Shuffle, $PS(i)$ (for "perfect shuffle"). Here data from processor $i$ is moved to processor $2i \bmod n - 1$. The inverse operation $PS^{-1}(i)$ is also defined (it moves data from processor $i$ to processor $j$, where $i \equiv 2j \pmod{n - 1}$). Most of the time these operations will be applied in parallel to all processors — the parallel operations will be denoted $PS$ and $PS^{-1}$, respectively.
2. Exchange, $EX(i)$. Here data from processor $i$ is moved to processor $i + 1$ if $i$ is even and $i - 1$ if $i$ is odd.

We will consider the effects of these operations. Suppose $b(i)$ is the binary representation of the number $i$.

PROPOSITION 5.1. Suppose processor $i$ has data in it and:

1. $PS(i)$ will send that data to processor $j$, or;
2. $EX(i)$ will send that data to processor $j'$.

Then:

1. $b(j)$ is the result of cyclically permuting $b(i)$ one bit to the left;
2. $b(j')$ is the same as $b(i)$ except that the low order bit is different.

PROOF. Recall that we are considering $n$ to be a power of 2. Statement b is clear. Statement a follows from considering how $j$ is computed: $i$ is multiplied by 2 and reduced $\pmod{n-1}$. Multiplication by 2 shifts the binary representation of a number one bit to the left. If the high-order bit is 1 it becomes equal to $n$ after multiplication by 2 — this result is congruent to 1 $\pmod{n - 1}$. $\square$

Our main result concerning the shuffle-exchange network is:

LEMMA 5.2. Every normal algorithm that runs on a degree-$k$ butterfly network (that has $k2^k$ processors) in $t$ time units can run on a $2^k$-processor shuffle-exchange network in $O(t)$ time.

PROOF. We assume that the processors of the shuffle-exchange network can carry out the same operations as the processors of the butterfly. The only thing that remains to be proved is that the data-movements of the butterfly can be simulated on the shuffle-exchange network. We will carry out this simulation (of data-movements) in such a way that the processors of the shuffle-exchange network correspond to the columns of the butterfly. In other words we will associate all processors of column $i$ of the butterfly with processor $i$ in the shuffle-exchange network. That it is reasonable to associate all processors in a column of the butterfly with a single processor of the shuffle-exchange network follows from the fact that the algorithm we are simulating is normal — only one rank of processors is active at any given time.

Recall how the processors of the butterfly were connected together. Processor $d_{r,i}$ ($r$ is the rank) was connected to processors $d_{r-1,i}$ and $d_{r-1,i'}$, where $i$ and $i'$ differ (in their binary representations) only in the $r-1^{st}$ bit from the left. Let us simulate the procedure of moving data from $d_{r,i}$ to $d_{r-1,i'}$:

Perform $PS^r$ on all processors in parallel. Proposition 5.1 implies that this will cyclically left-shift the binary representation of $i$ by $r$-positions. The $r-1^{st}$ bit from the left will wind up in the low-order position. EX will alter this value (whatever it is) and $PS^{-r}$ will right-shift this address so that the result will be in processor $i'$.

We have (in some sense) simulated the copy operation from $d_{r,i}$ to $d_{r-1,i'}$: by $PS^{-r} \circ EX(PS^r(i)) \circ PS^r$. The inverse copy operation (from $d_{r-1,i'}$ to $d_{r,i}$) is clearly simulated by $(PS^{-r} \circ EX(PS^r(i)) \circ PS^r)^{-1} = PS^{-r} \circ EX(PS^r(i)) \circ PS^r$.

*Incidentally — these composite operations must be read from right to left.*

There are several obvious drawbacks to these procedures: we must carry out $O(\lg n)$ steps (to do $PS^r$ and $PS^{-r}$) each time; and we must compute $PS^r(i)$ inside each EX. These problems are both solved as follows: Note that after doing the copy from $d_{r,i}$ to $d_{r-1,i'}$: the data will be in rank $r-1$ — consequently the next step will be an operation of the form $PS^{1-r} \circ EX(PS^{r-1}(i')) \circ PS^{r-1}$ — and the composite will have steps of the form $PS^{r-1} \circ PS^{-r} = PS^{-1}$. Consequently the simulations aren't so time-consuming if we compose successive operations and cancel out terms that are inverses of each other. In fact, with this in mind, we can define:

- **Simulation of** $d_{r,i} \rightarrow d_{r-1,i'}$:$PS^{-1} \circ EX(PS^r(i))$.
- **Simulation of** $d_{r,i} \rightarrow d_{r-1,i}$:$PS^{-1}$ (here we have represented movement of $d_{r,i} \rightarrow d_{r-1,i}$ by $PS^{-r} \circ PS^r$ before canceling).

Here we have lost the simple correspondence between columns of the butterfly and processors of the shuffle-exchange network. Now processor $d_{r,i}$ in the butterfly corresponds to processor $PS^r(i)$ of the shuffle-exchange network — here we are

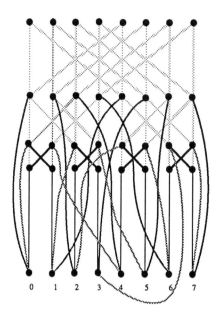

FIGURE III.6

regarding PS as an operation performed upon numbers (like squaring) as well as a data-movement command on a computer. This correspondence is illustrated by figure III.6.

Here the bottom row represents the processors of the shuffle-exchange computer and the top rows represent the butterfly computer (with its interconnections drawn very lightly). The correspondence between ranks 1, 2 and 3 of the butterfly and the shuffle-exchange computer are indicated by curved lines. The top row of the butterfly corresponds to the shuffle-exchange computer in exactly the same way as the bottom row so no lines have been drawn in for it.

It is easy to keep track of this varying correspondence and solve the second problem mentioned above by doing the following: in each step manipulate a list (data to be moved, column #). The first item is the data upon which we are performing calculations. The second is the column number of the butterfly that the data is supposed to be in. In each step we only carry out the PS and $PS^{-1}$ operations on the column numbers so that they reflect this varying correspondence. Our final simulation program can be written as:

- **Simulation of** $d_{r,i} \rightarrow d_{r-1,i'}$:$PS^{-1} \circ$ (column # i) — the EX-operation is only carried out on the data portion of the lists — the $PS^{-1}$ is carried out on both portions.
- **Simulation of** $d_{r,i} \rightarrow d_{r-1,i}$:$PS^{-1}$ — carried out on both portions of whole lists.
- **Simulation of** $d_{r,i} \rightarrow d_{r+1,i'}$:$EX$(column # $i'$) $\circ$ PS — the EX-operation is only carried out on the data portion of the lists — the PS is carried out on both

portions. Note that we most carry out the EX-operation on the processor whose column# field is $i'$ rather than $i$. It is easy for a processor to determine whether it has column# $i'$, however. The processors with column numbers $i$ and $i'$ will be adjacent along an EX-communication line before the PS-operation is carried out. So our simulation program has the processors check this; set flags; perform PS, and then carry out EX-operations if the flags are set.

• **Simulation of** $d_{r,i} \rightarrow d_{r+1,i}$:PS — carried out on both portions of whole lists.

□

This immediately implies that we can implement the DESCEND and ASCEND algorithms on the shuffle-exchange computer — we "port" the Butterfly versions of these algorithms (3.2 and 3.3 on page 74) to the shuffle-exchange computer.

ALGORITHM 5.3. **DESCEND Algorithm on the Shuffle-Exchange computer.** Suppose the input data is $T[0], \ldots, T[n-1]$, with $n = 2^k$ and we have a shuffle-exchange computer with $n$ processors. The first elements of these lists will be called their *data portions*. In each phase of the Shuffle-Exchange version of the DESCEND algorithm (2.1 on page 68) perform the following operations:

**do in parallel** (all processors)
   $L_1 \leftarrow \{T[i], i\}$ in processor $i$
**for** $j \leftarrow k - 1$ **downto** 0 **do**
   **do in parallel** (all processors)
      **Perform** PS
      $L_2 \leftarrow L_1$
      **Perform** EX upon the
         data-portions of $L_2$ in each processor.
      Each processor now contains two lists:
         $L_1 = $ old $\{T[m], m\}, L_2 = $ old $\{T[m + 2^j], m\}$
      **Compute** $\mathrm{OPER}(m, j; T[m], T[m + 2^j])$

ALGORITHM 5.4. **ASCEND Algorithm on the Shuffle-Exchange computer.** Suppose the input data is $T[0], \ldots, T[n-1]$, with $n = 2^k$ and we have a shuffle-exchange computer with $n$ processors. The first elements of these lists will be called their *data portions*. In each phase of the Shuffle-Exchange version of the ASCEND algorithm (2.2 on page 68) perform the following operations:

**do in parallel** (all processors)
   $L_1 \leftarrow \{T[i], i\}$ in processor $i$
**for** $j \leftarrow 0$ **to** $k - 1$ **do**
   **do in parallel** (all processors)
      $L_2 \leftarrow L_1$
      **Perform** EX upon the

data-portions of $L_2$ in each processor.

**Perform** $PS^{-1}$ in each processor, with both lists

Each processor now contains two lists:

$L_1 = $ old $\{T[m], m\}, L_2 = $ old $\{T[m + 2^j], m\}$

**Compute** $OPER(m, j; T[m], T[m + 2^j])$

COROLLARY 5.5. An algorithm for an CRCW-unbounded computer that executes in $\alpha$-steps using $n$ processors and $\beta$ memory locations can be simulated on an $n$-processor shuffle-exchange computer in $O(\alpha\lceil\beta/n\rceil \lg^2 n)$-time. Each processor of the shuffle-exchange computer must have at least $(\beta/n) + 3$ memory locations of its own.

Here we are assuming that $n$ is a power of 2.

**5.1. Discussion and further reading.** the shuffle-exchange network is also frequently called the DeBruijn network. This network, and variations on it, are widely used in existing parallel computers.

- The NYU Ultra Computer project is current building Ultra III which will be based on a shuffle-exchange network. See [134] for some additional material on this type of machine.
- Many new computer systems under development will use this design, including:
  1. the Triton project at the University of Karlsruhe. Here, according to Dr. Christian Herter[3]:

     "... to select a topology with very low average diameter because the average diameter is directly responsible for the expected latency of the network. The best network class we found are DeBruijn nets also known as perfect shuffle (which is nearly equivalent to shuffle exchange). Those networks have an average diameter very close to the theoretical limit. Each node has indegree two and outdegree two (constant for all network sizes) the (max.) diameter is still log N the average diameter is approximately log N - 1.7."

     This machine is intended to be a SIMD-MIMD hybrid with that initial would have 256 processors. See [69] for more information on this project.
  2. the Cedar Machine, begin developed at the Center for Supercomputing Research and Development, University of Illinois at Urbana Champaign. See [53].

---

[3]Private communication.

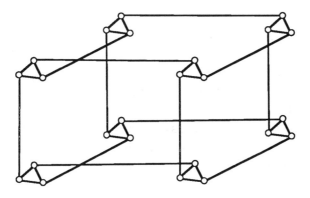

FIGURE III.7. The cube-connected cycles architecture

**Exercises:**

1. Devise a version of Pascal to run on the shuffle exchange computer, along the lines of butterfly pascal (page 77).

## 6. Cube-Connected Cycles

One of the limiting factors in designing a computer like the butterfly computer is the number of connections between processors. Clearly, if processors are only allowed to have two communication lines coming out of them the whole network will form a tree or ring. Since both of these networks only allow a limited amount of data to flow through them, it is clear that networks with high data flow must have at least *three* communication lines coming out of most of the processors. The butterfly network is close to this optimum since it has four lines through most processors. It is possible to design a network that has the same asymptotic performance as the butterfly with precisely three communications lines coming out of each processor. This is called the *cube-connected cycles* (or CCC) network. Figure III.7 shows how the processors are interconnected — as usual the nodes represent processors and the edges represent communication lines. These were first described by Preparata and Vuillemin in [124]

In general a $k$-dimensional CCC is a $k$-dimensional cube with the vertices replaced with loops of $k$ processors. In a $k$-dimensional cube every vertex has $k$ edges connected to it but in the $k$-dimensional CCC only one of the processors in a loop connects to each of the $k$ dimensions so that each processor has only three communication lines connected to it — two connecting it to other processors in its loop and one connecting it to another face of the cube.

If we identify rank 0 processors in the butterfly with rank $k$ processors we get a graph that turns out to contain the $k$-dimensional CCC network as a subgraph. In addition, every data movement in the butterfly network can be simulated in the embedded CCC, except that some movements in the butterfly require two steps in the CCC since it lacks some of the edges of the butterfly.

In a CCC with $n$ vertices, we will assume that $n = 2^k$, and that $k$ is of the form $k = r + 2^r$. Each vertex has a $k$-bit address, which is regarded as a pair of numbers $(\alpha, \ell)$, where $\alpha$ is a $k - r$ bit number and $\ell$ is an $r$ bit number. Each vertex has three connections to it:

- $F$ — the **forward** connection. $F(\alpha, \ell)$ is connected to $B(\alpha, (\ell + 1) \bmod 2^r)$;
- $B$ — the **back** connection. $B(\alpha, \ell)$ is connected to $F(\alpha, (\ell - 1) \bmod 2^r)$.
- $L$ — the **lateral** connection. $L(\alpha, \ell)$ is connected to $L(\alpha + \epsilon 2^\ell, \ell)$, where

$$\epsilon = \begin{cases} 1 & \text{if the } \ell^{\text{th}} \text{ bit of } \alpha \text{ is } 0 \\ -1 & \text{if the } \ell^{\text{th}} \text{ bit of } \alpha \text{ is } 1 \end{cases}$$

The forward and back connections connect the vertices within cycles, and the lateral connections connect different cycles, within the large-scale hypercube. If we shrink each cycle to a single vertex, the result is a hypercube, and the lateral connections are the only ones that remain.

In fact Zvi Galil and Wolfgang Paul show that the cube-connected cycles network could form the basis of a general-purpose parallel computer that could efficiently simulate all other network-computers in [52].

We will present several versions of the DESCEND and ASCEND algorithms. The first will closely mimic the implementations of these algorithms for the butterfly computer. Each *loop* the the CCC network will have a single data-item, and this will be cyclically shifted through the network in the course of the algorithm. This algorithm only processes $2^{k-r}$ data-items, but does it in $O(k - r)$ steps. We define the loop-operations:

DEFINITION 6.1. Define the following data-movement operations on the CCC network:

- **F-LOOP** is the permutation that transmits all data within the cycles of the CCC in the *forward* direction, i.e., through the $F(\alpha, \ell)$ port.
- **B-LOOP** is the permutation that transmits all data within the cycles of the CCC in the *forward* direction, i.e., through the $B(\alpha, \ell)$ port.

In the following algorithm, we assume that processor $i$ has two memory-locations allocated: $T[i]$ and $T'[i]$

ALGORITHM 6.2. **Wasteful version of the DESCEND algorithm** The input consists of $2^{k-r}$ data items stores, respectively, in $D[0], \ldots, D[2^{k-r} - 1]$.

**do in parallel** (for $0 \leq i \leq 2^{k-r} - 1$)
$T[i2^r + k - r - 1] \leftarrow D[i]$
**for** $j \leftarrow k - r - 1$ **downto** $0$ **do**
   **do in parallel** (for all processors)
      Transmit $T[*]$ along lateral lines, and
        store received data in $T'[*]$
      ( *if the $^{th}j$ bit of i is 0*
        $T'[i2^r + j] = old\ value\ of\ T[[i2^r + 2^{j+r} + j]$
        *of* $T[i], T[i + 2^j])$
      ( *if the $^{th}j$ bit of i is 1*
        $T'[i2^r + j] = old\ value\ of\ T[[i2^r - 2^{j+r} + j]$
        *of* $T[i], T[i + 2^j])$
      **Compute** $\text{OPER}(i, j; T[i], T[i + 2^j])$
   **Perform** B-LOOP
**endfor**

Note that processor number $i2^r + k - r - 1$ is the processor with coordinates $(i, k - r - 1)$.

ALGORITHM 6.3. **Wasteful version of the ASCEND algorithm** The input consists of $2^{k-r}$ data items stores, respectively, in $D[0], \ldots, D[2^{k-r} - 1]$.

**do in parallel** (for $0 \leq i \leq 2^{k-r} - 1$)
$T[i2^r] \leftarrow D[i]$
**for** $j \leftarrow 0$ **to** $k - r - 1$ **do**
   **do in parallel** (for all processors)
      Transmit $T[*]$ along lateral lines, and store received data in $T'[*]$
      ( *if the $^{th}j$ bit of i is 0*
        $T'[i2^r + j] = old\ value\ of\ T[[i2^r + 2^{j+r} + j]$
        *of* $T[i], T[i + 2^j])$
      ( *if the $^{th}j$ bit of i is 1*
        $T'[i2^r + j] = old\ value\ of\ T[[i2^r - 2^{j+r} + j]$
        *of* $T[i], T[i + 2^j])$
      **Compute** $\text{OPER}(i, j; T[i], T[i + 2^j])$
   **Perform** F-LOOP
**endfor**

Now we will present the implementations of DESCEND and ASCEND of Preparata and Vuillemin in [125]. Their implementations utilize *all* of the processors in each time-step. Their algorithms for the $n$-processor CCC operate upon $n$ data-items and execute in the same asymptotic time as the generic versions ($O(z \cdot \lg n)$, where $z$ is the execution-time of $\text{OPER}(a, b; U, V)$).

We begin with the DESCEND algorithm. To simplify the discussion slightly, we will initially assume that we are only interested in carrying out the first $k - r$ iterations of the DESCEND algorithm — so the **for**-loop only has the subscript going from $k - 1$ to $k - r$. We can carry out the these iterations of the algorithm in a way that is somewhat like the the "wasteful" algorithm, except that:

- Now we have more data to work with — the "wasteful" algorithm only had data in a single processor in each loop of the CCC.
- We must process each data-item at the proper time. This means that in the first few steps of the algorithm, we cannot do anything with *most* of the data-items. For instance, in the first step, the only data item that can be processed in each loop is the one that the "wasteful" algorithm *would* have handled — namely, the one with $\ell$-coordinate equal to $k - 1$. In the second step this data-item will have been shifted into the processor with $\ell$-coordinate equal to $k - 2$, but another data item will have shifted into the processor with coordinate $k - 1$. That second data-item will now be ready to be processed in the first iteration of the DESCEND algorithm.

  In this manner, each step of the algorithm will "introduce" a new element of the loop to the *first iteration* of the DESCEND algorithm. This will continue until the $k + 1^{\text{st}}$, at which time the element that started out being in the processor with $\ell$ coordinate $k - 2$ will be in position to start the algorithm (i.e., it will be in the processor with $\ell$ coordinate $k - 1$). Now we will need another $k - r$ steps to complete the computations for all of the data-items.
- We must have some procedure for "turning off" most of the processors in some steps.

ALGORITHM 6.4. If we have an $n$-processor CCC network and $n$ data-items $T[0]$, $\ldots, T[n - 1]$, where $T[i]$ is stored in processor $i$ (in the numbering scheme described above). Recall that $k = r + 2^r$.

**do in parallel** (for all processors)
**for** $i \leftarrow 2^r - 1$ **downto** $-2^r$ **do**
   **do in parallel** (for all processors whose $\alpha$ coordinate
        satisfies $0 \leq \alpha < 2^{k-r}$)
     and whose $\ell$-coordinate satisfies
        $\max(i, 0) \leq \ell < \min(2^r, 2^r + i))$
   *(if* $\text{bit}_\ell(\alpha) = 0$ *then processor* $(\alpha, \ell)$ *contains*
       $T[\alpha 2^r + \ell]$
   *and if* $\text{bit}_\ell(\alpha) = 1$ *then processor* $(\alpha, \ell)$ *contains*
       $T[\alpha 2^r + \ell + 2^{\ell+r}])$
   All processors transmit their value of $T[*]$ along a lateral
      communication line

Each processor $(\alpha, \ell)$ computes $\mathrm{OPER}(a, b; U, V)$
where: $a = \alpha 2^r + ((i - \ell - 1) \bmod 2^r)$
$b = \ell + r$
$U = T[\alpha 2^r + \ell]$
$V = T[\alpha 2^r + \ell + 2^{\ell+r}]$

B-LOOP
**endfor**

This handles iterations $k - 1$ to $k - r$. The remaining iterations do not involve any communication between distinct cycles in the CCC network.

We will discuss how the remaining iterations of the algorithm are performed later.

We will analyze how this algorithm executes. In the first iteration, $i = 2^r - 1$, and the only processors, $(\alpha, \ell)$, that can be active must satisfy the conditions

$$0 \le \alpha < 2^{k-r}$$
$$\max(i, 0) = 2^r - 1 \le \ell < \min(2^r, 2^r + i) = 2^r$$

so $\ell = 2^r - 1$. Data-item $T[\alpha 2^r + 2^r - 1]$ is stored in processor $(\alpha, 2^r - 1)$, for all $\alpha$ satisfying the condition above. In this step

$$\begin{aligned}
a =& \alpha 2^r + (((2^r - 1) - \ell - 1) \bmod 2^r) \\
=& \alpha 2^r + (-1 \bmod 2^r) \\
=& \alpha 2^r + 2^r - 1 \\
b =& \ell + r = k - 1 \text{ (recalling that } r + 2^r = k)
\end{aligned}$$

The lateral communication-lines in these processors connect processor $(\alpha, 2^r - 1)$ to $(\alpha', 2^r - 1)$, where $|\alpha - \alpha'| = 2^{2^r - 1} = 2^{k-r-1}$. These two processors contain *data-items* $T[\alpha 2^r + 2^r - 1]$ and $T[\alpha' 2^r + 2^r - 1]$, where

$$\begin{aligned}
|(\alpha 2^r + 2^r - 1) - (\alpha' 2^r + 2^r - 1)| =& 2^r |(\alpha - \alpha')| \\
=& 2^r \cdot 2^{k-r-1} = 2^{k-1}
\end{aligned}$$

Consequently, in the *first iteration*, the algorithm does *exactly* what it should — it performs $\mathrm{OPER}(a, b; U, V)$, where

$$\begin{aligned}
a =& \alpha 2^r + \ell \\
b =& k - 1 \\
U =& a \text{ if } \mathrm{bit}_{k-1}(a) = 0, \text{ in which case} \\
V =& a + 2^{k-1}
\end{aligned}$$

In the next step, $\ell$ can only take on the values $2^r - 1$ and $2^r - 2$. Our original data that was in processor $(\alpha, 2^r - 1)$ has been shifted into processor $(\alpha, 2^r - 2)$. In this case

$$
\begin{aligned}
a &= \alpha 2^r + (((2^r - 2) - \ell - 1) \bmod 2^r) \\
&= \alpha 2^r + (-1 \bmod 2^r) \\
&= \alpha 2^r + 2^r - 1 \\
b &= \ell + r = k - 2
\end{aligned}
$$

Note that the quantity $a$ remains unchanged — as it *should*. This is because $a$ represents the index of that data-item being processed, and this hasn't changed. The original data-item was shifted into a new processor (with a lower $\ell$-coordinate), but the formula for $a$ compensates for that. It is also not difficult to verify that the correct version of $\mathrm{OPER}(a, b; U, V)$ is performed in this step.

This iteration of the algorithm has *two* active processors in each cycle of the CCC network. It processes all data-items of the form $T[\alpha 2^r + 2^r - 1]$ and $T[\alpha 2^r + 2^r - 2]$ — a new data-item has entered into the "pipeline" in each cycle of the network. This is what happens for the first $2^r - 1$ iterations of the algorithm — new elements enter into the algorithm. At iteration $2^r - 1$ — when $i = 0$, the first data-items to enter the entire algorithm (namely the ones of the form $T[\alpha 2^r + 2^r - 1]$, for all possible values of $\alpha$) are *completely processed*. We must continue the algorithm for an additional $2^r$ steps to process the data-items that entered the algorithm late.

The **if**-statement that imposes the condition $\max(i, 0) \leq \ell < \min(2^r, 2^r + i)$ controls which processors are allowed to be active in any step.

We have described how to implement the first $k - r - 1$ iterations of the DESCEND algorithm. Now we must implement the remaining $r$ iterations. In these iterations, the pairs of data-items to be processed in each call of $\mathrm{OPER}(a, b; U, V)$ *both* lie in the same loop of the CCC network, so that no *lateral* moves of data are involved (in the sense of the definition on page 90). Consider the $i^{\text{th}}$ step of the last $r$ iterations of the DESCEND algorithm. It involves the computation:

> if $(\mathrm{bit}_{r-i}(j) = 0$ then
>     $\mathrm{OPER}(j, r - i; T[j], T[j + 2^{r-i}])$

It is completely straightforward to implement this on a linear array of processors (and we can think of the processors within the same cycle of the CCC as a linear array) — we:

1. Move data-item $T[j + 2^{r-i}]$ to processor $j$, for all $j$ with the property that $\mathrm{bit}_{r-i}(j) = 0$. This is a parallel data-movement, and moves each selected data-item a distance of $2^{r-i}$. The execution-time is, thus, proportional to $2^{r-i}$.

2. For every $j$ such that $\mathrm{bit}_{r-i}(j) = 0$, processor $j$ now contains $T[j]$ and $T[j + 2^{r-i}]$. It is in a position to compute $\mathrm{OPER}(j, r - i; T[j], T[j + 2^{r-i}])$. It performs the computation in this step.

3. Now the new values of $T[j]$ and $T[j + 2^{r-i}]$ are in processor $j$. We send $T[j + 2^{r-i}]$ back to processor $j + 2^{r-i}$ — this is exactly the reverse of step 1 above. The execution-time is also $2^{r-i}$.

The *total* execution-time is thus

$$T = \sum_{i=0}^{r} 2^{r-i} = 2^{r+1} - 1 = 2(k-r) - 1$$

We can combine all of this together to get out master DESCEND algorithm for the CCC network:

ALGORITHM 6.5. If we have an $n$-processor CCC network and $n$ data-items $T[0]$, ..., $T[n-1]$, where $T[i]$ is stored in processor $i$ (in the numbering scheme described above). Recall that $k = r + 2^r$.

**do in parallel** (for all processors)
**for** $i \leftarrow 2^r - 1$ **downto** $-2^r$ **do**
    **do in parallel** (for all processors whose $\alpha$ coordinate
        satisfies $0 \le \alpha < 2^{k-r}$)
        and whose $\ell$-coordinate satisfies
        $\max(i, 0) \le \ell < \min(2^r, 2^r + i))$
    (*if* $\text{bit}_\ell(\alpha) = 0$ *then processor* $(\alpha, \ell)$ *contains*
        $T[\alpha 2^r + \ell]$
        *and if* $\text{bit}_\ell(\alpha) = 1$ *then processor* $(\alpha, \ell)$ *contains*
        $T[\alpha 2^r + \ell + 2^{\ell+r}])$
    All processors transmit their value of $T[*]$ along a lateral
        communication line
    Each processor $(\alpha, \ell)$ computes $\text{OPER}(a, b; U, V)$
        where: $a = \alpha 2^r + ((i - \ell - 1) \bmod 2^r)$
            $b = \ell + r$
            $U = T[\alpha 2^r + \ell]$
            $V = T[\alpha 2^r + \ell + 2^{\ell+r}]$
  B-LOOP
**endfor**


**for** $i \leftarrow r$ **downto** $0$ **do**
    **for** all processors $j$ such that $\text{bit}_i(j) = 1$
        transmit $T[j]$ to processor $j - 2^i$
    **endfor**
    **for** all processors $j$ such that $\text{bit}_i(j) = 0$
    **Compute** $\text{OPER}(j, i; T[j], T[j + 2^i])$
    (*Both of these data-items are now in processor* $j$)

**for** all processors $j$ such that $bit_i(j) = 0$
  transmit $T[j + 2^i]$ to processor $j + 2^i$
 **endfor**
**endfor**

The "transmit" operations are completely straightforward — they simply involve a sequence of steps in which data is sent to the appropriate neighboring processor.

The corresponding ASCEND algorithm is:

ALGORITHM 6.6. If we have an $n$-processor CCC network and $n$ data-items $T[0]$, $\ldots, T[n-1]$, where $T[i]$ is stored in processor $i$ (in the numbering scheme described above).

**for** $i \leftarrow r$ **downto** 0 **do**
 **for** all processors $j$ such that $bit_i(j) = 1$
  transmit $T[j]$ to processor $j - 2^i$
 **endfor**
 **for** all processors $j$ such that $bit_i(j) = 0$
 **Compute** $\mathrm{OPER}(j, i; T[j], T[j + 2^i])$
 *(Both of these data-items are now in processor $j$)*
 **for** all processors $j$ such that $bit_i(j) = 0$
  transmit $T[j + 2^i]$ to processor $j + 2^i$
 **endfor**
**endfor**


**do in parallel** (for all processors)
**for** $i \leftarrow -2^r$ **to** $2^r - 1$ **do**
 **do in parallel** (for all processors whose $\alpha$ coordinate
   satisfies $0 \le \alpha < 2^{k-r}$)
  and whose $\ell$-coordinate satisfies
   $\max(i, 0) \le \ell < \min(2^r, 2^r + i)$
 *(if* $\mathrm{bit}_\ell(\alpha) = 0$ *then processor* $(\alpha, \ell)$ *contains*
  $T[\alpha 2^r + \ell]$
 *and if* $\mathrm{bit}_\ell(\alpha) = 1$ *then processor* $(\alpha, \ell)$ *contains*
  $T[\alpha 2^r + \ell + 2^{\ell+r}])$
 All processors transmit their value of $T[*]$ along a lateral
  communication line
 Each processor $(\alpha, \ell)$ computes $\mathrm{OPER}(a, b; U, V)$
  where: $a = \alpha 2^r + ((i - \ell - 1) \bmod 2^r)$
    $b = \ell + r$
    $U = T[\alpha 2^r + \ell]$
    $V = T[\alpha 2^r + \ell + 2^{\ell+r}]$

F-LOOP
**endfor**

Although these algorithms look considerably more complex than the corresponding algorithms for the Butterfly and the shuffle-exchange network, their execution-time is comparable.

**Exercises:**

1. Consider a modification of the butterfly network, where we introduce a new edge in each column that connects the top and bottom vertex — we will call this the m-Butterfly network. It is similar, but not identical to, the form of the Butterfly that identifies the top and bottom rows. Show that the CCC network with a given value of $r$ is *isomorphic* to an m-Butterfly network with $2^r$ rows.
   a. Show that the wasteful DESCEND and ASCEND algorithms (6.2 and 6.3 on page 91) of the CCC map into algorithms 3.2 and 3.3 on page 74 for the Butterfly network, under this isomorphism.
   b. Map the master DESCEND and ASCEND algorithms for the CCC into an algorithm for the m-Butterfly network under this isomorphism

2. Implement the shift-operation (2.8 on page 71) on the CCC network. Describe all of the data-movements.

3. Implement the Bitonic sort algorithm (originally defined in § 3 on page 22) on the CCC network.

## 7. Dataflow Computers

Recall the definition of computation network in 7.1 on page 48. Dataflow computers represent an interesting approach to parallel computation in which a computation network for a problem is directly implemented by the hardware. The processors of a dataflow computer perform the arithmetic operations of the program and directly correspond to the vertices of the computation network. A program for a dataflow computer consists of a kind of symbolic representation of a computation network.

As one might think, many complex and interesting issues arise in connection with the design of dataflow computers — for instance:

- The architecture must reconfigure itself somehow during execution of a program.
- The vertices in the hardware-implemented computation network must perform various auxiliary functions, such as the queuing of data that has been received along input lines before the other data needed to perform a computation. For instance, if a processor is to add two numbers, and one of the numbers has arrived and the other hasn't, the processor must hold the number that is available and wait for the second number.

There is a great deal of research being conducted on the development of dataflow computers. At M.I.T. a group is working on the Monsoon Project — see [13]. In Japan a group is developing the Sigma-1 computer — see [136], [135].

## 8. The Granularity Problem

In all of the simulation algorithms described above we assumed that the unbounded parallel computer had an amount of memory that was proportional to the number of processors (this factor of proportionality is the term $\beta$ that appears so prominently). Suppose we want to simulate an unbounded parallel computer with a large amount of RAM — much larger than the number of processors. Also suppose that the processors in the network used to carry out the simulation together have enough memory between them to accomplish this. We then encounter the so-called *Granularity Problem* — how is this distributed memory to be efficiently accessed?

This is an interesting problem that has been considered before in the context of distributed databases — suppose a database is broken up into many pieces in distinct computers that are networked together. Suppose people use all of the computers in the network to access the database. Under some (possibly rare) circumstances it is possible that all of the users want data that is be located on the one computer in the network and response time will slow down tremendously because each individual computer in the network can only handle a small number of requests for data at a time. The question arises: Is is possible to organize the data in the database (possibly with multiple copies of data items) so that access to data is always fast regardless of how users request it?

The term granularity comes from the fact that the number of processors is much lower than the amount of memory available so that each processor has a sizable chunk of local memory that must be accessed by all of the other processors. This was an open question until recently. Work of Upfal and Wigderson solved this problem in a very satisfactory way. Although this entire chapter has made references to graphs to some extent (for instance all of the networks we have considered are graphs), the present section will use slightly more graph theory. We will make a few definitions:

DEFINITION 8.1. A *graph* (or *undirected graph*) is a pair $(V, E)$, where $V$ is a set of objects called the *vertices* of the graphs and $E$ is a set of ordered-pairs of vertices, called

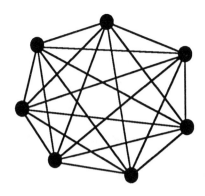

FIGURE III.8. A complete graph on 7 vertices

the *edges* of the graph. In addition, we will assume the *symmetry condition*

If a pair $(v_1, v_2) \in E$, then $(v_2, v_1) \in E$ also.

A *directed graph* (or *digraph*) is defined like an undirected graph except that the symmetry condition is dropped.

Throughout this section we will assume that our network computer has processors connected in a *complete graph*. This is a graph in which every vertex is connected to every other vertex. Figure III.8 shows a complete graph.

If not it turns out that a complete graph network computer can be efficiently simulated by the models presented above.

The main result is the following:

THEOREM 8.2. A program step on an EREW computer with $n$ processors and RAM bounded by a polynomial in $n$ can be simulated by a complete graph computer in $O((\lg n)(\lg \lg n)^2)$ steps.

The idea of this result is as follows:

1. Problems occur when many processors try to access the data in a single processor's local memory. Solve this by keeping many copies of each data item randomly distributed in other processors' local memory. For the time being ignore the fact that this requires the memory of the whole computer to be increased considerably.

2. When processors want to access local memory of some other processor, let them randomly access one of the many redundant copies of this data, stored in some other processor's local memory. Since the multiple copies of data items are randomly distributed in other processor's local memory, and since the processors that read this data access randomly chosen copies of it, *chances are* that the bottleneck described above *won't occur. Most* of the time the many

processors that want to access the same local data items will *actually* access *distinct* copies of it.

We have, of course, ignored several important issues:

1. If multiple copies of data items are maintained, how do we insure that all copies are *current* — i.e., some copies may get updated as a result of a calculation and others might not.
2. Won't total memory have to be increased to an unacceptable degree?
3. The vague probabilistic arguments presented above don't, in themselves prove anything. How do we really know that the possibility of many processors trying to access the same copy of the same data item can be ignored?

The argument presented above is the basis for a randomized solution to the granularity problem developed by Upfal in [154]. Originally the randomness of the algorithm was due to the fact that the multiple copies of data items were distributed randomly among the other processors. Randomness was needed to guarantee that other processors trying to access the same data item would usually access distinct copies of it. In general, the reasoning went, if the pattern of distributing the copies of data items were regular most of the time the algorithm would work, but under some circumstances the processors might try to access the data in the same pattern in which it was distributed.

Basically, he showed that if only a limited number of copies of each data item are maintained ($\approx \lg n$) conflicts will still occur in accessing the data (i.e. processors will try to access the same copy of the same data item) but will be infrequent enough that expect access time will not increase unacceptably. The fact that a limited number of copies is used answers the object that the total size of the memory would have to be unacceptably increased. Since this is a *randomized algorithm*, Upfal rigorously computed the *expected*[4] execution time and showed that it was $O(\beta \lg^2 n)$ — i.e. that the vague intuitive reasoning used in the discussion following the theorem was essentially correct. The issue of some copies of data getting updated and others being out of date was addressed by maintaining a time stamp on each copy of the data and broadcasting update-information to all copies in a certain way that guaranteed that a certain minimum number of these copies were current at any given time. When a processor accessed a given data item it was required to access several copies of that data item, and this was done in such a way that the processor was guaranteed of getting at least one current copy. After accessing these multiple copies of the data item the processor would then simply use the copy with the latest time stamp.

After this Upfal and Wigderson (in [155]) were able the make the randomized algorithm *deterministic*. They were able to show that there exist fixed patterns for distributing data among the processors that allow the algorithm to work, even in the worst case.

---

[4]In randomized algorithms the expected execution time is the *weighted average* of all possible execution times, weighted by the *probabilities* that they occur.

This is the algorithm we will consider in this section.

Let $U$ denote the set of all simulated RAM locations in the EREW computer. The important idea here is that of an *organization scheme* $S$. An organization scheme consists of an assignment of sets $\Gamma(u)$ to every $u \in U$ — where $\Gamma(u)$ is the set of processors containing RAM location $u$ — with a protocol for execution of read/write instructions.

We will *actually* prove the following result:

THEOREM 8.3. There exists a constant $b_0 > 1$, such that for every $b \geq b_0$ and $c$ satisfying $b^c \geq m^2$, there exists a consistent scheme with efficiency $O(b[c(\lg c)^2 + \lg n \lg c])$.

1. Note that we get the time estimates in the previous result by setting $c$ proportional to $\lg n$. In fact, $c$ must be $\geq \log_b(m^2) = 2\log(m)/\log(b)$.

2. Here:

> $m$ is the number of memory locations in the RAM that is being simulated,
> $n$ is the number of processors,
> $b$ is a constant parameter that determines how many real memory locations are needed to accomplish the simulation. It must be $\geq 4$ but is otherwise *arbitrary*.

To get some idea of how this works suppose we are trying to simulate a megabyte of RAM on a computer with 1024 processors. Each PE must contain 1024 simulated memory locations following the straightforward algorithm in the previous chapter, and each simulation step might require $1024 \lg^2 n$-time. Using the present algorithm, $m = 220, n = 210$, and execution time is

$$O\left(b\frac{40}{\log b}\log^2\left(\frac{40}{\log b}\right) + 10\log\left(\frac{40}{\log b}\right)\right)$$

with $40/\log b$-megabytes of real memory needed to simulate the one megabyte of simulated memory. By varying $b$ we determine execution time of the simulation verses amount of real memory needed — where $b$ must be at least 4.

In our scheme every item $u \in U$ will have *exactly* $2c - 1$ copies. It follows that $\Gamma(u)$ is actually a set of $2c - 1$ values: $\{\gamma_1(u), \ldots, \gamma_{2c-1}(u)\}$. These $\gamma$-functions can be regarded as *hashing functions*, like those used in the sorting algorithm for the hypercube computer. Each copy of a data item is of the form <value,time_stamp>. The protocol for accessing data item $u$ at the $t^{\text{th}}$ instruction is as follows:

1. to *update* $u$, access *any* $c$ copies in $\Gamma(u)$, update their values and set their time-stamp to $t$;
2. to *read* $u$ access *any* $c$ copies in $\Gamma(u)$ and use the value of the copy with the latest time-stamp;

The algorithm maintains the following *invariant condition*:

**Every $u \in U$ has at least $c$ copies that agree in value and are time-stamped with the index of the last instruction in which a processor updated $u$.**

This follows from the fact that every two $c$-subsets of $\Gamma(u)$ have a *non-empty intersection* (because the size of $\Gamma(u)$ is $2c - 1$).

Processors will *help* each other to access these data items according to the protocol. It turns out to be efficient if at most $n/(2c - 1)$ data items are processed at a time. We consequently shall partition the set of processors into $k = n/(2c - 1)$ groups, each of size $2c - 1$. There will be $2c$ phases, and in each phase each group will work in parallel to satisfy the request of one of its members. The current distinguished member will broadcast its request (access $u_i$ or write $v_i$ into $u_i$) to the other members of its group. Each of them will repeatedly try to access a fixed distinct copy of $u_i$. After each step the processors in this group will check whether $u_i$ is still alive (i.e., $c$ of its copies haven't yet been accessed). When $c$ of a given data item's copies have been accessed the group will *stop working* on it — the copy with the *latest* time stamp is computed and sent to $P_i$.

Each of the first $2c - 1$ phases will have a time limit that may stop processing of the $k$ data items while some are still alive (i.e., haven't been fully processed). We will show, however, that at most $k/(2c - 1)$ of the original $k$ items will remain. These are distributed, using *sorting*, one to a group. The last phase (which has no time limit) processes these items.

8.4. Let $P_{(m-1)(2c-1)+i}$, $i = 1, \ldots, 2c - 1$ denote the processors in group $m$, $m = 1, \ldots, k$, $k = n/(2c - 1)$. The structure of the $j^{\text{th}}$ copy of data item $u$ is
$<\text{value}_j(u), \text{time\_stam+}_j(u)>$

Phase (i, time_limit):
{
$m \leftarrow \lceil \text{processor\_\#}/2c - 1 \rceil$
$f \leftarrow (m - 1)(2c - 1)$
  $P_{f+i}$ broadcast its request
    (read($u_{f+i}$)) or
      update($u_{f+i}, v_{f+i}$) to
        $P_{f+1}, \ldots, P_{f+2c-1}$;
        live($u_{f+i}$) $\leftarrow$true;
        count $\leftarrow 0$;
      **while** live($u_{f+i}$) and count $<$
      time_limit **do**
        count $=$ count $+ 1$;
        $P_{f+j}$
        tries to access copy$_j(u_{f+i})$;

```
        if permission_granted
          if read_request
            read
            <value_j(u_{f+i}),
            time_stamp_j(u_{f+i})>
            else
            /*update request */
            <value_j(u_{f+i}),
            time_stamp_j(u_{f+i}>
            =<v_{f+i},t>
          if < c copies of u_{f+i} are still alive
            live(u_{f+i})← false;
        endwhile
          if read_request then find and send
            to P_{f+i} the value with the latest
            time_stamp;
}
/* phase i;*/
```

ALGORITHM 8.5. Here is the top-level algorithm:
```
begin
    for i = 1 to 2c − 1 do run phase(i, log_η 4c);
        for a fixed h to be calculated later
            sort the k'2^k live requests
            and route them
        to the first processors in the k'
        first groups, one to each processor;
    run phase(1, lg n);
    endfor
end; end algorithm;
```

LEMMA 8.6. If the number of live items at the beginning of a phase is $w \leq k$ then after the first $s$ iterations of the while loop at most $2(1 − 1/b)^s w$ live copies remain.

The significant point here is that the number of live data-items is reduced by a factor. This implies that at most a logarithmic number of iterations are required to reduce the number to zero.

PROOF. At the beginning of a phase there are $w$ live items, and all of their copies are alive so there are a total of $(2c − 1)w$ live copies. By the lemma above after $s$ iterations the number of live copies is $2 (1 − 1/b)^s(2c − 1)w$. Since $|\Gamma'(u)| \geq c$ for each live

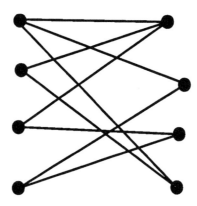

FIGURE III.9. A bipartite graph

item, these can be the live copies of at most $(1 - 1/b)^s(2c - 1)w/c \le 2(1 - 1/b)^s w$ live items. $\square$

COROLLARY 8.7. Let $\eta = (1 - 1/b)^{-1}$:

1. After the first $\log_\eta(4c - 2)$ iterations of the while loop, at most $k/(2c - 1)$ live items remain alive (so the last phase has to process at most $k$ requests);

2. After $\log_\eta 4c \le \log_\eta n$ iterations in a phase, no live items remain.

To complete the analysis note that each group needs to perform the following operations during each phase: *broadcast, maximum, summation* (testing whether $u_i$ is still alive). Also, before the last phase all the requests that are still alive are sorted.

It remains to give an efficient memory access protocol. In each iteration of the while loop in the algorithm the number of requests sent to each processor is equal to the number of live copies of live data items this processor contains. (Recall that a data item is called *live* if at least $c$ copies of it are live.) Since a processor can only process one data item at a time the number of copies processed in an iteration of the while loop is equal to the number of processors that contain live copies of live data items. We will, consequently, attempt to organize memory allocation in such a way as to maximize this number. We will use bipartite graphs to describe this memory allocation scheme.

DEFINITION 8.8. A *bipartite graph* is a graph with the property that its vertices can be partitioned into two sets in such a way that the endpoints of each edge lie in different sets. Such a graph is denoted $G(A, B, E)$, where $A$ and $B$ are the sets of vertices and $E$ is the set of edges.

Figure III.9 shows a bipartite graph.

In our memory allocation scheme $A$ we will consider bipartite graphs $G(U, N, E)$, where $U$ represents the set of $m$ shared memory items and $n$ is the set of processors.

In this scheme, if $u \in U$ is a data item then $\Gamma(u)$ is the *set* of vertices adjacent to $u$ in the bipartite graph being used.

LEMMA 8.9. For every $b \geq 4$, if $m \leq (b/(2e)^4)^{c/2}$ then there is a way to distribute the $2c - 1$ copies of the $m$ shared data items among the $n$ processors such that before the start of each iteration of the **while** loop $\Gamma \geq A(2c - 1)/b$. Here $\Gamma$ is the number of processors containing live copies of live data items and $A$ is the number of live data items.

We give a probabilistic proof of this result. Essentially we will compute the probability that an arbitrary bipartite graph has the required properties. If this probability is greater than 0 there must exist at least one good graph. In fact it turns out that the probability that a graph is good approaches 1 as $n$ goes to $\infty$ — this means that most arbitrary graphs will suit our purposes. Unfortunately, the proof is not constructive. The problem of verifying that a given graph is good turns out to be of exponential complexity.

It turns out that the bipartite graphs we want, are certain *expander graphs* — see § 4.3 in chapter VI (page 402). It is intriguing to compare the operation we are performing here with the *sorting-operation* in § 4.3.

PROOF. Consider the set $G_{m,n,c}$ of all bipartite graphs $G(U, N, E)$ with $|U| = m$, $|N| = n$, and with the degree of each vertex in $U$ equal to $2c - 1$.

We will say that a graph $G(U, N, E) \in G_{m,n,c}$ is *good* if for all possible choices of sets $\{\Gamma'(u): \Gamma'(u) \subseteq \Gamma(u), |\Gamma'(u)| \geq c$, for all $u \in U\}$ and for all $S \subseteq U$ such that $|S| \leq n/(2c - 1)$ the inequality $|\bigcup_{u \in U} \Gamma'(u)| \geq |S|(2c - 1)/b$ — here $S$ represents the set of live data items and $\bigcup_{u \in U} \Gamma'(u)$ represents the set of processors containing live copies of these data items. We will count the number of bipartite graphs in $G_{m,n,c}$ that aren't good or rather compute the *probability* that a graph isn't good. If a graph isn't good then there exists a choice $\{\Gamma'(u): \Gamma'(u) \subseteq \Gamma(u),$ such that $|\Gamma'(u)| \geq c$, for all $u \in U\}$ and a set $S \subseteq U$ such that $|S| \leq n/(2c - 1)$ and $|\Gamma'(u)| < |S|(2c - 1)/b$.

(2) $\quad$ Prob $[G \in G_{m,n,c}$ is not good]

$$\leq \sum_{g \leq \frac{n}{2c-1}} \binom{m}{g} \binom{n}{\frac{g}{b}(2c-1)} \binom{2c-1}{c}^g \left\{ \frac{g(2c-1)}{bn} \right\}^{gc}$$

Here $g$ is the size of the set $S$ and $g(2c-1)/b$ is the size of $|\bigcup_{u \in U} \Gamma'(u)|$. The formula has the following explanation:

1. the fourth factor is the probability that an edge coming out of a *fixed set* $S$ will hit a *fixed set* $|\bigcup_{u \in U} \Gamma'(u)|$ — the exponent $qc$ is the probability that *all* of the edges coming out of *all* of the elements of $S$ will have their other end in $|\bigcup_{u \in U} \Gamma'(u)|$.

The idea is that we imagine the graph $G$ as varying and this results in variations in the ends to the edges coming out of vertices of $S$.

2. the first factor is the number of ways of *filling out* the set $S$ to get the $m$ vertices of the graph;

3. the second factor is the number of ways of filling out the set $|\bigcup_{u \in U} \Gamma'(u)|$, whose size is $< |S|(2c-1)/b \le q(2c-1)/b$ to get the $n$ processor-vertices of the graph;

4. the third factor is the number of ways of adding edges to the graph to get the *original graph* in $G_{m,n,c}$ — we are choosing the edges that were deleted to get the subgraph connecting $S$ with $\Gamma'(u)$.

This can be *approximately evaluated* using Stirling's Formula, which states that $n!$ is asymptotically proportional to $n^{n-.5}e^{-n}$. We want to get an estimate that is $\ge$ the original.

DEFINITION 8.10.

$$\binom{m}{q} = \frac{m!}{q!(m-q)!}$$

We will use the estimate

(3)
$$\binom{m}{q} \le m^q q^{-q+1/2} e^q$$

*Claim: Formula (2) asymptotically behaves like* $o(1/n)$ *as* $n \to \infty$, *for* $b \ge 4$ *and*
$$m \le \left(\frac{b}{(2e)^4}\right)^{\frac{c}{2}}.$$

Since the formula *increases* with increasing $m$ we will assume $m$ has its maximal allowable value of $(b/(2e)^4)^{c/2}$. This implies that the first term is $\le ((b/(2e)^4)^{c/2})^g q^{-g+1/2}e^g = b^{gc/2} 2^{-2gc} e^{g-2gc} g^{-g+1/2}$. Now the third term is $((2c-1)!/c!(c-1)!)^g$, which can be estimated by $((2c-1)^{2c-1.5}/c^{c-.5}(c-1)^{c-1.5}))^g$, and this is

$$\le ((2c)^{2c-1.5}/c^{c-.5}c^{c-1.5})^g = ((2c)^{2c-1.5}/c^{2c-2})^g$$
$$= (2^{2c-1.5}c^{2c-1.5}/c^{2c-2})^g k = (2^{2c-1.5}c^{.5})^g \le c^{.5g}2^{2cg}$$

The product of the first and the third terms is therefore

$$\le b^{gc/2} 2^{-2gc} e^{g-2gc} g^{-g+1/2} c^{.5g} 2^{2cg}$$
$$= b^{gc/2} e^{g-2gc} g^{-g+1/2} c^{g/2}$$

Now the second term is

$$\le n^{g(2c-1)/b}(g(2c-1)/b)^{-(g(2c-1)/b)+1/2}e^{(g(2c-1)/b)}$$
$$\le ng^{c/2}(g(2c-1)/b)^{-(g(2c-1)/b)+1/2}e^{gc/2}$$

— here we have replaced $g(2c-1)/b$ first by $g(2c-1)/4$ (since we are only trying to get an upper bound for the term and 4 is the *smallest allowable* value for $b$), and then by $gc/2$. We can continue this process with the term raised to the 1/2 power to get

$$n^{gc/2}(g(2c-1)/b)^{-(g(2c-1)/b)}(gc/2)^{1/2}e^{gc/2}$$
$$= n^{gc/2}(g(2c-1))^{-(g(2c-1)/b)}b^{(g(2c-1)/b)}(gc/2)^{1/2}e^{gc/2}$$
$$= n^{gc/2}(g(2c-1))^{-(g(2c-1)/b)}b^{gc/2}(gc/2)^{1/2}e^{gc/2}$$

The product of the first three terms is

$$bgc/2e^{g-2gc}g^{-g+1/2}c^{g/2}n^{gc/2}(g(2c-1))^{-(g(2c-1)/b)}b^{gc/2}(gc/2)^{1/2}e^{gc/2}$$
$$= b^{gc}e^{g-3gc/2}g^{-g+1}c^{g+1/2}n^{gc/2}(g(2c-1))^{-(g(2c-1)/b)}$$

The product of all four terms is

$$b^{gc}e^{g-3gc/2}g^{-g+1}c^{g+1/2}n^{gc/2}(g(2c-1))^{-(g(2c-1)/b)}(g(2c-1)/bn)^{gc}$$
$$= e^{g-3gc/2}g^{-g+1}c^{g+1/2}n^{gc/2}(g(2c-1))^{-(g(2c-1)/b)}(g(2c-1)/n)^{gc}$$

Now note that the exponential term (on the left) dominates all of the polynomial terms so the expression is $\le g^{-g}c^g n^{gc/2}(g(2c-1))^{-(g(2c-1)/b)}(g(2c-1)/n)^{gc}$. In the sum the first term is $cn^{-c/2}(2c-1)^{c-((2c-1)/b)}$. This dominates the series because the factors of $g^{-g}$ and $(g(2c-1))k^{-(g(2c-1)/b)}$ *overwhelm* the factors $c^g n^{gc/2}$ — the last factor is bounded by 1. $\square$

**Exercises:**

1. The algorithm for the Granularity Problem requires the underlying network to be a *complete graph*. Can this algorithm be implemented on *other* types of networks? What would have to be changed?
2. Program the algorithm for the granularity problem in Butterfly Pascal.
3. find an embedding of the CCC-network in a butterfly network that has had its highest and lowest ranks identified.
4. Show how a network computer with processors connected in a complete graph can be efficiently simulated by a shuffle-connected computer.

# IV
# Examples of Existing Parallel Computers

## 1. Asynchronous Parallel Programming

In this section we will consider software for asynchronous parallel programs. These are usually run on MIMD computers with relatively few processors, although it is possible to develop such programs for *networks* of sequential computers. Many of the important concepts were first discussed in § 7.3.1 in chapter II.

**1.1. Portable programming packages.** In this section we will discuss a general programming system that has been implemented on many different parallel computers. Our main emphasis will be on the LINDA package. It is widely available striking in its simplicity and elegance. The original system, called LINDA, was developed by Scientific Computing Associates and they have ported it to many platforms. There are also sever free versions of LINDA:

1. a package called POSYBL that can be obtained via anonymous ftp from `ariadne.csi.forth.gr`.
2. and a package called Prolog-D-Linda (available by ftp from `ftp.cs.uwa.edu.au`, with a full research report). This package is particularly interesting because it is designed to be used in conjunction with a version of Prolog (SICSTUS prolog).

Linda is particularly interesting because it is available on parallel computers and on sets of networked uni-processor machines. The network version of LINDA effectively converts networked uni-processor machines into a kind of MIMD-parallel computer. It consists of:

- A set of commands that can be issued from a C or Pascal program.
- A preprocessor that scans the source program (with the embedded LINDA commands) and replaces them by calls to LINDA functions.

In order to describe the commands, we must first mention the LINDA concept of *tuple space*. Essentially, LINDA stores all data that is to be used by several processes in table created and maintained by LINDA itself, called tuple space. User-defined processes issue LINDA commands to put data into, or to take data out of tuple space. In particular, they are completely unable to share memory *explicitly*. LINDA controls access to tuple space in such a way that race-conditions are impossible — for instance the program on page 55 can't be written in LINDA[1].

A *tuple* is an ordered sequence of typed values like:

(1,zz,"cat",37)

or

(1,3.5e27)

An *anti* tuple is similar to a tuple, except that some of its entries are written as variable-names preceded by ?. Example:

(1,?x,?y,37)

A tuple is said to *match* an anti-tuple if they both have the same number of entries, and all of the quantities in the anti-tuple are exactly equal to the corresponding quantities in the tuple. For instance:

(1,?x,?y,37)

matches

(1,17,"cat",37)

but *doesn't* match

(1,17,37)

(wrong number of entries) or

(2,17,"cat",37)

(first entries don't agree).

The LINDA system defines the following operations upon tuples:

1. **out**(*tuple*). This inserts the tuple that appears as a parameter into tuple space. Example:

   **out**(1,zz,"cat",37)

---

[1]So LINDA is not a suitable system to use in the *beginning* of a course on concurrent programming!

Suppose $t$ and $a$ are, respectively, a tuple and an anti-tuple. The operation of *unifying* $t$ with $a$ is defined to consist of assigning to every element of the anti-tuple that is preceded by a ?, the corresponding value in the tuple. After these assignments are made, the anti-tuple and the tuple are *identical*.

2. **in**(*anti-tuple*) This attempts to find an actual tuple in tuple-space that *matches* the given anti-tuple. If no such matching tuple exists, the process that called **in** is suspended until a matching tuple appears in tuple space (as a result of another process doing an **out** operation). If a match *is* found
   a. the matching tuple is atomically removed from tuple-space
   b. the anti-tuple is unified with the matching tuple. This results in the assignment of values to variables in the anti-tuple that has been preceded by a ?.

The **in** statement represents the primary method of receiving data from other processes in a system that employs LINDA. Here is an example: We have a process that has a variable defined in it named zz. The statement:

**in**(1,?zz)

tests whether a tuple exists of the form (1,x) in tuple-space. If none exists, then the calling process is blocked. If such a tuple is found then it is *atomically* removed from tuple-space (in other words, if several processes "simultaneously" attempt such an operation, only one of them succeeds) and the assignment zz←x is made.

3. **eval**(*special-tuple*) Here *special-tuple* is a kind of tuple that contains one or more *function-calls*. LINDA created separate processes for these function-calls. This is the primary way to create processes in the LINDA system. Example:

**eval**(compeq(pid))

4. **rd**(*anti-tuple*) Similar to **in**(*anti-tuple*), but it doesn't remove the matching tuple from tuple-space.

5. **inp**(*anti-tuple*) Similar to **in**(*anti-tuple*), but it doesn't suspend the calling process if no matching tuple is found. It simply returns a boolean value (an integer equal to 1 or 0) depending on whether a matching tuple was found.

6. **rdp**(*anti-tuple*) Similar to **inp**(*anti-tuple*), but it doesn't remove a matching tuple from tuple-space.

Although LINDA is very simple and elegant, it has the unfortunate shortcoming that is lacks anything analogous to semaphore *sets* as defined on page 62.

Here is a sample LINDA program that implements the generic DESCEND algorithm (see 2.1 on page 68):

```
#define TRUE 1
#define K  /* the power of 2 defining the size of the
```

```
                * input−data. */
#define N  /* 2^k*/
int bit_test[K];
real_main(argc, argv) /* The main program in LINDA must be called
                * 'real_main' ––– the LINDA system has its own
                * main program that calls this. */
int   argc;
char **argv;
{
  struct DATA_ITEM {/* declaration */}

  struct DATA_ITEM T[N];
  int i,k;
  for (i=0; i<N; ++i) out(T[i],i,K); /* Place the data into
                        * tuple−space */
  for (i=0; i<K; i++) bit_test[i]=1<<i; /* Initialize bit_test array. */
  for (k=0; k<K, k++)
    for (i=0; i<N; ++i) eval(worker(i,k)); /* Start processes */

}


void worker(x,k)
int x,k;
{
int k;
struct DATA_ITEM data1,data2;
  if ((bit_test[k]&x) == 0) then
    {
    in(?data1,x,k);
    in(?data2,x+bit_test[k],k);
    OPER(x,k,&data1,&data2);
    out(data1,x,k−1);
    out(data2,x+bit_test[k],k−1);
    }
}
```

In this program, we are assuming that the procedure **OPER**(x,k,*data1,*data2); is declared like:

**void OPER**(x,k,p1,p2); **int** x,k; **struct** DATA_ITEM *p1, *p2;

and is defined to perform the computations of $OPER(m, j; T[m], T[m + 2^j])$. This program is not necessarily practical:

1. The time required to start a process or even to carry out he **in** and **out** operations may exceed the time required to simply execute the algorithm sequentially.
2. Only half of the processes that are created ever execute.

Nevertheless, this illustrates the simplicity of using the parallel programming constructs provided by the LINDA system. Substantial improvements in the speed of LINDA (or even a hardware implementation) would negate these two objections. In addition, a practical algorithm can be developed that is based upon the one presented above — it would have each process do much more work in executing the DESCEND algorithm so that the time required for the real computations would dominate the overall computing time. For instance we can have each worker routine perform the $OPER(m, j; T[m], T[m + 2^j])$ computation for $G$ different values of $m$ rather than just 1. Suppose:

$$p_t = \text{Time required to create a process with \textbf{eval}.}$$

$$i_t = \text{Time required to \textbf{in} a tuple.}$$

$$o_t = \text{Time required to \textbf{out} a tuple.}$$

$$r_t = \text{Time required to perform } OPER(m, j; T[m], T[m + 2^j]).$$

Then the time required to perform the whole algorithm, when each worker procedure works on $G$ distinct rows of the T-array is approximately:

$$nkp_t/G + k(i_t + o_t) + kGr_t$$

Here, we are assuming that, if each worker process handles $G$ different sets of data, we can get away with creating $1/G$ of the processes we would have had to otherwise. We are also assuming that the times listed above *dominate* all of the execution-time — i.e., the time required to increment counters, perform **if**-statements, etc., is negligible. The overall execution time is a minimum when the derivative of this with respect to $G$ is zero or:

$$-nkp_t/G^2 + kr_t = 0$$

$$nkp_t = G^2 kr_t$$

$$G = \sqrt{\frac{np_t}{r_t}}$$

The resulting execution-time is

$$nkp_t/G + k(i_t + o_t) = \frac{nkp_t}{\sqrt{\frac{np_t}{r_t}}} + kr_t\sqrt{\frac{np_t}{r_t}} + k(i_t + o_t)$$
$$= k\sqrt{np_t r_t} + k\sqrt{np_t r_t} + k(i_t + o_t)$$
$$= k\left(2\sqrt{np_t r_t} + i_t + o_t\right)$$

One good feature of the LINDA system is that it facilitates the implementation of calibrated-algorithms, as described in 7.10 on page 62.

ALGORITHM 1.1. Suppose we have a calibrated algorithm for performing computation on an EREW-SIMD computer. Then we can implement this algorithm on a MIMD computer in LINDA as follows:

1.  Create a LINDA process for each processor in the SIMD algorithm (using **eval**).
2.  Add all input-data for the SIMD algorithm to the tuple-space via **out**-statements of the form

    **out**(addr,0,data);

    where addr is the address of the data (in the SIMD-computer), 0 represents the $0^{th}$ time step, and the data is the data to be stored in this memory location (in the SIMD-computer). In general, the middle entry in the **out**-statement is a *time-stamp*.
3.  In program step $i$, in LINDA process $p$ (which is simulating a processor in the SIMD computer) we want to read a data-item from address $a$. The fact that the algorithm was *calibrated* implies that we *know* that this data-item was written in program-step $i' = f(i, a, p)$, where $i' < i$, and we We perform

    **rd**($a,i'$,data);
4.  In program step $i$, in LINDA process $p$, we want to store a data-item. We perform

    **out**($a,i$,data);

## 2. SIMD Programming: the Connection Machine

**2.1. Generalities.** In this section we will consider the Connection Machine manufactured by Thinking Machines Corporation. Aside from the inherent interest in this particular piece of hardware, it will be instructive to see how a number of important issues in the theory of parallel algorithms are addressed in this system.

We first consider some general features of the architecture of this system. The computer (model CM-2) is physically implemented as a 12-dimensional hypercube, each node of which has 16 processors. This is a SIMD machine in which programs are stored on a front end computer. The front end is a conventional computer that is hard-wired into the Connection Machine. As of this writing (March, 1990) the types of machines that can be front end for the Connection Machine are: DEC VAX's; Symbolics Lisp machines; Sun machines. In a manner of speaking the Connection Machine may be viewed as a kind of coprocessor (!) for the front end — even though its raw computational power is usually many thousands of times that of the front end. The front end essentially "feeds" the Connection Machine its program, one instruction at a time. Strictly speaking, it broadcasts the instructions to all of the processors of the Connection Machine though the hardware connection between them. Each processor executes the instructions, using data from:

1. the front end (which acts like common RAM shared by all processors)
2. its own local RAM (8K);
3. local RAM of other processors.

The first operation is relatively slow if many processors try to access distinct data items from the front end at the time time — the front end can, however, broadcast data to all processors very fast. This turns out to mean that it is advantageous to store data that must be the same for all processors, in the front end. Copying of data between processors (statement 3 above) follows two basic patterns:

1. **Grid communication** This makes use of the fact that the processors in the machine can be organized (at run-time) into arrays with rapid communication between neighboring processors. Under these conditions, the Connection Machine behaves like a mesh-connected computer. Grid communication is the fastest form of communication between processors.
2. **General communication** This uses an algorithm somewhat like the randomized sorting algorithm of Valiant and Brebner (see [157] and §4.1 in chapter II of this book). This sorting algorithm is used when data must be moved between distinct nodes of the 12-dimensional hypercube (recall that each node has 16 processors). The sorting algorithm is built into hardware and essentially transparent to the user. Movement of data between processors is slower than access of local memory, or receiving broadcast data from the front end. Since this data movement uses a *randomized* algorithm, the time it takes varies — but is usually reasonably fast. In this case the Connection Machine can be viewed as a P-RAM machine, or *unbounded parallel computer*.

You can think of the memory as being organized into hierarchies:

- local memory — this is memory contained in the individual processors. This memory can be accessed very rapidly, in parallel.
- front-end — this is memory contained in the front-end computer, that also contains the *program* for the connection machine. This memory can also be accessed rapidly, but only a limited amount of parallelism can be used. The

front-end computer can
- read data from a single individual processor in the Connection Machine, in one program-step;
- perform a *census-operation* on all active processors. See § 2.2 on page 117 for a discussion of census-operations.
- "broadcast" data from the front-end to all of the processors in a single program-step. This is a fast operation, since it uses the same mechanism used for broadcasting the instructions to the Connection Machine.

- non-local — here processors access local memory of other processors. This is a parallel memory access that is somewhat slower than the other two operations. It still uses a single program-step — program steps on the Connection Machine do not all use the same amount of time.

This division into hierarchies turns out to fit in well with programming practices used in high-level programming languages, since modular programs make use of local memory in procedures — and this turns out to be local memory of processors.

I/O can be performed by the front-end, or the Connection Machine itself. The latter form of I/O make use of a special piece of hardware called a Data Vault — essentially disk drives with a high data rate (40 megabytes/second).

The Connection Machine is supplied with several programming languages, including:

**Paris** essentially the assembly language of the Connection Machine (it stands for Parallel Instruction Set). It is meant to be used in conjunction with a high-level language that runs on the front-end computer. This is usually C. You would, for instance, write a C program that contained Paris instructions embedded in it.

**\*Lisp** this is a version of Common Lisp with parallel programming features added to it. These extra commands are in a close correspondence with PARIS instructions — causing \*Lisp to be regarded as a "low-level" language (!). There is also a form of lisp called CM-Lisp that is somewhat higher-level than \*Lisp.

**CM-Fortran** this is a version of Fortran with parallel programming primitives.

**C\*** This language a superset of ANSI C. We will focus on this language in some detail. Since it is fairly typical of a language used on SIMD machines we will use it as a kind of *pseudocode* for presenting parallel algorithms in the sequel. There are two *completely different* versions of this language. The old versions (< 6.0) were very elegant but also inefficient to the point of being essentially *unusable*. The new versions (with version-numbers ≥ 6.0) are lower-level languages[2] that are very powerful and efficient — they rival C/Paris in these

---

[2]As this statement implies, these two versions have almost *nothing* in common but the name, C\*, and

respects. We will focus on this language in the remainder of this chapter.

**2.2. Algorithm-Design Considerations.** In many respects designing algorithms for the Connection Machine is like designing algorithms for a PRAM computer. The processors have their own local RAM, and they don't share it with other processors (at least in a straightforward way), but modular programming techniques make *direct* sharing of memory unnecessary in most circumstances.

One important difference between the Connection Machine and a PRAM machine is that two operations have different relative timings:

- *Census* operations take constant time on a Connection Machines and $O(\lg n)$-time on a PRAM computer. Census operations are operations that involving accumulating data over all of the active processors. "Accumulating data" could amount to:
    1. taking a (sum, product, logical OR, etc.) of data items over all active processors.
    2. counting the number of active processors.

    On a PRAM computer, census operations are implemented via an algorithm like the one on page 69 or in § 1 of chapter VI. Although the computations are not faster on the Connection Machine, the census operations are effectively built into the hardware, the time spent doing them is essentially incorporated into the machine cycles.
- Parallel data movement-operations require *unit-time* on a PRAM computer, but may require longer on a Connection Machine. Certain basic movement-operations are built into the hardware and require unit time, but general operations[3] can require much longer. See section 2.4.8 for more details.

**2.3. The C\* Programming Language.** In addition to the usual constructs of ANSI C, it also has ways of describing code to be executed in parallel. Before one can execute parallel code, it is necessary to describe the layout of the processing elements. This is done via the **shape** command. It allocates processors in an *array* and assigns a name to the set of processors allocated. All code to be executed on these processors must contain a reference to the name that was assigned. We illustrate these ideas with a sample program — it implements the parallel algorithm for forming cumulative sums of integers:

```
#include <stdio.h>
#define MACHINE_SIZE 8192
shape [MACHINE_SIZE]sum;
```

---

the fact that they are loosely based on C.

[3]Such as a parallel movement of data form processor $i$ to processor $\sigma(i)$, with $\sigma$ an arbitrary permutation

```
int:sum n;          /* each processor will hold an integer value */
int:sum PE_number;   /* processors must determine their identity. */
int:current lower(int);
void add(int);
int:current lower (int iteration)
  {
    int next_iter=iteration+1;
    int:current PE_num_reduced = (PE_number >>
          next_iter)<<next_iter;
    return PE_num_reduced + (1<<iteration) − 1;
  }
void add(int iteration)
  {
    where (lower(iteration)<PE_number)
       n+=[lower(iteration)]n;
  }
void main()  /* code not prefaced by a shape name
          is executed sequentially on the front end. */
  {
    int i;
    with (sum) PE_number= pcoord(0);
    printf("Enter 8 numbers to be added in parallel:\n");
    for (i=0;i<8;i++)
      {
        int temp;
        scanf("%d",&temp);
        [i]n=temp;
      }
    for (i=0;i<=2;i++)
      with(sum) add(i);
        /* the call to add is executed in parallel
         by all processors in the shape named sum*/
    printf("The cumulative sums are:\n");
    for (i=0;i<8;i++)
      printf("Processor %d, value =%d\n",
          [i]PE_number,[i]n);
  }
```

The first aspect of this program that strikes the reader is the statement: **shape** [MA-CHINE_SIZE]sum;. This allocates an *array* of processors (called a *shape*) of size equal to MACHINE_SIZE and assigns the name sum to it. Our array was one-dimensional, but you can allocate an array of processors with up to 32 dimensions[4]. The dimension is actually significant in that transmission of data to neighboring processors can be done faster than to random processors. This language construct mirrors the fact that the hardware of the Connection Machine can be dynamically reconfigured into arrays of processors. Of course transmission of data between processors can also be done at *random* (so that the Connection Machine can be viewed as a CREW computer) — this motion is optimized for certain configurations of processors. If you are writing a program in which the data movement doesn't fit into any obvious array-pattern it is usually best to simply use a one-dimensional array and do random-access data movement (data-movement commands can contain subscripts to denote target-processors).

All statements that use this array of processors must be within the scope of a statement **with**(sum).

The statement **int**:sum n; allocates an integer variable in each processor of the shape sum. Note that the shape name follows the data type and that they are separated by a colon. In C* there are *two* basic classes of data: *scalar* and *parallel*. Scalar data is allocated on the front-end computer in the usual fashion (for the C-language), and parallel data is allocated on the processors of the Connection Machine.

You may have noticed the declaration: **int**:**current** lower(int); and wonder where the shape **current** was declared. This is an example of a pre-defined and *reserved* shape name. Another such predefined name is **physical** — it represents all processors physically present on a given Connection Machine. The word **current** simply represents all *currently-active* processors. The fact that the procedure lower has such a shape declaration in it implies that it is *parallel code* to be executed on the processors of the Connection Machine, rather than the *front end*. The word **current** is a kind of *generic term* for a shape that can be used to declare a procedure to be parallel.

Note that the *parameter* to this procedure has *no* shape declaration — this means it represents *scalar* data allocated on the *front end* computer and *broadcast* to the set of active processors whenever it is needed. Note that the procedure allocates data on the *front end* and on the processors — the second allocation is done via **int**:**current** PE_num_reduced....

Consider the next procedure:

```
void add(int iteration)
{
  where (lower(iteration)<PE_number)
    n+=[lower(iteration)]n;
```

---

[4]Currently, (i.e., as of release 6.0 of C*) each dimension of a shape must be a power of 2 and the total number of processors allocated must be a power of 2 times the number of processors physically present in the machine

```
}
```

This block of code is executed sequentially on the front end computer. On the other hand, it *calls* parallel code in the **where**-statement. A **where** statement evaluates a logical condition on the set of active processors and executes its consequent on the processors that satisfy the test. In a manner of speaking, it represents a parallel-if statement.

Now we look at the main program. It is also executed on the front end. The first unusual piece of code to catch the eye is:

```
with (sum) PE_number = pcoord(0);
```

At most one shape can be active at any given time. When a shape is activated all processors in the shape become active and parallel code can be executed on them. Note that **where** and **with**-statements can be *nested*.

Nested **where** statements simply *decrease* the number of active processors — each successive **where**-statement is executed on the processors that remained active from the previous **where** statements. This is a *block* structured statement in the sense that, when the program *exits* from the **where** statement the context that existed prior to it is restored. In other words, when one leaves a **where** block, all processors that were turned off by the logical condition are turned back on.

Nested **with** statements simply change the current shape (and *restore* the previously active shape when the program leaves the inner **with** statement).

Note the expression **pcoord**(0). This is a predefined parallel procedure (i.e. it runs on all active processors) that returns the coordinate of a processor in a given shape-array. These coordinates are numbered from 0 so the only coordinate that exists in a one-dimensional array like sum is the $0^{th}$. This statement assigned a unique number (in parallel) to each processor in the shape sum. Here is a list of the pre-defined functions that can be used to access shapes:

1. **pcoord**(*number*) — defined above.
2. **rankof**(*shape*) — returns the *rank* of the given shape — this is the number of dimensions defined for it. Example: **rankof**(sum) is equal to 1.
3. **positionsof**(*shape*) — returns the total number of positions of a given shape — essentially the number of processors allocated to it. **positionsof**(sum) is 8192.
4. **dimof**(*shape, number*) — returns the range of a given dimension of a shape.

Next, we examine the code:

```
for (i=0;i<8;i++)
{
```

```
int temp;
scanf("%d",&temp);
[i]n=temp;
}
```

This reads 8 numbers typed in by the user and stores them into 8 processors. The scanf procedure is the standard C-procedure so that it reads data into the *front-end* computer. We read the data into a temporary variable names temp and then assign it into the variable n declared in the processors of sum. Note the statement [i]n=temp; — this illustrates how one refers to data in parallel processors. The data temp is assigned to the $i^{th}$ copy of n. Such statements can be used *in* parallel code as well — this is how one does *random access* data movement on a Connection Machine. With a multi-dimension shape of processor, each subscript is enclosed in its own pair of square brackets: [2][3][1]z.

There is also a special notation for referencing *neighboring* processors in a *shape*. Suppose p is a parallel variable in a one-dimensional shape. Then the notation [**pcoord**(0)+1]p refers to the value of p in the *next-higher* entry in this shape. The statement

[**pcoord**(0)+1]p=p;

sends data in each processor in this shape to the next higher processor in the shape — in other words it does a *shift operation*. The statement

**with** (this_shape)
z=[**pcoord**(0)+1][**pcoord**(1)−1]z;

does a two-dimensional shift operation.

In such cases, there is a shorthand for referring to the *current position* or *current value* of some index in a shape — this is the *dot notation*. The expression 'pcoord(i)', where i is the current index, is represented by a *dot*. The data-movements described above can be replaced by:

[.+1]p=p;

and

**with** (this_shape) z=[.+1][.−1]z;

The next sample program illustrates the use of a two-dimensional shape. It implements the Levialdi Counting Algorithm, discussed in the Introduction. First, it generates a random array of pixels using the **rand** function. Then it implements the Levialdi algorithm and reports on the number of connected components found.

```
#include <stdio.h>
#include <stdlib.h>
shape [64][128]pix;            /* image, represented as an array of
                                * pixels */
int:pix val, nval;             /* 1 or 0, representing "on" or "off"
                                * pixel */
int        total = 0;     /* total number of clusters found
                                * thus far */
int:current h(int:current);
int:pix h(int:pix);
void        update();
int         isolated();
void        print_array();
void
main()
{
    int        i, j, on_pixels;      /* number of 1 pixels in
                                      * image */
    unsigned int   seed;  /* seed value passed to random number
                                * generator */
    int        density;/* density (% of 1's) in original
                                * image (0-100) */
    printf("Enter density and seed:  ");
    scanf("%d %u", &density, &seed);
    srand(seed);
    for (i = 0; i < 64; i++)
        for (j = 0; j < 64; j++)
            [i][j] val = ((i == 0) || (i >= 63) ||
                    (j == 0) || (j >= 63)) ? 0 :
                  (rand() % 100 < density);
    printf("Array Initialized\n");
    print_array();
    on_pixels = 1;
    i = 0;
    with(pix)
```

```
        while (on_pixels > 0) {
        int        t = 0;
        update();
        i++;
        t += val;
        if (i < 2)
                print_array();
        on_pixels = t;
        printf(
        "Iteration %.3d -- %.4d pixels on, %.3d clusters found\n",
            i, on_pixels, total);
    }
    printf("Final tally:    %d clusters found \n",
        total);
}
int:current h(int:current x)
{
    return (x >= 1);
}
void update()
{
    total += isolated();
    where((pcoord(0) > 0)
        & (pcoord(0) < 63)
        & (pcoord(1) > 0)
        & (pcoord(1) < 63)) {
      val = h(h([.][. − 1] val
            +[.][.] val
            +[. + 1][.] val − 1)
          + h([.][.] val
            +[. + 1][. − 1] val − 1));
    }
}
int isolated()
{
    int        t = 0;
    int:current iso;
    iso = (val == 1)
        & ([. − 1][.] val == 0)
        & ([. + 1][.] val == 0)
```

```
                    & ([.][. − 1] val == 0)
                    & ([.][. + 1] val == 0)
                    & ([. − 1][. − 1] val == 0)
                    & ([. + 1][. − 1] val == 0)
                    & ([. − 1][. + 1] val == 0)
                    & ([. + 1][. + 1] val == 0);
           t += iso;
           return t;
}
void
print_array()
{
       int         i, j;
       for (i = 0; i < 15; i++) {
            for (j = 0; j < 64; j++)
                 printf("%d",[i][j] val);
            printf("\n");
       }
       printf("\n\n");
}
```

2.3.1. *Running a C\* program.* To run a C\* program:

1. Give it a name of the form 'name.cs';
2. Compile it via a command of the form:

       cs  name.cs

   there are a number of options that can be supplied to the compiler. These closely correspond to the options for the ordinary C compiler. If no option regarding the name of the output file is supplied, the default name is 'a.out' (as with ordinary C).

3. Execute it via a command:

       a.out

   This attaches to the Connection Machine, and executes a.out.

## 2.4. Semantics of C\*.

2.4.1. *Shapes and parallel allocation of data.* We will begin this section by discussing the precise meaning of the **with** statement and its effects on **where** statements. As has been said before, the **with** statement determines the *current* shape. In general, the current shape determines which data can *actively* participate in computations. This rule has the following exceptions:

1. You can declare a parallel variable that is not of the current shape. This variables cannot be initialized, since that would involve its participation in active computations.
2. A parallel variable that is not of the current shape can be operated upon if it is left-indexed by a scalar or scalars — since the result of left-indexing such a parallel variable is effectively a scalar. In other words, you can execute statements like [2][3]z=2; even though z is not of the current shape.
3. The result of left-indexing a variable not of the current shape by a parallel variable that *is* of the current shape is regarded as being of the current shape. In other words, the following code is valid:

    > **shape** [128][64]s1;
    > **shape** [8192]s2;
    > **int**:s2 a,b;
    > **int**:s1 z;
    > **with** (s2) [a][b]z=2;

    since the expression [a][b]z is regarded as being of the shape s2. This is essentially due to the fact that the set of *values* it takes on are *indexed* by the shape s2 — since a and b are indexed by s1.
4. It is possible to apply **dimof** and **shapeof** and the address-of operator, &, to a parallel variable that is not of the current shape. This is due to the fact that these are implemented by the compiler, and don't involve actual parallel computations.
5. You can right-index a parallel array that is not of the current shape with a *scalar expression*.
6. You can use the "dot" operator to select a field of a parallel structure or union that is not of the current shape — provided that field is not another structure, union, or array.

2.4.2. *Action of the **with**-statement.* The **with**-statement *activates* a given shape and makes it the *current* shape. The concept of current shape is central to the C* language. In order to understand it, we must consider the problem of describing SIMD code. Since each processor executes the same instruction, it suffices to describe what *one processor* is doing in each program step. We consider a single *generic* processor and its actions. This *generic* processor will be called the *current* processor. The term "current processor" does not mean that only one processor is "current" at any given time — it merely selects one processor out of all of the processors and uses it as an *example* of how the parallel program executes on all of the processors.

The *current shape* defines:

- how the current processor is accessed in program-code. If the current shape is $k$ dimensional, the current processor is a member of a $k$-dimensional array.
- the functions **pcoord**($i$).

The next example illustrates how variables that are *not* of the current shape can be accessed in parallel code. This code-fragment will multiply a vector v by a matrix A:

```
shape [64][128]twodim;
shape [8192]onedim;
int:twodim A;
int:onedim v,prod;
void main()
{
   with(onedim) prod=0; /* Zero out the v-vector. */
   with(twodim) /* Make the shape 'twodim' current. */
   {
      [pcoord(0)]prod+=A*[pcoord(1)]v;
   /* This code accesses 'prod' and 'v', in
    * spite of the fact that they are not of the
    * current shape. The fact that 'twodim' is
    * the current shape defines the 'pcoord(0)'
    * 'pcoord(1)' */


}
```

The use of the shape twodim is not *essential* here — it merely makes it more convenient to code the array-access of A. We could have written the code above without using this shape. The processors in the shape twodim are the *same* processors as those in the shape *onedim* (not simply the same *number* of processors). Here is a version of the code-fragment above that only uses the onedim-shape:

```
shape [64][128]twodim;
shape [8192]onedim;
int:twodim A;
int:onedim v,prod;
void main()
{
   with(onedim) prod=0; /* Zero out the v-vector. */
   with(onedim) /* Continue to use the shape 'onedim'. */
   {
   [pcoord(0)/128]prod+=[pcoord(0)/128][pcoord(0) % 128]A*v;
```

```
    }
}
```

Here, we use the expression [**pcoord**(0)/128][**pcoord**(0) % 128]A to refer to elements of A. We assume that the shape twodim is numbered in a row-major format.

The second code-fragment carries out the same activity as the first. The use of the shape twodim is not entirely essential, but ensures:

- The code-fragment

   [**pcoord**(0)]prod+=A*[**pcoord**(1)]v;

  is *simpler,* and
- It executes faster — this is due to features of the hardware of the Connection Machine. Shapes are implemented in hardware and a with-statement has a direct influence on the Paris code that is generated by the C* compiler.

2.4.3. *Action of the* **where**-*statement.* There are several important considerations involving the execution of the **where**-statement. In computations that take place within the scope of a **where**-statement, the active processors that are subject to the conditions in the **where**-statement are the ones whose coordinates are exactly equal to {**pcoord**(0),...,**pcoord**($n$)}. This may seem to be *obvious*, but has some interesting consequences that are not apparent at first glance. For instance, in the code:

```
shape [8192]s;
int:s z;
where(z > 2) [.+2]z=z;
```

a processor [i+2]z will receive the value of [i]z if *that value* is $> 2$ — the original value of [i+2]z (and, therefore, the whether processor number $i + 2$ was active) is irrelevant. Processor [.+2] in this statements is simply a *passive receiver* of data — it doesn't have to be *active* to receive the data from processor [.]z. On the other hand, processor [.] must be active in order to *send* the data.

Suppose the processors had z-values given by:

```
[0]z=1;
[1]z=3;
[2]z=3;
[3]z=0;
[4]z=1;
[5]z=4;
[6]z=2;
[7]z=0;
```

then the code fragment given above will result in the processors having the values:

$$[0]z=1;$$
$$[1]z=3;$$
$$[2]z=3;$$
$$[3]z=3;$$
$$[4]z=3;$$
$$[5]z=4;$$
$$[6]z=2;$$
$$[7]z=4;$$

When processors are inactive in a **where**-statement they mark time in a certain sense — they execute the code in a manner that causes no changes in data in their memory or registers. They can be passive receivers of data from other processors, however.

Other processors can also *read* data from an inactive processor's local memory. For instance the effect of the **where** statement:

**where**(z>2)
z=[.−1]z;

is to cause the original data in the parallel variable z to be transformed into:

$$[0]z=1;$$
$$[1]z=1;$$
$$[2]z=3;$$
$$[3]z=0;$$
$$[4]z=1;$$
$$[5]z=1;$$
$$[6]z=2;$$
$$[7]z=0;$$

In order to understand more complex code like

**where**(cond)
[.+3]z=[.+2]z;

it is necessary to pretend the existence of a new variable temp in active processors:

```
where(cond)
    {
    int:shapeof(z) temp;
    temp=[.+2]z;
    [.+3]z=temp;
    }
```

a given data-movement will take place if and only if the condition in cond is satisfied in the current processor. In this case *both* [.+2]z and [.+3]z play a *passive* role — the processor associated with one is a storage location for parallel data and the other is a passive receiver for parallel data.

These **where**-statements can be coupled with an optional **else** clause. The **else**-clause that follows a **where**-statement is executed on the *complement* of the set of processors that were active in the **where**-statement.

```
    where (c) s1;
        else s2;
```

Statement s2 is executed on all processors that *do not* satisfy the condition c.

There is also the **everywhere**-statement, that explicitly *turns on* all of the processors of the current shape.

Ordinary arithmetic operations and assignments in C* have essentially the same significance as in C except when scalar and parallel variables are combined in the same statement. In this case the following two rules are applied:

1. **Replication Rule:** A scalar value is automatically replicated where necessary to form a parallel value. This means that in a mixed expression like:

```
    int a;
    int:someshape b,c;
    b=a+c;
```

the value of a is converted into a parallel variable before it is substituted into the expression. This means it is *broadcast* from the front-end computer.

This is also used in the Levialdi Counting Program in several places — for instance:

```
    int:current h(int:current x)
    {
```

```
        return (x >= 1);
    }
```

2. **As-If-Serial Rule:** A parallel operator is executed for all active processors as if in some serial order. This means that in an expression:

```
    int a;
    int:someshape b;
    a=b;
```

a will get the value of *one* of the copies of b — *which* one cannot be determined before execution. Code like the following:

```
    int a;
    int:someshape b;
    a += b;
```

will assign to a the sum of *all* of the values of b. This is accomplished via an efficient parallel algorithm (implemented at a low-level) like that used to add numbers in the sample program given above. Note that the basic computations given in the sample program were unnecessary, in the light of this rule — the main computation could simply have been written as

```
    int total;
    with(sum) total += n;
```

This is used in several places in the Levialdi Counting Program:

```
with(pix)
    while (on_pixels > 0) {
    int        t = 0;
    update();
    i++;
    t += val;
    if (i < 2)
        print_array();
    on_pixels = t;
    printf(
```

```
"Iteration %.3d -- %.4d pixels on, %.3d clusters found\n",
    i, on_pixels, total);
}
```

2.4.4. *Parallel Types and Procedures.* 1. **Scalar data**. This is declared exactly as in ANSI C.

2. **Shapes** Here is an example:

> **shape** *descriptor* shapename;

where *descriptor* is a list of 0 or more symbols like [*number*] specifying the size of each dimension of the array of processors. If no descriptor is given, the shape is regarded as having been only *partially* specified – its specification must be completed (as described below) before it can be used.

When shapes are defined *in* a procedure, they are dynamically allocated on entry to the procedure and de-allocated on exit. Shapes can also be *explicitly* dynamically allocated and freed. This is done with the **allocate_shape** command. This is a procedure whose:

> 1. first parameter is a *pointer* to the shape being created;
> 2. second parameter is the *rank* of the shape (the number of dimensions);
> 3. the remaining parameters are the size of each dimension of the shape.

For instance, the shape sum in the first sample program could have been dynamically allocated via the code:

> **shape** sum;
> sum=**allocate_shape**(&sum,1,8192);

The shape pix in the Levialdi Counting Program could have been created by:

> **shape** pix;
> pix=**allocate_shape**(&pix,2,64,128);

Shapes can be explicitly deallocated by calling the function **deallocate_shape** and passing a pointer to the shape. Example: **deallocate_shape**(&pix);. You have to include the header file <stdlib.h> when you use this function (you do not need this header file to use **allocate_shape**).

3. **Parallel data** Here is a typical parallel type declaration:

> typename:shapename variable;

**4. Scalar pointer to a shape** In C*, shapes can be regarded as *data items* and it is possible to declare a scalar pointer to a shape. Example:

   **shape** *a;

Having done this we can make an assignment like

   a=&sum;

and execute code like

   **with**(*a)...

A function related to this is the **shapeof**-function. It takes a parallel variable as its parameter and returns the *shape* of the parameter.

**5. Scalar pointers to parallel data.** Sample:

   typename:shapename *variable;

Here variable becomes a pointer to *all* instances of a given type of parallel data. It becomes possible to write

   [3]*variable.data

Typecasting can be done with parallel data in C*. For example, it is possible to write a=(int:sum)b;.

6. type procedure_name (type1 par1, type2 par2,
         ...,typen parn);

Note that the preferred syntax of C* looks a little like Pascal. All procedures are supposed to be declared (or prototyped), as well as being defined (i.e. coded). The *declaration* of the procedure above would be:

7. type procedure_name (type1, type2,
         ...,typen);

— in other words, you list the type of data returned by the procedure, the name of the procedure, and the types of the parameters. This is a rule that is not strictly enforced by present-day C*, but may be in the future. Example:

```
int:sum lower (int iteration) /* DEFINITION of
                  'lower' procedure */
 {
  int next_iter=iteration+1;
  int:sum PE_num_reduced = (PE_number >>
        next_iter)<<next_iter;
  return PE_num_reduced + (1<<iteration) − 1;
 }
void add(int iteration) /* DEFINITION of 'add'
              procedure */
 {
```

```
    where (lower(iteration)<PE_number)
        n+=PE[lower(iteration)].n;
}
```

In the remainder of this discussion of C* we will always follow the preferred syntax.

8. **Overloaded function-definitions.** In C* it is possible to define scalar and parallel versions of the same function or procedure. In other words, one can declare procedures with the same name with with different data types for the parameters. It is possible to have different versions of a procedure for different *shapes* of input parameters. This is already done with many of the standard C functions. For instance, it is clear the the standard library <math.h> couldn't be used in the usual way for parallel data: the standard functions only compute single values. On the other hand, one can code C* programs as if many of these functions existed. The C* compiler automatically resolves these function-calls in the proper way when parallel variables are supplied as parameters. In fact, many of these function-calls are implemented on the Connection Machine as *machine language* instructions, so the C* compiler simply generates the appropriate assembly language (actually, PARIS) instructions to perform these computations.

In order to overload a function you must use its name in an **overload**-statement. This must be followed by the declarations of all of the versions of the function, and the definitions of these versions (i.e., the actual code).

2.4.5. *Special Parallel Operators.* The following special operators are defined in C* for doing parallel computations:

- <?= Sets the target to the *minimum* of the values of the parallel variable on the right;
- >?= Sets the target to the *maximum* of the values of the parallel variable on the right;
- ,= Sets the target to an *arbitrary selection* of the values of the parallel variable on the right.
- <? This is a function of two variables (scalar or parallel) that returns the minimum of the variables.
- >? This is a function of two variables (scalar or parallel) that returns the maximum of the variables.
- (?:) This is very much like the standard C operator-version of the if-statement. It is coded as (*var?stmt1*:*stmt2*). If *var* is a scalar variable it behaves *exactly* like the corresponding C statement. If *var* is a *parallel* variable it behaves like: **where**(*var*)     *stmt1*;     **else** *stmt2*;

C* also defines a new data type: **bool**. This can be used for flags and is advantageous, since it is implemented as a single bit and the Connection Machine can access single bits.

2.4.6. *Sample programs.* We begin with sample programs to compute Julia sets and the Mandelbrot set. These are important fractal sets used in many areas of pure and applied mathematics and computer science including data compression and image processing. Computing these sets requires a enormous amount of CPU activity and lends itself to parallelization. We begin by quickly defining these sets. These computations are interesting in that they require essentially *no* communication between processors.

Let $p \in \mathbb{C}$ be some complex number. The Julia set with parameter $p$ is defined to be the set of points $z \in \mathbb{C}$ with the property that the sequence of points $z_i$ remains bounded. Here $z_0 = z$ and for all $i$ $z_{i+1} = z_i^2 + p$. Note that if $z$ starts out sufficiently large, $\|z_i\| \to \infty$ as $i \to \infty$ since the numbers are being squared each time. In our program we will calculate $z_{100}$ for each initial value of $z$ and reject initial values of $z$ that give a value of $z_{100}$ whose absolute value is $> 5$ since adding $p$ to such a number doesn't have a chance of canceling it out[5]. We will assign a different processor to each point that is tested. Here is the program:

```
#include <stdio.h>
shape[64][128] plane;
float:plane r, c, r1, c1, x, y, mag;
int:plane injulia, row, column;
void
main()
{
     float        p_r = 0.0, p_c = 0.0;
     int          i, j;
     with(plane) {
          r = pcoord(0);      /* Determine row of processor. */
          c = pcoord(1);      /* Determine column of processor. */
          x = r / 16.0 − 2.0;   /* Determine x−coordinate
                              * (real part of point that
                              * this processor will
                              * handle.  We subtract 2.0
                              * so that x−coordinates will
                              * range from −2 to +2.  */
          y = c / 32.0 − 2.0;   /* Determine y−coordinate
                              * (complex part of point
                              * that this processor will
                              * handle.  We subtract 2.0
```

---

[5]This is not obvious, incidentally. It turns out that the values of $p$ that produce a nonempty Julia set have an absolute value $< 2$.

```
                         * so that y-coordinates will
                         * range from -2 to +2.  */
      r = x;
      c = y;
      injulia = 1;    /* Initially assume that all points
                       * are in the Julia set.  */
      mag = 0.;       /* Initially set the absolute value
                       * of the z(i)'s to 0.  We will
                       * calculate these absolute values
                       * and reject all points for which
                       * they get too large. */
}
printf("Enter real part of par:\n");
scanf("%g", &p_r);
printf("Enter complex part of par:\n");
scanf("%g", &p_c);
/*
 * Compute the first 100 z's for each point of the selected
 * region of the complex plane.
 */
for (i = 0; i < 100; i++) {
    with(plane) {
        /*
         * We only work with points still thought to
         * be in the Julia set.
         */
        where(injulia == 1) {
            r1 = r * r - c * c + p_r;
            c1 = 2.0 * r * c + p_c;
            r = r1;
            c = c1;
            mag = (r * r + c * c);
            /*
             * This is the square of the absolute
             * value of the point.
             */
        }
        where(mag > 5.0) injulia = 0;
        /*
         * Eliminate these points as candidates for
```

```
                    * membership in the Julia set.
                    */
                }
            }
            /*
            * Now we print out the array of 0's and 1's for half of the
            * plane in question.  A more serious program for computing a
            * Julia set might work with more data points and display the
            * set in a graphic image, rather than simply printing out a
            * boolean array.
            */
            for (i = 0; i < 64; i++) {
                for (j = 0; j < 64; j++)
                    printf("%d",[i][j] injulia);
                printf("\n");
            }
        }
```

Mandelbrot sets are very similar to Julia sets, except that, in computing $z_{i+1}$ from $z_i$, we add the *original value* of $z$ instead of a fixed parameter $p$. In fact, it is not hard to see that a given point is *in* the Mandelbrot set if and only if the Julia set generated using this point as its parameter is *nonempty*. This only requires a minor change in the program:

```
#include <stdio.h>
shape [64][128]plane;
float:plane r, c, r1, c1, x, y, mag;
int:lane inmand, row, column;
void main()
{
    int        i, j;
    with(plane) {
        r = pcoord(0);     /* Determine row of processor. */
        c = pcoord(1);     /* Determine column of processor. */
        x = r / 16.0 − 2.0;    /* Determine x−coordinate
                           * (real part of point that
                           * this processor will
                           * handle.  We subtract 2.0
                           * so that x−coordinates will
                           * range from −2 to +2.  */
```

```
y = c / 32.0 − 2.0;    /* Determine y−coordinate
                        * (complex part of point
                        * that this processor will
                        * handle.  We subtract 2.0
                        * so that y−coordinates will
                        * range from −2 to +2.  */
r = x;
c = y;
inmand = 1;    /* Initially assume that all points
                * are in the Mandelbrot set.  */
mag = 0.;      /* Initially set the absolute value
                * of the z(i)'s to 0.  We will
                * calculate these absolute values
                * and reject all points for which
                * they get too large. */
}
/*
 * Compute the first 100 z's for each point of the selected
 * region of the complex plane.
 */
for (i = 0; i < 100; i++) {
    with(plane) {
        /*
         * We only work with points still thought to
         * be in the Mandelbrot set.
         */
        where(inmand == 1) {
            r1 = r * r − c * c + x;
            c1 = 2.0 * r * c + y;
            r = r1;
            c = c1;
            mag = (r * r + c * c);
            /*
             * This is the square of the absolute
             * value of the point.
             */
        }
        where(mag > 5.0) inmand = 0;
        /*
         * Eliminate these points as candidates for
```

```
              * membership in the Mandelbrot set.
              */
         }
    }
    /*
     * Now we print out the array of 0's and 1's for half of the
     * plane in question.  A more serious program for computing a
     * Mandelbrot set might work with more data points and
     * display the set in a graphic image, rather than simply
     * printing out a boolean array.
     */
    for (i = 0; i < 64; i++) {
        for (j = 0; j < 64; j++)
            printf("%d",[i][j] inmand);
        printf("\n");
    }
}
```

2.4.7. *Pointers.* In this section we will discuss the use of pointers in C\*. As was mentioned above, C\* contains all of the usual types of C-pointers. In addition, we have:

- scalar pointers to shapes;
- scalar pointers to parallel variables.

It is interesting that there is no *simple* way to have a parallel pointer to a parallel variable. Although the CM-2 Connection Machine has indirect addressing, the manner in which floating point data is processed requires that indirect addressing be done in a rather restrictive and non-intuitive way[6]. The designer of the current version of C\* has simply opted to not implement this in C\*. It can be done in PARIS (the assembly language), but it is not straightforward (although it is *fast*). We can "fake" parallel pointers to parallel variables by making the *arrays* into *arrays of processors* — i.e., storing individual elements in separate processors. It is easy and straightforward to refer to sets of processors via parallel variables. The disadvantage is that this is slower than true indirect addressing (which only involves the local memory of individual processors).

A scalar pointer to a shape can be coded as:

**shape \*ptr;**

This can be accessed as one might expect:

**with(\*ptr)...**

---

[6]The address must be "shuffled" in a certain way

As mentioned before, shapes can be dynamically allocated via the **allocate_shape** command. In order to use this command, the user needs to include the file <stdlib.h>. **allocate_shape** is a procedure whose:

1. first parameter is a *pointer* to the shape being created;
2. second parameter is the *rank* of the shape (the number of dimensions);
3. the remaining parameters are the size of each dimension of the shape.

The shape pix in the Levialdi Counting Program could have been create by:

**shape** pix;
pix=**allocate_shape**(&pix,2,64,128);

Dynamically allocated shapes can be freed via the **deallocate_shape** command.

Scalar pointers to parallel variables point to *all instances* of these variables in their respective shape. For instance, if we write:

```
shape [8192]s;
int:s a;
int:s *b;
b=&a;
```

causes a and *b to refer to the same parallel data-item. Parallel data can be dynamically allocated via the **palloc** function. In order to use it one must include the file <stdlib.h>. The parameters to this function are:

1. The shape in which the parallel variable lives;
2. The size of the parallel data item in **bits**. This is computed via the **boolsizeof** function.

Example:

```
#include <stdlib.h>
shape [8192]s;
int:s *p;
p=palloc(s, boolsizeof(int:s));
```

Dynamically allocated parallel variables can be freed via the **pfree** function. This only takes one parameter — the pointer to the parallel data-item.

2.4.8. *Subtleties of Communication.* A number of topics were glossed over in the preceding sections, in an effort to present a simple account of C*. We will discuss these topics here. The alert reader may have had a few questions, about communication:

- How are *general* and *grid-based* communications handled in C* — assuming that one has the ability to program these things in C*?

- Although statements like a+=b, in which a is scalar and b is parallel, adds up values of b into a, how is this accomplished if a is a *parallel* variable? A number of algorithms (e.g., matrix multiplication) will require this capability.
- How is data passed between different shapes?

The C* language was designed with the idea of providing all of the capability of assembly language in a higher-level language (like the C language, originally) and addresses all of these topics.

We first note that the *default* forms of communication (implied whenever one uses left-indexing of processors) is:

1. *general* communication, if the left-indexing involves data-items or numbers;
2. *grid communication*, if the left-indexing involves *dot notation* with fixed displacements. This means the numbers added to the dots in the left-indexes are the same for all processors. For instance a=[.+1][.−1]a; satisfies this condition. In addition, the use of the **pcoord** function to compute index-values results in grid communication, if the displacements are the same for all processors. For instance, the compiler is smart enough to determine that

a=[**pcoord**(0)+1][**pcoord**(1)−1]a;

involves grid communication, but

a=[**pcoord**(1)+1][**pcoord**(0)−1]a;

does not, since the data movement is different for different processors.

Even in general communication, there are some interesting distinctions to be made. Suppose a is some parallel data-item. We will call the following code-fragment

a = [2]a

a parallel *get* statement, and the following

[7]a = a

a parallel *send* statement. In the first case data enters the *current* processor from some other processor, and in the second, data from the current processor is plugged into another processor.

It turns out that *send* operations are *twice* as fast as the parallel *get* operations. Essentially, in a get operation, the current processor has to send a message to the other processor requesting that it send the data.

It is also important to be aware of the distinction between get and send operations when determining the effect of *inactive processors*. The get and send operations are only carried out by the active processors. Consider the code:

```
shape [8192]s;
int:s dest,source;
int index=4;
```

```
where(source < 30)
    dest = [index]source;
```

here, the processors carrying out the action are those of dest — they are getting data from source. It follows that this code executes by testing the value of source in every processor, then then plugging [4]source into dest in processors in which source is < 30. In code like

```
shape [8192]s;
int:s dest,source;
int index=4;
where(source < 30)
    [.+1]source = source;
```

the processors in [.+1]source are passive receivers of data. Even when they are inactive, data will be sent to them by the next lower numbered processor unless *that processor* is also inactive.

2.4.9. *Collisions.* In PARIS, there are commands for data-movement on the Connection Machine that resolve collisions of data in various ways. The C* compiler is smart enough to generate appropriate code for handling these collisions in many situations — in other words the compiler can determine the programmer's intent and generate the proper PARIS instructions.

1. if the data movement involves a simple assignment and there are collisions, a random instance of the colliding data is selected;
2. if a reduction-assignment operation is used for the assignment, the appropriate operation is carried out. Table IV.1 lists the valid reduction-assignment operations in C*.
   These reduction-operations are carried out using an efficient algorithm like the algorithm for adding 8 numbers on page 69. Reduction-operations take $O(\lg n)$-time to operate on $n$ numbers. See §1 chapter VI for more information on the algorithm used.

These considerations are essential in programming some kinds of algorithms. For instance, matrix multiplication (at least when one uses the $O(\lg n)$ time algorithm) requires the use of reduction operations:

```
#include <stdio.h>
#include <stdlib.h>
shape [64][128]mats;
```

| Reduction-Assignment Operations in C* | | |
|---|---|---|
| Operation | Effect | Value if no active processors |
| += | Sum | 0 |
| -= | Negative sum | 0 |
| ^= | Exclusive OR | 0 |
| \|= | OR | 0 |
| <?= | MIN | Maximum possible value |
| >?= | MAX | Minimum possible value |

TABLE IV.1

```
void matmpy(int, int, int, int:current*,
    int:current*, int:current*); /* Function prototype. */


/*  This is a routine for multiplying an
    m x n matrix (*mat1) by an n x k matrix (*mat2)
    to get an m x k matrix (*outmat)
*/
void matmpy(int m, int n, int k,
        int:current *mat1,
        int:current *mat2,
        int:current *outmat)
{
    shape [32][32][32]tempsh; /* A temporary shape used
                * to hold intermediate
                * results –– products
                * that are combined
                * in the reduction
                * operations.
                */
    int:tempsh tempv;
    *outmat = 0;
    with(tempsh)
    {
    bool:tempsh region = (pcoord(0) < m) &
            (pcoord(1) < n)
            & (pcoord(2) < k);
                /* This is the most efficient
                * way to structure a where
                * statement with complex
```

```
                    * conditions.
                    */
        where(region) {
            tempv =[pcoord(0)][pcoord(1)] (*mat1) *
                    [pcoord(1)][pcoord(2)] (*mat2);
            [pcoord(0)][pcoord(2)](*outmat) += tempv;
        /* Parallel reduction operations here ——
          via the '+='—operation. */
        }
    }
}
void main()
{
    int n = 3;
    int m = 3;
    int k = 3;
    int:mats a, b, c;
    int        i, j;
    with(mats) {
        a = 0;
        b = 0;
        c = 0;
       /* Test data. */
        [0][0]a = 1;
        [0][1]a = 2;
        [0][2]a = 3;
        [1][0]a = 4;
        [1][1]a = 5;
        [1][2]a = 6;
        [2][0]a = 7;
        [2][1]a = 8;
        [2][2]a = 9;
        b = a;
        matmpy(3, 3, 3, &a, &b, &c);
        for (i = 0; i < n; i++)
            for (j = 0; j < k; j++)
                printf("   %d\n",[i][j]c);
    }
}
```

Note that the matrices are passed to the procedure as pointers to parallel integers. The use of pointers (or pass-by-value) is generally the most efficient way to pass parallel data. C programmers may briefly wonder why it is necessary to take pointers at all — in C matrices are normally stored as pointers to the first element. The answer is that these "arrays" are not arrays in the usual sense (as regarded by the C programming language). They are *parallel integers* — the *indexing* of these elements is done by the parallel programming constructs of C* rather than the array-constructs of the C programming language. In other words these are regarded as *integers* rather than *arrays* of integers. They store the same kind of information as an array of integers because they are *parallel* integers.

Note the size of the shape **shape** [32][32][32]tempsh; — this is the fundamental limitation of the $O(\lg n)$-time algorithm for matrix multiplication. A more practical program would dynamically allocate tempsh to have dimensions that approximate $n$, $m$, and $k$ if possible. Such a program would also switch to a slower algorithm that didn't require as many processors if $n \times m \times k$ was much larger than the available number of processors. For instance, it is possible to simply compute the $m \times k$ elements of *outmat in parallel — this is an $O(m)$-time algorithm that requires $m \times k$ processors.

**Exercises:**

1. How would the CM-2 be classified, according to the Händler scheme described on page 19?
2. Write the type of program for matrix multiplication described above. It should dynamically allocate (and free) tempsh to make its dimensions approximate $n$, $m$, and $k$ (recall that each dimension of a shape must be a power of 2), and switch to a slower algorithm if the number of processors required is too great.
3. Suppose we want to program a parallel algorithm that does *nested parallelism*. In other words it has statements like:

   **for** $i = 1$ **to** $k$ **do in parallel**
       int $t$;
       **for** $j = 1$ **to** $\ell$ **do in parallel**
           $t \leftarrow \min\{j|A[i,j] = 2\}$

   How do we implement this in C*?
   Note that before we can find the minimum, we must set $t$ to MAXINT, since the semantics of the <?= operation imply that the value of $t$ is compared with the values of pcoord(1) when the minimum is taken. If $t$ started out being < all of the existing values of pcoord(1) that satisfy the criterion above, $t$ would be unchanged.

**2.5. A Critique of C\*.** In this section we will discuss some of the good and bad features of the C\* language. Much of this material is taken from [64], by Hatcher, Philippsen, and Tichy.

We do this for several reasons:

- Analysis of these features shows how the hardware influences the design of the software.
- This type of analysis is important in pointing the way to the development of newer parallel programming languages.

The development of parallel languages cannot proceed in a purely abstract fashion. Some of the issues discussed here were not known until the C\* language was actually developed.

If one knows PARIS (the "assembly language" of the CM-1 and 2), the first flaw of C\* that comes to mind is the fact that most PARIS instructions are not implemented in C\*. This is probably due to the fact that people working with the Connection Machine originally programmed mostly in PARIS. There was the general belief that one had to do so in order to get good performance out of the machine. The designers of C\* may have been felt that the the people who wanted to extract maximal performance from the CM-series by using some of the more obscure PARIS instructions, would simply program in PARIS from the beginning. Consequently, they may have felt that there was no compelling reason to implement these instructions in C\*. This position is also amplified by the fact that PARIS is a very large assembly language that is generally more "user-friendly" than many other assembly languages[7] — in other words, PARIS was designed with the idea that "ordinary users" might program in it (rather than just system programmers).

We have already seen the problem of parallel pointers and indirect addresses — see § 2.4.7 on page 138. This limitation was originally due to certain limitations in the CM-2 hardware. In spite of the hardware limitations, the C\* language could have provided parallel pointers — it would have been possible to implement in software, but wasn't a high enough priority.

The use of left indices for parallel variables is a non-orthogonal feature of C\*. Parallel scalar variables are *logically* arrays of data. The only difference between these arrays and right-indexed arrays is that the data-items are stored in separate processors. Since the left-indexed data is often accessed and handled in the same way as right-indexed arrays, we should be able to use the same notation for them. In other words, the criticism is that C\* requires us to be more aware of the hardware (i.e. which array-elements are in separate processors) than the logic of a problem requires. The issue is further confused slightly by the fact that the CM-1 and 2 frequently allocate *virtual processors*. This means that our knowledge of how data-items are distributed among

---

[7]Some people might disagree with this statement!

the processors (as indicated by the left-indices) is frequently an *illusion* in any case.

In [64] Hatcher, Philippsen, and Tichy argue that the statements for declaring data in C* are non-orthogonal. They point out that *type* of a data-item should determine the kinds of operations that can be performed upon it — and that C* mixes the concepts of data-type and array size. In C* the *shape* of a parallel variable is regarded as an aspect of its *data-type*, where it is really just a kind of array declaration. For instance, in Pascal (or C) one first declares a data-type, and then can declare an array of data of that type. The authors mentioned above insist that these are two logically independent operations that C* forces a user to combine. Again the C* semantics are clearly dictated by the hardware, but Hatcher, Philippsen, and Tichy point out that a high-level language should insulate the user from the hardware as much as possible.

In the same vein Hatcher, Philippsen, and Tichy argue that C* requires users to declare the same procedure in more different ways than are logically necessary. For instance, if we have a procedure for performing a simple computation on an integer (say) and we want to execute this procedure in parallel, we must define a version of the procedure that takes a parallel variable as its parameter. This also results in logical inconsistency in parallel computations.

For instance, suppose we have a procedure for computing absolute value of an integer abs (). In order to compute absolute values in parallel we must define a version of this procedure that takes a suitable parallel integer as its parameter:

> **int:current** abs(**int:current**);

Now suppose we use this procedure in a parallel block of code:

```
int:someshape c,d;
where(somecondition)
   {
   int:someshape a;
   a=c+d;
   e=abs(a);
   }
```

Now we examine how this code executes. The add-operation a=c+d is performed in parallel, upon parallel data. When the code reacher the statement e=abs (a), the code *logically* becomes sequential. That is, it performs a *single* function-call with a single data-item (albeit, a parallel one). Hatcher, Philippsen, and Tichy argue that it should be possible to simply have a single definition of the abs-function for scalar variables, and have that function executed in parallel upon all of the *components* of the parallel variables involved. They argue that this would be logically more consistent than what C* does, because the semantics of the abs-function would be the same as for addition-operations.

In defense of C*, one should keep in mind that it *is* based upon C, and the C programming language is somewhat lower-level than languages like Pascal or Lisp[8]. One could argue that it is better than working in PARIS, and one shouldn't expect to be completely insulated from the hardware in a language based upon C.

An international group of researchers have developed a programming language that attempts to correct these perceived shortcomings of C* — see [68], [67], and [69]. This language is based upon Modula-2, and is called Modula*. We discuss this language in the next section.

## 3. Programming a MIMD-SIMD Hybrid Computer: Modula*

**3.1. Introduction.** In this section we will discuss a new parallel programming language based upon Modula-2 in somewhat the same way that C* was based upon the C programming language.

The semantics of this language are somewhat simpler and clearer than C*, and it is well-suited to both SIMD and MIMD parallel programming. This language has been developed in conjunction with a MIMD-SIMD hybrid computer — the Triton system, being developed at the University of Karlsruhe. They have implemented Modula* on the CM-2 Connection Machine, and the MasPar MP-1, and intend to implement it upon their MIMD-SIMD hybrid computer, the Triton/1, when it is complete. The discussion in this section follows the treatment of Modula* in [68], by Christian Herter and Walter Tichy.

**3.2. Data declaration.**

3.2.1. *Elementary data types.* The elementary data-types of Modula* are the same as those of Modula-2. They closely resemble basic data-types in Pascal. The most striking difference (at first glance) is that the language is case-sensitive and all keywords are capitalized. We have:

- **INTEGER.**
- **CARDINAL.** These are *unsigned* integers.
- **CHAR.** Character data.
- **BOOLEAN.** Boolean data.
- **REAL.** Floating point data.
- **SIGNAL.** This is a special data-type used for synchronizing processes. A **SIGNAL** is essentially a *semaphore*. These data-items are used in Modula* somewhat more than in Modula-2, since this is the main synchronization primitive for asynchronous parallel processes.

---

[8]It is sometimes called "high-level assembly language".

3.2.2. *Parallel data types.* Unlike C*, Modula* has no special parallel data type. Parallel data is indistinguishable from *array* data, except that one can specify certain aspects of parallel allocation.

In Modula-2 array-types are defined with statements like:

**ARRAY** *RangeDescriptor* **OF** *type*;

Here *RangeDescriptor* describes the range of subscript-values, and *type* is the data-type of the elements of the array. Example:

**VAR**
x:**ARRAY** [1..10] **OF INTEGER**;

This is very much like Pascal, except that keywords must be capitalized. This construct is generalized to higher dimensions in a way reminiscent of C:

**VAR**
y:**ARRAY** [1..10], [-3..27] **OF INTEGER**;

In Modula*, arrays declared this way will be parallel, *by default*. If we want to control how this allocation is done, Modula* allows an *optional* additional parameter, called the *allocator* that follows the *RangeDescriptor* for each dimension:

**ARRAY** *RangeDescriptor* [ *allocator*] **OF** *type*;

If an allocator is missing it is assumed to be the same as the next allocator to its *right*. The possible values for this allocator are:

1. **SPREAD**. This is the *default* allocator. It means that values of the array are allocated in ranges of consecutive values in the separate processors. This means that if we want to allocate $n$ array elements among $p$ processors, segments of size $\lceil n/p \rceil$ are allocated to each processor. This allocator guarantees that array elements with consecutive subscript-values are allocated, as much as possible, to the same processors. In a one-dimensional array (from 0 to $n - 1$, say), element $i$ is stored in location $i \bmod p$ in processor $\lceil i/p \rceil$.

2. **CYCLE**. This allocator assigns elements of the array in a round-robin fashion among the available processors. As a result, array-elements with consecutive subscript-values are almost always put in different processors. In a one-dimensional array (from 0 to $n - 1$), element $i$ is put in location $\lceil i/p \rceil$ of processor $i \bmod p$.

   In both of the preceding cases, if more than one subscript in a sequence uses the same allocator, all of the consecutive subscripts with the same allocator are combined into a single one-dimensional index that is processed using **SPREAD** or **CYCLE**, as the case may be. For instance in the following declaration:

   **VAR**
   y:**ARRAY** [1..10] **CYCLE** [-2..27] **CYCLE OF INTEGER**;

the two subscripts are combined into a single subscript that runs from 0 to 299, and the elements are distributed by the **CYCLE** distribution, using this one subscript.

The following two allocators correspond to **SPREAD** and **CYCLE**, respectively, but inhibit the process of combining several consecutive subscripts with the same allocator into a single subscript:

3. **SBLOCK**. this corresponds to **SPREAD**.
4. **CBLOCK**. this corresponds to **CYCLE**.
5. **LOCAL**. This forces all elements of the array whose subscripts only differ in an index defined to be **LOCAL**, to be in the same processor.

**Exercises:**

1. If we have a computer with 6 processors (numbered from 0 to 5), describe how the following array is allocated among the processors:

   **VAR**
       **y:ARRAY [1..10] CYCLE [-2..27] CYCLE OF INTEGER;**

2. If we have a computer with 6 processors (numbered from 0 to 5), describe how the following array is allocated among the processors:

   **VAR**
       **y:ARRAY [1..10] CBLOCK [-2..27] CBLOCK OF INTEGER;**

**3.3. The FORALL Statement.** In this section we will describe the command that initiates parallel execution of code-segments. There is a single statement that suffices to handle asynchronous and synchronous parallel execution, and can consequently, be used for MIMD or SIMD programs.

3.1. The basic syntax is:

$$\text{FORALL identifier:rangeType IN} \left\{ \begin{array}{c} \text{PARALLEL} \\ \text{SYNC} \end{array} \right\}$$

    statementSequence
**END**

Here:

1. identifier is a variable that becomes available inside the block — its scope is the **FORALL**-block. This identifier takes on a unique value in each of the threads of control that are created by **FORALL**-statement.
2. rangeType is a data-type that defines a range of values — it is like that used to define subscript-ranges in arrays. It define the number of parallel threads

created by the **FORALL**-statement. One such thread is created for each value in the rangeType, and in each such thread, the identifier identifier takes on the corresponding value.

3. statementSequence is the code that is executed in parallel.

Example:

**FORALL x:[1..8] IN PARALLEL**
    y:=f(x)+1;
**END**

This statement creates 8 processes, in a 1-1 correspondence with the numbers 1 through 8. In process $i$, the variable x takes the value $i$.

The two synchronization-primitives, **PARALLEL** and **SYNC**, determine how the parallel threads are executed. The **PARALLEL** option specifies *asynchronous* processes — this is MIMD-parallel execution. The **SYNC** option specifies synchronous execution of the threads — essentially SIMD-parallel execution.

Two standard Modula-2 commands **SEND** and **WAIT** are provided for explicit synchronization of asynchronous processes. The *syntax* of these functions is given by:

    SEND(**VAR** x:**SIGNAL**);
    WAIT(**VAR** x:**SIGNAL**);

The sequence of events involved in using these functions is:

1. A variable of type SIGNAL is declared globally;
2. When a process calls SEND or WAIT with this variable as its parameter, the variable is initialized. Later, when the matching function is called this value is accessed.

3.3.1. *The asynchronous case.* In the asynchronous case, the **FORALL**-statement terminates when the last process in the *StatementSequence* terminates. No assumptions are made about the order of execution of the statements in the asynchronous case, except where they are *explicitly* synchronized via the **SEND** and **WAIT** statements. The statements in statementSequence may refer to their private value of identifier or any global variable. If two processes attempt to store data to the same global memory-location, the result is undefined, but equal to one of the values that were stored to it.

3.3.2. *The synchronous case.* This is the case that corresponds to execution on a SIMD computer. This case is like the asynchronous case except that the statements in the body of the construct are executed synchronously. There are a number of details to be worked out at this point.

**Sequence** A sequence of statements of the form

$$S_1; S_2; \ldots S_k$$

is executed synchronously by a set of processes by synchronously executing each of the $S_i$ in such a way that $S_{i+1}$ is not begun until statement $S_i$ is completed. This can be regarded as a recursive definition.

**Assignment** An assignment statement of the form

L:=E

is executed synchronously by a set rangeType of processes as follows.

1. All processes in rangeType evaluate the designator $L$ synchronously, yielding |rangeType|, results, each designating a variable.
2. All processors in rangeType evaluate expression $E$ synchronously, yielding |rangeType|, values.
3. All processes in rangeType store their values computed in the second step into their respective variables computed in the first step (in arbitrary order).

Then the assignment terminates. If the third step results in several values being stored into the same memory location, one of the store-operations is successful — and which of them succeeds is indeterminate.

**expressions** These are executed somewhat like the assignment statement above. The set of processes rangeType apply the operators to the operands in the same order and in unison. The order of application is determined by the precedence-rules of Modula-2.

**If statement** An if statement executes somewhat like a where statement in C* — see page 120 in the previous section. The following statement

IF e1 THEN t1
    ELSIF e2 THEN t2
    . . .
    ELSE tk
END

is executed synchronously as follows:

1. All processes evaluate expression e1 synchronously.
2. Those processes for which e1 evaluates to TRUE then execute t1 synchronously, while the other processes evaluate e2 synchronously.
3. Those processes whose evaluation of e2 results in TRUE then execute t2 synchronously, and so on.

The synchronous execution of the if statement terminates when the last non-empty subset of rangeType terminates.

**CASE statement** Syntax:

CASE e OF
    c1: s1
    . . .

```
    ck: sk
ELSE s0
END
```

(here the ELSE clause is optional).

1. All processes in rangeType synchronously evaluate expression e.
2. Each process selects the case whose label matches the computed value of e. The set of processes rangeType gets partitioned into $k$ (or $k + 1$, if the **ELSE** clause appears). The processes in each set synchronously execute the statements that correspond to the case that they have selected. This is very much like the switch statement in the *old* version of C*. (In the current version of C*, the switch statement is used only for sequential execution).

**LOOP** statement. This is one of the standard looping constructs in Modula-2.

```
LOOP
    t1
END
```

Looping continues until the **EXIT** statement is executed. This type of statement is executed synchronously in Modula* by a set of processors as follows:

- As long as there is one active process within the scope of the LOOP statement, the active processes execute statement t1 synchronously.
- An **EXIT** statement is executed synchronously by a set of processes by marking those processes *inactive* with respect to the smallest enclosing **LOOP** statement.

**WHILE statement** Syntax:

```
WHILE condition DO
    t1
END
```

In Modula* each iteration of the loop involves:

1. all active processes execute the code in condition and evaluate it.
2. processes whose evaluation resulted in FALSE are marked inactive.
3. all other processes remain active and synchronously execute the code in t1.

The execution of the **WHILE** statement continues until there are no active processes.

**FORALL statement** One of the interesting aspects of Modula* is that parallel code can be *nested*. Consider the following **FORALL** statement:

```
FORALL D: U
    t1
```

> . . .
>     tk
> **END**

This is executed synchronously by a set rangeType, of processes, as follows:

1. Each process $p$ ∈rangeType computes $U_p$, the set of elements in the range given by U. Processes may compute different values for $U_p$, if U has identifier (the index created by the enclosing FORALL statement that identifies the process — see 3.1 on page 149), as a parameter.
2. Each process $p$ creates a set, $S_p$, of processes with $|S_p| = U_p$ and supplies each process with a constant, $D$, bound to a unique value of $U_p$.
3. Each process, $p$, initiates the execution of the statement sequence t1 through tk by $S_p$. Thus there are |rangeType| sets sets of processes, each executing the sequence of statements independently of the others. Each such set of processes executes its statements synchronously or asynchronously, depending on the option chosen in this **FORALL** statement.

**WAIT and SEND** When a set of processes synchronously executes a **WAIT** command, if *any* process in the set blocks on this command, then *all* of them do. This rule allows synchronization to be maintained.

**Procedure call** The set of processes rangeType synchronously executes a procedure call of the form

> P(x1,...,xk);

as follows:

1. Each process in rangeType creates an activation of procedure P.
2. All processes in rangeType evaluate the actual parameters x1,...,xk synchronously in the same order (not necessarily left to right) and substitute them for the corresponding formal parameters in their respective activations of P.
3. All processes execute the body of P synchronously in the environment given by their respective activations.
4. The procedure call terminates when all processes in rangeType have been designated inactive with respect to the call.

Note: there is an implied return statement at the end of each procedure. In an explicit return statement like

> RETURN [E];

if the expression E is supplied all processes synchronously evaluate it and use this value as the result of the function-call. After evaluating E (or if no expression is supplied) all processes in rangeType are marked inactive with respect to the most recent procedure call.

**3.4. Sample Programs.** Here are some sample programs. Our standard example is the program for adding up $n$ integers. This example was first introduced in the Introduction (on page 69), and a implementation in C* appeared in § 2.3 (on page 117). It is interesting to compare these two programs:

```
VAR V: ARRAY [0..N-1] OF REAL;
VAR stride:  ARRAY [0..N-1] OF CARDINAL;
BEGIN
   FORALL i : [0..N-1] IN SYNC
      stride[i] := 1;
      WHILE stride[i] < N DO
        IF ((i MOD (stride[i] * 2)) = 0) AND ((i - stride[i]) >= 0) THEN
          V[i] := V[i] + V[i - stride[i]]
        END;
        stride[i] := stride[i] * 2
      END
   END  (* sum in V[N] *)
END
```

Note that we must explicitly declare a parallel variable stride[i] and its use is indexed by the process-indexing variable i.

Modula* does not take advantage of the hardware support for performing operations like the above summation in a single statement.

We can also perform these calculations asynchronously:

```
VAR V: ARRAY [0..N-1] OF REAL;
VAR stride:  CARDINAL;
BEGIN
     stride := 1;
     WHILE stride < N DO
   FORALL i : [0..N-1] IN PARALLEL
      IF ((i MOD (stride * 2)) = 0) AND ((i - stride) >= 0) THEN
        V[i] := V[i] + V[i - stride]
      END;
      stride := stride * 2
   END
   END (* sum in V[N-1] *)
END
```

**3.5. A Critique of Modula*.** The language Modula* responds to all of the points raised by Tichy, Philippsen, and Hatcher in [64] and it might, consequently, be regarded as a "Second generation" parallel programming language (with C* a first generation language). This is particularly true in light of the probable trend toward the development of hybrid MIMD-SIMD computers like the CM-5 and the Triton/1, and the fact that Modula* handles asynchronous as well as synchronous parallel code.

The simple and uniform semantics of the parallel code in Modula* makes programs very portable, and even makes Modula* a good language for describing parallel algorithms in the abstract sense[9]

One suggestion: some operations can be performed very quickly and efficiently on a parallel computer, but the implementation of the algorithm is highly dependent upon the architecture and the precise layout of the data. It would be nice if Modula* supported these operations.

For instance the census-operations like:

1. Forming a sum or product of a series of numbers.
2. Performing versions of the generic ASCEND or DESCEND algorithms in chapter III (see page 67).

Many architectures have special hardware support for such operations — for instance the Connection Machines support census-operations. The material in chapter III shows that such hardware support is at least *possible* for the ASCEND and DESCEND algorithms. I feel that it might be desirable to add ASCEND and DESCEND operations to the Modula* language, where the user could supply the $OPER(*, *; *, *)$-function. Failing this, it might be desirable to include a generic census operation in the language.

The alternative to implementing these generic parallel operations would be to write the Modula* compiler to automatically recognize these constructs in a program. This does not appear to be promising — such a compiler would constitute an "automatically parallelizing" compiler (at least to some extent[10]).

---

[9]This is reminiscent of the Algol language, which was used as much as a *publication language* or *pseudocode* for describing algorithms, as an actual programming language.

[10]The important point is that we are requiring the compiler to recognize the programmer's intent, rather than to simply translate the program. The experience with such compilers has been discouraging.

# V
# Numerical Algorithms

In this chapter we will develop SIMD algorithms for solving several types of numerical problems.

## 1. Linear algebra

**1.1. Matrix-multiplication.** In this section we will discuss algorithms for performing various matrix operations. We will assume that all matrices are $n \times n$, unless otherwise stated. It will be straightforward to generalize these algorithms to non-square matrices where applicable. We will also assume that we have a CREW parallel computer at our disposal, unless otherwise stated. We also get:

PROPOSITION 1.1. Two $n \times n$ matrices $A$ and $B$ can be multiplied in $O(\lg n)$-time using $O(n^3)$ processors.

PROOF. The idea here is that we form the $n^3$ products $A_{ij}B_{jk}$ and take $O(\lg n)$ steps to sum over $j$. $\square$

Since there exist algorithms for matrix multiplication that require fewer than $n^3$ multiplications (the best current asymptotic estimate, as of 1991, is $n^{2.376}$ multiplications — see [33]) we can generalize the above to:

COROLLARY 1.2. If multiplication of $n \times n$ matrices can be accomplished with $M(n)$ multiplications then it can be done in $O(\lg n)$time using $M(n)$ processors.

This algorithm is efficiently implemented by the sample program on page 141. We present an algorithm for that due to Reif and Pan which inverts an $n \times n$-matrix in $O(\lg^2 n)$-time using $M(n)$ processors — see [121]. Recall that $M(n)$ is the number of multiplications needed to multiply two $n \times n$ matrices.

**1.2. Systems of linear equations.** In this section we will study parallel algorithms for solving systems of linear equations

$$Ax = b$$

where $A$ is an $n \times n$ matrix, and $b$ is an $n$-dimensional vector.

We will concentrate upon *iterative* methods for solving such equations. There are a number of such iterative methods available:

- The Jacobi Method.
- The JOR method — a variation on the Jacobi method.
- The Gauss-Seidel Method.
- The SOR method.

These general procedures build upon each other. The last method is the one we will explore in some detail, since a variant of it is suitable for implementation on a parallel computer. All of these methods make the basic assumption that the *largest* elements of the matrix $A$ are located on the main diagonal. Although this assumption may seem rather restrictive

- This turns out to be a natural assumption for many of the applications of systems of linear equations. One important application involves numerical solutions of *partial differential equations,* and the matrices that arise in this way are mostly zero. See § 5 for more information on this application.
- Matrices *not* dominated by their diagonal entries can sometimes be transformed into this format by permuting rows and columns.

In 1976, Csanky found NC-parallel algorithms for computing *determinants* and *inverses* of matrices — see [38], and § 1.5 on page 193. Determinants of matrices are defined in 1.5 on page 159. Csanky's algorithm for the inverse of a matrix wasn't numerically stable and in 1985, Pan and Reif found an improved algorithm for this — see [121] and § 1.3 on page 182. This algorithm is an important illustration of the fact that the best parallel algorithm for a problem is often entirely different from the best sequential algorithms. The standard sequential algorithm for inverting a matrix, Gaussian elimination, does not lend itself to parallelization because the process of choosing pivot point is *P-complete* — see [158]. It is very likely that there doesn't *exist* a fast parallel algorithm for inverting a matrix based upon the more traditional approach of *Gaussian elimination*. See the discussion on page 47 for more information.

§ 1.3 discusses methods that work for arbitrary invertible matrices. These methods require more processors than the iterative methods discussed in the other sections, but are of some theoretical interest.

1.2.1. *Generalities on vectors and matrices.* Recall that a matrix is a linear transformation on a vector-space.

DEFINITION 1.3. Let $A$ be an $n \times n$ matrix. $A$ will be called

1. *lower triangular* if $A_{i,j} = 0$ whenever $i \geq j$
2. *upper triangular* if $A_{i,j} = 0$ whenever $i \leq j$.

Note that this definition implies that upper and lower triangular matrices have zeroes on the main diagonal. Many authors define these terms in such a way that the matrix is permitted to have nonzero entries on the main diagonal.

DEFINITION 1.4. Let $L = \{n_1, \ldots, n_k\}$ be the result of permuting the list of integers $\{1, \ldots, k\}$. The *parity* of the permutation, $\wp(n_1, \ldots, n_k)$ is $\pm 1$, and is computed via:

1. For each $n_i$ in $L$, count the number of values $n_j$ appearing to the right of $n_i$ in $L$ (i.e. $j > i$) such that $n_i > n_j$. Let this count be $c_i$.
2. Form the total $c = \sum_{i=1}^{k} c_i$, and define $\wp(n_1, \ldots, n_k) = (-1)^c$.

The number $c$ roughly measures the extent to which the permutation alters the normal sequence of the integers $\{1, \ldots, k\}$. Here are a few examples:

Suppose the permutation is $\{3, 1, 2, 4, 5\}$. Then $c_1 = 4$ because the first element in this sequence is $3$ and there are $4$ smaller elements to the right of it in the sequence. It follows that the parity of this permutation is $+1$. Permutations with a parity of $+1$ are commonly called *even permutations* and ones with a parity of $-1$ are called *odd permutations*.

DEFINITION 1.5. If $A$ is an $n \times n$ matrix, the *determinant* of $A$ is defined to be the

$$\det(A) = \sum_{\substack{i_1, \ldots, i_n \\ \text{all distinct}}} \wp(i_1, \ldots, i_n) A_{1, i_1} \cdots A_{n, i_n}$$

where $\wp(i_1, \ldots, i_n)$ is the parity of the permutation $\begin{pmatrix} 1 & \cdots & n \\ i_1 & \cdots & i_n \end{pmatrix}$.

Here are some basic properties of determinants of matrices:

PROPOSITION 1.6. If $A$ and $B$ are $n \times n$ matrices then:

1. $\det(A \cdot B) = \det(A) \cdot \det(B)$;
2. The linear transformation represented by $A$ is invertible if and only if $\det(A) \neq 0$.
3. If $A$ is a lower or upper triangular matrix and $D$ is a diagonal matrix, then $\det(D + A) = \det(D)$.

Vector-spaces can be equipped with measures of the *size* of vectors: these are called *norms* of vectors. Our measure of the size of a *matrix* will be called the *norm* of a *matrix* – it will be closely associated with a norm of a vector[1]. Essentially, the norm of a matrix

---

[1] Although there exist norms of matrices that are *not* induced by norms of vectors, we will not use them in the present discussion

will measure the extent to which that matrix "stretches" vectors in the vector-space — where *stretching* is measured with respect to a norm of vectors.

DEFINITION 1.7. Let $V$ be a vector space. A *norm* on $V$ is a function $\| * \|: V \to \mathbb{R}$, with the following properties:

1. $\|v\| = 0$, if and only if the vector $v$ is 0;
2. for all $v, w \in V$, $\|v + w\| \leq \|v\| + \|w\|$;
3. for all $v \in V, c \in \mathbb{C}$, $\|c \cdot v\| = |c| \cdot \|v\|$;
   A function $\| * \|:$ Matrices over $V \to \mathbb{R}$ is called a *matrix norm* if it satisfies conditions like 1, 2, and 3 above and, in addition:
4. for all matrices $X$ and $Y$ over $V$, $\|XY\| \leq \|X\| \cdot \|Y\|$.

DEFINITION 1.8. Given a vector norm $\| * \|$ on a vector space we can define the *associated matrix norm* as follows: $\|M\| = \max_{v \neq 0} \|Mv\|/\|v\|$.

1. The matrix norm inherits properties 1, 2, and 3 from the vector norm. Property 4 results from the fact that the vector $v$ that gives rise to the maximum value for $\|Xv\|$ or $\|Yv\|$ might not also give the maximum value of $\|XYv\|$.

2. Property 3 of the vector norm implies that we can define the matrix norm via: $\|X\| = \max_{v, \|v\|=1} \|Xv\|$;

Here are three fairly standard vector norms:

DEFINITION 1.9.    1. $\|v\|_1 = \sum_i |v_i|$;
2. $\|v\|_\infty = \max_i |v_i|$;
3. $\|v\|_2 = \sqrt{\sum_i |v_i|^2}$;

The three vector norms give rise to corresponding matrix norms.

DEFINITION 1.10.    1. A square matrix will be called a *diagonal matrix* if all entries not on the main diagonal are 0;
2. Given a matrix $A$, the *Hermitian transpose* of $A$, denoted $A^H$, is defined by $(A^H)_{ij} = A^*_{ji}$, where $*$ denotes the *complex conjugate*;
3. $\det(\lambda \cdot I - A) = 0$ is a polynomial in $\lambda$ called the *characteristic polynomial* of $A$.
4. A number $\lambda \in \mathbb{C}$; is called an *eigenvalue* of $A$ if there exists a vector $v \neq 0$ such that $Av = \lambda v$. This vector is called the *eigenvector* corresponding to $\lambda$; An alternate definition of eigenvalues of $A$ is: $\lambda$ is an *eigenvalue* of $A$ if and only if the matrix $\lambda \cdot I - A$ is *not invertible*. This leads to an equation for computing eigenvalues (at least, for *finite-dimensional* matrices): eigenvalues are roots of the characteristic polynomial. In this definition, an eigenvalue can occur more than once — for instance, if the characteristic polynomial has a factor of $(\lambda - t)^k$, we say that the original matrix has $k$ eigenvalues equal to $t$.

5. If $\lambda$ is an eigenvalue of a matrix $A$, then the *eigenspace* associated with $\lambda$ is the space of vectors, $v$, satisfying $(\lambda \cdot I - A)v = 0$. If $A$ has only one eigenvalue equal to $\lambda$ the eigenspace associated to $\lambda$ is 1 dimensional — it consists of all scalar multiples of the eigenvector associated with $\lambda$.

6. The *nullspace* of a matrix $A$ is the space of all vectors $v$ such that $Av = 0$. It is the same as the eigenspace of 0 (regarded as an eigenvalue of $A$).

7. $\rho(A)$, the *spectral radius* of $A$ is defined to be the maximum magnitude of the eigenvalues of $A$.

8. A matrix $A$ is called *Hermitian* if $A = A^H$.

9. The *condition number* of a matrix is defined by cond $A = \|A\|_2 \cdot \|A^{-1}\|_2 \geq \|I\| = 1$ if $A$ is nonsingular, $\infty$ otherwise.

It is not difficult to see that the eigenvalues of $A^H$ are the *complex conjugates* of those of $A$ — consequently, if $A$ is *Hermitian* its eigenvalues are *real*.

Note that eigenvectors are not nearly as uniquely determined as eigenvalues — for instance any scalar multiple of an eigenvector associated with a given eigenvalue is also an eigenvector associated with that eigenvalue.

Although computation of eigenvalues and eigenvectors of matrices is somewhat difficult in general, some special classes of matrices have symmetries that simplify the problem. For instance, the eigenvalues of a *diagonal matrix* are just the values that occur in the main diagonal. For a more interesting example of computation of eigenvalues and eigenvectors of a matrix see § 2.4 on page 218 and § 5.1.3 on page 275.

PROPOSITION 1.11. If $\| * \|$ is any norm and $A$ is any matrix, then:

1. $\rho(A) \leq \|A\|$.
2. $\|A^k\| \to 0$ as $k \to \infty$ if and only if $\rho(A) < 1$.
3. $\det(A) = \prod_{i=1}^{n} \lambda_i$, where the $\lambda_i$ are the eigenvalues of $A$. Here, given values of eigenvalues may occur more than once (as roots of the characteristic polynomial).

PROOF. First statement: Let $\lambda$ be the largest eigenvalue of $A$. Then $\rho(A) = |\lambda|$, and $Av = \lambda v$, where $v$ is the eigenvector corresponding to $\lambda$. But $\|Av\| \leq \|A\| \cdot \|v\|$ by 1.8, and $Av = \lambda v$ and $\|Av\| = |\lambda| \cdot \|v\| = \rho(A)\|v\|$. We get $\rho(A)\|v\| \leq \|A\| \cdot \|v\|$ and the conclusion follows upon dividing by $\|v\|$.

Second statement: Suppose $\rho(A) \geq 1$. Then $\rho(A)^k \leq \rho(A^k) \geq 1$ for all $k$ and this means that $\|A^k\| \geq 1$ for all $k$. On the other hand, if $\|A^k\| \to 0$ as $k \to \infty$, then $\rho(A^k) \to 0$ as $k \to \infty$. The fact that $\rho(A)^k \leq \rho(A^k)$ and the fact that powers of numbers $\geq 1$ are all $\geq 1$ imply the conclusion.

Third statement: This follows from the general fact that the constant-term of a polynomial is equal to its value at zero and the product of the roots. We apply this general fact to the characteristic polynomial $\det(A - \lambda \cdot I) = 0$. □

DEFINITION 1.12. Two $n \times n$ matrices $A$ and $B$ will be said to be *similar* if there exists a third invertible $n \times n$ $C$ such that $A = CBC^{-1}$.

Two matrices that are similar are equivalent in some sense. Suppose:
- $A$ is a transformation of a vector-space with basis vectors $\{b_i\}$, $i = 1, \ldots, n$.
- $B$ is a transformation of the same vector-space with basis vectors $\{b_i'\}$.
- $C$ is the matrix whose columns are the result of expressing the $\{b_i'\}$ in terms of the $\{b_i\}$.

Then the result of writing the $B$-transformation in terms of the basis $\{b_i\}$ is the $A$ matrix (if $A = CBC^{-1}$). In other words, similar matrices represent the *same transformation* — in different coordinate-systems. It make sense that:

LEMMA 1.13. Let $A$ and $B$ be similar $n \times n$ matrices with $A = CBC^{-1}$ for some invertible matrix $C$. Then $A$ and $B$ have the same eigenvalues and spectral radii.

PROOF. The statement about the eigenvalues implies the one about spectral radius. Let $\lambda$ be an eigenvalue of $A$ with corresponding eigenvector $V$ (see 1.10 on page 160). Then

$$Av = \lambda v$$

Now replace $A$ by $CBC^{-1}$ in this formula to get

$$CBC^{-1}v = \lambda v$$

and multiply (on the left) by $C^{-1}$ to get

$$BC^{-1}v = \lambda C^{-1}v$$

This implies that $\lambda$ is an eigenvalue of $B$ with corresponding eigenvector $C^{-1}v$. A similar argument implies that every eigenvalue of $B$ also occurs as an eigenvalue of $A$. $\square$

LEMMA 1.14. Let $\lambda$ be an eigenvalue of a nonsingular matrix $W$. Then $1/\lambda$ is an eigenvalue of $W^{-1}$.

PROOF. Note that $\lambda \neq 0$ because the matrix is nonsingular. Let $v$ be an associated eigenvector so $Wv = \lambda v$. Then $Wv \leq 0$ and $W^{-1}(Wv) = v = (1/\lambda)\lambda v = (1/\lambda)Wv$. $\square$

LEMMA 1.15. $\|W\|_2 = \|W^H\|_2$; $\|W\|_2 = \rho(W)$ if $W = W^H$.

PROOF. Recall the definition of the *inner product*: $(v, w) = \sum_{i=1}^n v_i^* w_i$, where $v$ and $w$ are $n$-dimensional vectors. Then:
1. $\|v\|_2^2 = (v, v)$
2. $(v, w) = (w, v)^*$
3. $(v, Aw) = (A^H v, w)$

These three basic properties of inner product imply the first statement, since

$$\|W\|_2 = \sqrt{\max_{v \neq 0}(Wv, Wv)/(v,v)}$$

$$= \sqrt{\max_{v \neq 0}(Wv, Wv)/(v,v)}$$

$$= \sqrt{\max_{v \neq 0}|(v, W^H W v)|/(v,v)}$$

$$= \sqrt{\max_{v \neq 0}|(W^H W v, v)|/(v,v)}$$

Suppose $V$ is the vector-space upon which $W$ acts: $W\colon V \to V$. The second statement follows from the fact that we can find a *basis* for $V$ composed of eigenvectors of $W$ (here, we use the term eigenvector in the loosest sense: the nullspace of $W$ is generated by the eigenvectors of 0. This is a well-known result that is based upon the fact that eigenvectors of distinct eigenvalues are linearly independent, and a count of the dimensions of the nullspaces of $W - \lambda \cdot I$ shows that the vector-space generated by the eigenspaces of all of the eigenvectors is all of $V$. Suppose that $v = \sum_{i=1}^n c_i e_i$, where the $e_i$ are eigenvectors of $W$. Then

$$(Wv, Wv) = \sum_{i=1}^n (Wc_ie_i, Wc_ie_i)$$

$$= \sum_{i=1}^n (\lambda_i c_i e_i, \lambda_i c_i e_i)$$

$$= \sum_{i=1}^n |\lambda_i|^2 |c_i|^2$$

so $(Wv, Wv)$ is a *weighted average* of the eigenvalues of $W$. It follows that the maximum value of $(Wv, Wv)/(v,v)$ occurs when $v$ is an eigenvector, and that value is the square of an eigenvalue. $\square$

DEFINITION 1.16. A matrix that can be represented as $W^H W$ for some matrix $W$ will be called *Hermitian positive semidefinite*. A nonsingular Hermitian positive semidefinite matrix is called *Hermitian positive definite*.

LEMMA 1.17. All eigenvalues of a Hermitian positive semidefinite matrix are nonnegative.

It turns out that the 1-norm and the $\infty$-norm of matrices are very easy to compute:

PROPOSITION 1.18. The matrix norm associated with the vector norm $\| * \|_\infty$ is given by $\|M\|_\infty = \max_i \sum_{j=1}^n |M_{ij}|$.

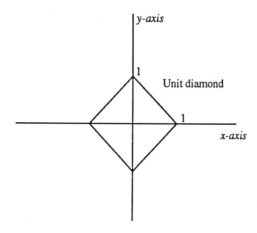

FIGURE V.1. A 2-dimensional diamond

PROOF. We must maximize $\max_i \sum_j M_{ij} \cdot v_j$, subject to the requirement that all of the $v_j$ are between $-1$ and $1$. If is clear that we can maximize one of these quantities, say the $i^{\text{th}}$, by setting:

$$v_j = \begin{cases} +1, & \text{if } M_{ij} \text{ is positive;} \\ -1, & \text{if } M_{ij} \text{ is negative.} \end{cases}$$

and this will result in a *total value* of $\sum_j |M_{ij}|$ for the $i^{\text{th}}$ row. The norm of the matrix is just the *maximum* of these row-sums.  □

PROPOSITION 1.19. The 1-norm, $\|M\|_1$, of a matrix $M$ is given by

$$\max_j \sum_{i=1}^{n} |M_{ij}|$$

PROOF. This is somewhat more difficult to show than the case of the $\infty$-norm. We must compute the maximum of $\sum_i |M_{ij} \cdot v_j|$ subject to the requirement that $\sum_j |v_j| = 1$. The *crucial step* involves showing:

**Claim:** The maximum occurs when *all but one* of the $v_j$ is **zero**. The remaining nonzero $v_j$ *must* be $+1$ or $-1$.

We will give a *heuristic proof* of this claim. With a little work the argument can be made completely rigorous. Essentially, we must consider the geometric significance of the 1-norm of vectors, and the shape of "spheres" with respect to this norm. The set of all vectors $v$ with the property that $\|v\|_1 = 1$ forms a polyhedron centered at the origin. We will call this a *diamond*. Figure V.1 shows a 2-dimensional diamond.

The 1-norm of a vector can be regarded as the radius (in the sense of 1-norms) of the smallest diamond centered at the origin, that encloses the vector. With this in mind,

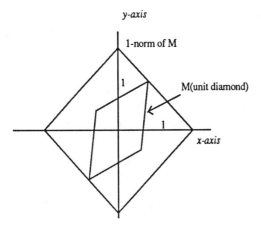

FIGURE V.2

we can define the 1-norm of a matrix $M$ as the radius of the smallest diamond that encloses $M(unit\ diamond)$, as in figure V.2.

Note that the radius of a diamond centered at the origin is easy to measure — it is just the *x-intercept* or the *y-intercept*. The *heuristic part* of the argument is to note that the smallest enclosing diamond *always* intersects $M$(unit diamond) at a *vertex*. This implies the claim, however, because vertices of the unit diamond are precisely the points with one coordinate equal to $+1$ or $-1$ and all other coordinates 0.

Given this claim, the proposition follows quickly. If the $j^{\text{th}}$ component of $v$ is 1 and all other components are 0, then $\|Mv\|_1 = \sum_i |M_{ij}|$. The 1-norm of $M$ is computed with the value of $j$ that *maximizes* this.  $\square$

Unfortunately, there is no *simple* formula for the 2-norm of a matrix in terms of the *entries* of the matrix. The 2-norm of a *Hermitian positive semidefinite matrix* (defined below) turns out to be equal to the largest *eigenvalue* of the matrix. We do have the following result that relates these quantities, and the spectral radius:

LEMMA 1.20. Let $W = (w_{ij})$. Then $\|W^{\mathrm{H}}W\|_2 = \rho(W^{\mathrm{H}}W) = \|W\|_2^2 \le \|W^{\mathrm{H}}W\|_1 \le \max_i \sum_j |w_{ij}| \max_j \sum_i |w_{ij}| \le n\|W^{\mathrm{H}}W\|_2$.

PROOF. It is not hard to see that $\|W^{\mathrm{H}}W\|_1 \le \max_i \sum_j |w_{ij}| \max_j \sum_i |w_{ij}|$ since $\|W^{\mathrm{H}}W\|_1 \le \|W^{\mathrm{H}}\|_1 \cdot \|W\|_1$ and $\|W^{\mathrm{H}}\|_1 = \|W\|_\infty$, by the explicit formulas given for these norms. The remaining statements follow from 1.41 on page 188.  $\square$

We also have the following interesting properties of the spectral radius:

**Exercises:**

1. Let $A$ be a square matrix. Show that if $\lambda_1$ and $\lambda_2$ are two eigenvalues of $A$ with associated eigenvectors $v_1$ and $v_2$ and $\lambda_1 = \lambda_2$, then any linear combination of $v_1$ and $v_2$ is a valid eigenvector of $\lambda_1$ and $\lambda_2$.

2. Let $A$ be a square matrix and let $\lambda_1, \ldots, \lambda_k$ be eigenvalues with associated eigenvectors $v_1, \ldots, v_k$. In addition, suppose that $\lambda_i \neq \lambda_j$ for all $i, j$ such that $i \neq j$, and that all of the $v_i$ are nonzero. Show that the set of vectors $\{v_1, \ldots, v_k\}$ are linearly independent — i.e. the only way a linear combination

$$\sum_{i=1}^{k} \alpha_i v_i$$

(where the $\alpha_i$ are scalars) can equal *zero* is for all of the coefficients, $\alpha_i$, to be zero.

The two exercises above show that the numerical values of eigenvalues determine many of the properties of the associated eigenvectors: if the eigenvalues are equal, the eigenvectors are strongly related to each other, and if the eigenvalues are distinct, the eigenvectors are independent.

3. Show that the two exercises above imply that an $n \times n$ matrix cannot have more than $n$ distinct eigenvalues.

4. Compute the 1-norm and the $\infty$-norm of the matrix

$$\begin{pmatrix} 1 & 2 & 3 \\ 0 & -1 & 0 \\ 2 & -4 & 8 \end{pmatrix}$$

5. Compute the 1-norm and the $\infty$-norm of the matrix

$$\begin{pmatrix} 0 & 1/4 & 1/4 & 1/4 & 1/4 \\ 1/4 & 0 & -1/4 & -1/4 & 1/4 \\ -1/4 & -1/4 & 0 & 1/4 & 1/4 \\ -1/4 & -1/4 & -1/4 & 0 & 1/4 \\ -1/4 & -1/4 & -1/4 & 1/4 & 0 \end{pmatrix}$$

6. compute the eigenvalues and eigenvectors of the matrix

$$A = \begin{pmatrix} 1 & 2 & 3 \\ 2 & -1 & 1 \\ 3 & 0 & 0 \end{pmatrix}$$

Use this computation to compute its spectral radius.

7. Let $A$ be the matrix

$$\begin{pmatrix} 1 & 2 \\ 3 & -1 \end{pmatrix}$$

Compute the 2-norm of $A$ directly.

8. Suppose that $A$ is an $n \times n$ matrix with eigenvectors $\{v_i\}$ and corresponding eigenvalues $\{\lambda_i\}$. If $I$ is a $n \times n$ identity matrix and $\alpha$ and $\beta$ are constants show that the eigenvectors of $\alpha A + \beta I$ are also $\{v_i\}$ and the corresponding eigenvalues are $\alpha \lambda_i + \beta$.

1.2.2. *The Jacobi Method.* The most basic problem that we will want to solve is a system of linear equations:

(4) $$Ax = b$$

where $A$ is a given $n \times n$ matrix, $x = (x_0, \ldots, x_{n-1})$ is the set of unknowns and $b = (b_0, \ldots, b_{n-1})$ is a given vector of dimension $n$. Our method for solving such problems will make use of the matrix $D(A)$ composed of the diagonal elements of $A$ (which we now assume are *nonvanishing*):

(5) $$D(A) = \begin{pmatrix} A_{0,0} & 0 & 0 & \cdots & 0 \\ 0 & A_{2,2} & 0 & \cdots & 0 \\ \vdots & \vdots & \vdots & \ddots & \vdots \\ 0 & 0 & 0 & \cdots & A_{n-1,n-1} \end{pmatrix}$$

As remarked above, the traditional methods (i.e., Gaussian elimination) for solving this do not lend themselves to easy parallelization. We will, consequently, explore *iterative* methods for solving this problem. The iterative methods we discuss requires that the $D$-matrix is invertible. This is equivalent to the requirement that all of the diagonal elements are nonzero. Assuming that this condition is satisfied, we can rewrite equation (4) in the form

$$D(A)^{-1}Ax = D(A)^{-1}b$$
$$x + (D(A)^{-1}A - I)x = D(A)^{-1}b$$
(6) $$x = (I - D(A)^{-1}A)x + D(A)^{-1}b$$

where $I$ is an $n \times n$ identity matrix. We will be interested in the properties of the matrix $Z(A) = I - D(A)^{-1}A$. The basic iterative method for solving equation (4) is to
1. Guess at a solution $u^{(0)} = (r_0, \ldots, r_{n-1})$;
2. Forming a sequence of vectors $\{u^{(0)}, u^{(1)}, \ldots\}$, where $u^{(i+1)} = Z(A)u^{(i)} + D(A)^{-1}b$.

Now suppose that the sequence $\{u^{(0)}, u^{(1)}, \dots\}$ converges to some vector $\bar{u}$. The fact that $\bar{u}$ is the *limit* of this sequence implies that $\bar{u} = Z(A)\bar{u} + D(A)^{-1}b$ — or that $\bar{u}$ is a *solution* of the original system of equations (4). This general method of solving systems of linear equations is called the *Jacobi method* or the *Relaxation Method*. The term "relaxation method" came about as a result of an application of linear algebra to numerical solutions of partial differential equations – see the discussion on page 263.

We must, consequently, be able to say whether, and when the sequence $\{u^{(0)}, u^{(1)}, \dots\}$ converges. We will use the material in the preceding section on norms of vectors and matrices for this purpose.

PROPOSITION 1.21. Suppose $A$ is an $n \times n$ matrix with the property that all of its diagonal elements are nonvanishing. The Jacobi algorithm for solving the linear system

$$Ax = b$$

converges to the same value regardless of starting point $(u^{(0)})$ if and only if $\rho(Z(A)) = \mu < 1$, where $\rho(Z(A))$ is the spectral radius defined in 1.10 on page 160.

Note that this result also gives us some idea of *how fast* the algorithm converges.

PROOF. Suppose $\bar{u}$ is an exact solution to the original linear system. Then equation (6) on page 167 implies that:

$$(7) \qquad\qquad \bar{u} = (I - D(A)^{-1}A)\bar{u} + D(A)^{-1}b$$

Since $\rho(Z(A)) = \mu < 1$ it follows that $\|Z(A)^k\| \to 0$ as $k \to \infty$ for any matrix-norm $\| * \|$ — see 1.11 on page 161. We will compute the amount of *error* that exists at any given stage of the iteration. The equation of the iteration is

$$u^{(i+1)} = (I - D(A)^{-1}A)u^{(i)} + D(A)^{-1}b$$

and if we subtract this from equation (7) above we get

$$
\begin{aligned}
\bar{u} - u^{(i+1)} =& (I - D(A)^{-1}A)\bar{u} + D(A)^{-1}b \\
& - (I - D(A)^{-1}A)u^{(i)} + D(A)^{-1}b \\
=& (I - D(A)^{-1}A)(\bar{u} - u^{(i)}) \\
=& Z(A)(\bar{u} - u^{(i)})
\end{aligned}
$$

$$(8)$$

The upshot of this is that each iteration of the algorithm has the effect of multiplying the error by $Z(A)$. A simple inductive argument shows that at the end of the $i^{\text{th}}$ iteration

$$(9) \qquad\qquad \bar{u} - u^{(i+1)} = Z(A)^i(\bar{u} - u^{(0)})$$

The conclusion follows from 1.11 on page 161, which implies that $Z(A)^i \to 0$ as $i \to \infty$ if and only if $\rho(Z(A)) < 1$. Clearly, if $Z(A)^i \to 0$ the error will be killed off as $i \to \infty$ regardless of how large it was initially. $\square$

The following corollary give us an estimate of the rate of convergence.

COROLLARY 1.22. The conditions in the proposition above are satisfied if $\|Z(A)\| = \tau < 1$ for any matrix-norm $\| * \|$. If this condition is satisfied then

$$\|\bar{u} - u^{(i)}\| \leq \tau^{i-1}\|\bar{u} - u^{(0)}\|$$

where $\bar{u}$ is an exact solution of the original linear system.

PROOF. This is a direct application of equation (9) above:

$$\bar{u} - u^{(i+1)} = Z(A)^i(\bar{u} - u^{(0)})$$
$$\|\bar{u} - u^{(i+1)}\| = \|Z(A)^i(\bar{u} - u^{(0)})\|$$
$$\leq \|Z(A)^i\| \cdot \|(\bar{u} - u^{(0)})\|$$
$$\leq \|Z(A)\|^i \cdot \|(\bar{u} - u^{(0)})\|$$
$$= \tau^i\|(\bar{u} - u^{(0)})\|$$

$\square$

We conclude this section with an example. Let

$$A = \begin{pmatrix} 4 & 1 & 1 \\ 1 & 4 & -1 \\ 1 & 1 & 4 \end{pmatrix}$$

and

$$b = \begin{pmatrix} 1 \\ 2 \\ 3 \end{pmatrix}$$

Then

$$D(A) = \begin{pmatrix} 4 & 0 & 0 \\ 0 & 4 & 0 \\ 0 & 0 & 4 \end{pmatrix}$$

and

$$Z(A) = \begin{pmatrix} 0 & -1/4 & -1/4 \\ -1/4 & 0 & 1/4 \\ -1/4 & -1/4 & 0 \end{pmatrix}$$

so our iteration-step is:

$$u^{(i+1)} = \begin{pmatrix} 0 & -1/4 & -1/4 \\ -1/4 & 0 & 1/4 \\ 1/4 & -1/4 & 0 \end{pmatrix} u^{(i)} + \begin{pmatrix} 1/4 \\ 1/2 \\ 3/4 \end{pmatrix}$$

It is not hard to see that $\|Z(A)\|_1 = 1/2$ so the Jacobi method converges for any initial starting point. Set $u^{(0)} = 0$. We get:

$$u^{(1)} = \begin{pmatrix} \dfrac{1}{4} \\ \dfrac{1}{2} \\ \dfrac{3}{4} \end{pmatrix}$$

and

$$u^{(2)} = \begin{pmatrix} -\dfrac{1}{16} \\ \dfrac{5}{8} \\ \dfrac{9}{16} \end{pmatrix}$$

$$u^{(3)} = \begin{pmatrix} -\dfrac{3}{64} \\ \dfrac{21}{32} \\ \dfrac{39}{64} \end{pmatrix}$$

Further computations show that the iteration converges to the solution

$$x = \begin{pmatrix} -\dfrac{1}{15} \\ \dfrac{2}{3} \\ \dfrac{3}{5} \end{pmatrix}$$

**Exercises:**

1. Apply the Jacobi method to the system

$$Ax = b$$

where

$$A = \begin{pmatrix} 4 & 0 & 1 \\ 1 & 4 & -1 \\ 1 & 1 & 4 \end{pmatrix}$$

and

$$b = \begin{pmatrix} 1 \\ 2 \\ 3 \end{pmatrix}$$

2. Compute $D(A)$ and $Z(A)$ when

$$A = \begin{pmatrix} 4 & 1 & 1 \\ 2 & 5 & 1 \\ 0 & -3 & 10 \end{pmatrix}$$

Can the Jacobi method be used to solve linear systems of the form

$$Ax = b?$$

1.2.3. *The JOR method.* Now we will consider a variation on the Jacobi method. JOR, in this context stands for Jacobi Over-Relaxation. Essentially the JOR method attempts to speed up the convergence of the Jacobi method by modifying it slightly. Given the linear system:

$$Ax = b$$

where $A$ is a given $n \times n$ matrix whose diagonal elements are nonvanishing and $b$ is a given $n$-dimensional vector. The Jacobi method solves it by computing the sequence

$$u^{(i+1)} = Z(A)u^{(i)} + D(A)^{-1}b$$

where $D(A)$ is the matrix of diagonal elements of $A$ and $Z(A) = I - D(A)^{-1}A$. In other words, we solve the system of equations by "moving from $u^{(i)}$ toward $u^{(i+1)}$. the basic idea of the JOR method is that:

> If motion in this direction leads to a solution, maybe moving *further* in this direction at each step leads to the solution *faster*.

We, consequently, replace $u^{(i)}$ not by $u^{(i+1)}$ as defined above, but by $(1 - \omega)u^{(i)} + \omega u^{(i+1)}$. when $\omega = 1$ we get the Jacobi method exactly. The number $\omega$ is called the *relaxation coefficient*. Actual *over*relaxation occurs when $\omega > 1$. Our basic iteration-step is:

$$
\begin{aligned}
(10) \qquad u^{(i+1)} &= (\omega Z(A))u^{(i)} + \omega D(A)^{-1}b + (1 - \omega)u^{(i)} \\
&= (\omega Z(A) + I(1 - \omega))u^{(i)} + \omega D(A)^{-1}b \\
&= (\omega(I - D(A)^{-1}A + I(1 - \omega))u^{(i)} + \omega D(A)^{-1}b \\
&= (I - \omega D(A)^{-1}A)u^{(i)} + \omega D(A)^{-1}b
\end{aligned}
$$

Note that, if this iteration converges, then the limit is still a solution of the original system of linear equations. In other words a solution of

$$ x = (I - \omega D(A)^{-1}A)x + \omega D(A)^{-1}b $$

is a solution of $x = x - \omega D(A)^{-1}Ax + \omega D(A)^{-1}b$ and $0 = -\omega D(A)^{-1}Ax + \omega D(A)^{-1}b$. Dividing by $\omega$ and multiplying by $D(A)$ gives $Ax = b$.

The rate of convergence of this method is can be determined in the same way as in the Jacobi method:

COROLLARY 1.23. The conditions in the proposition above are satisfied if $\|I - \omega D(A)^{-1}A\| = \tau < 1$ for any matrix-norm $\| * \|$. In this case

$$ \|\bar{u} - u^{(i)}\| \le \tau^{i-1}\|\bar{u} - u^{(0)}\| $$

where $\bar{u}$ is an exact solution of the original linear system.

PROOF. The proof of this is almost identical to that of 1.22 on page 169. □

We must, consequently, adjust $\omega$ in order to make $\|I - \omega D(A)^{-1}A\| < 1$ and to minimize $\omega\tau^{i-1}\|D(A)^{-1}Au^{(0)} + D(A)^{-1}b\|$.

PROPOSITION 1.24. If the Jacobi method converges, then the JOR method converges for $0 < \omega \le 1$.

PROOF. This is a result of the fact that the eigenvalues of $(I - \omega D(A)^{-1}A)$ are closely related to those of $Z(A)$ because $(I - \omega D(A)^{-1}A) = \omega Z(A) + (1 - \omega)I$ If $\theta_j$ is an eigenvalue of $(I - \omega D(A)^{-1}A)$ then $\theta_j = \omega\lambda_j + 1 - \omega$, where $\lambda_j$ is an eigenvalue of $Z(A)$. If $\lambda_j = re^{it}$, then

$$
\begin{aligned}
|\theta_j|^2 &= \omega^2 r^2 + 2\omega r \cos t(1 - \omega) + (1 - \omega)^2 \\
&\le (\omega r + 1 - \omega)^2 < 1
\end{aligned}
$$

□

1.2.4. *The SOR and Consistently Ordered Methods.* We can combine the iterative meth-ods described above with the Gauss-Seidel method. The Gauss-Seidel method per-forms iteration as described above with one important difference:

> In the computation of $u^{(i+1)}$ from $u^{(i)}$ in equation (10), computed values for $u^{(i+1)}$ are substituted for values in $u^{(i)}$ *as soon as they are available* during the course of the computation.

In other words, assume we are computing $u^{(i+1)}$ *sequentially* by computing $u_1^{(i+1)}$, $u_2^{(i+1)}$, and so forth. The regular Jacobi method or the JOR method involves perform-ing these computations in a straightforward way. The Gauss-Seidel method involves computing $u_1^{(i+1)}$, and immediately setting $u_1^{(i)} \leftarrow u_1^{(i+1)}$ before doing any other com-putations. When we reach the point of computing $u_2^{(i+1)}$, it will *already* contain the computed value of $u_1^{(i+1)}$. This technique is easily implemented on a *sequential com-puter*, but it is not clear how to implement it in parallel.

The combination of the Gauss-Seidel method and overrelaxation is called the *SOR method*. The term SOR means *successive overrelaxation*. Experience and theoretical results show that it almost always converges faster than the JOR method.

In order to write down an equation for this iteration-scheme, we have to consider what it means to use $u_j^{(i+1)}$ for $0 \leq j < k$ when we are computing the $k^{\text{th}}$ entry of $u_k^{(i+1)}$. We are essentially multiplying $u^{(i)}$ by a matrix $(Z(A))$ in order to get $u^{(i+1)}$. When computing the $k^{\text{th}}$ entry of $u^{(i+1)}$, the entries of the matrix $Z(A)$ that enter into this entry are the entries whose *column*-number is strictly less than their *row*-number. In other words, they are the lower triangular entries of the matrix. It amounts to using the *following* iteration-scheme:

(11) $$u^{(i+1)} = \omega(L(A)u^{(i+1)} + U(A)u^{(i)} + D(A)^{-1}b) + (1 - \omega)u^{(i)}$$

Here $L(A)$ is the lower-triangular portion of $Z(A)$ and $U(A)$ is the upper-triangular portion.

(12) $$u^{(i+1)} = \mathcal{L}_\omega u^{(i)} + (1 - \omega L(A))^{-1}\omega D(A)^{-1}b$$

where $\mathcal{L}_\omega = (I - \omega L)^{-1}(\omega U + (1 - \omega)I)$. As before, we have the following criterion for the convergence of the SOR method:

THEOREM 1.25. The SOR iteration-scheme for solving the linear system

$$Au = b$$

(where $A$ has nonvanishing diagonal elements) converges if $\|\mathcal{L}_\omega\| = \tau < 1$ for any matrix-norm $\| * \|$. In this case

$$\|\bar{u} - u^{(i)}\| \leq \tau^{i-1}\|\bar{u} - u^{(0)}\|$$

where $\bar{u}$ is an exact solution of the original linear system.

PROOF. As before, the proof of this is almost identical to that of 1.22 on page 169.  □

THEOREM 1.26. (See [78]) The spectral radius of $\mathcal{L}_\omega$ satisfies

$$\rho(\mathcal{L}_\omega) \geq |\omega - 1|$$

In addition, if the SOR method converges, then

$$0 < \omega < 2$$

PROOF. This follows from the fact that the determinant of a matrix is equal to the product of the eigenvalues — see 1.11 on page 161.

$$\begin{aligned}
\det \mathcal{L}_\omega &= \det((I - \omega L)^{-1}(\omega U + (1 - \omega)I)) \\
&= (\det(I - \omega L))^{-1} \det(\omega U + (1 - \omega)I) \\
&= \det(\omega U + (1 - \omega)I) \\
&= (1 - \omega)^n
\end{aligned}$$

since the determinant of a matrix that is the sum of a lower or upper triangular matrix and the identity matrix is 1. It follows that:

$$\rho(\mathcal{L}_\omega) \geq (|\omega - 1|^n)^{1/n} = |1 - \omega|$$

□

Experiment and theory show that this method tends to converge twice as rapidly as the basic Jacobi method — 1.29 on page 176 computes the spectral radius $\rho(\mathcal{L}_\omega)$ if the matrix $A$ satisfies a condition to be described below.

Unfortunately, the SOR method as presented above doesn't lend itself to parallelization. Fortunately, it is possible to modify the SOR method in a way that does lend itself to parallelization.

DEFINITION 1.27. Let $A$ be an $n \times n$ matrix. Then:

1. two integers $0 \leq i, j \leq n$ are *associated* if $A_{i,j} \neq 0$ or $A_{j,i} \neq 0$;
2. Let $\Sigma = \{1, \ldots, n\}$, the set of numbers from 1 to $n$ and let $S_1, S_2, \ldots, S_k$ be disjoint subsets of $\Sigma$ such that

$$\bigcup_{i=1}^{k} S_i = \Sigma$$

Then the partition $S_1, S_2, \ldots, S_k$ of $\Sigma$ is
  a. an *ordering* of $A$ if
      i. $1 \in S_1$ and for any $i, j$ contained in the *same* set $S_t$, $i$ and $j$ are *not* associated — i.e., $A_{i,j} = A_{j,i} = 0$.

       ii. If $j$ is the lowest number in $\Sigma \setminus S_i$, then $j \in S_{i+1}$ for all $1 \le i < k$.
   b. a *consistent ordering* of $A$ if for any pair of associated integers $0 \le i, j \le n$,
      such that $i \in S_t$,
- $j \in S_{t+1}$ if $j > i$;
- $j \in S_{t-1}$ if $j < i$;

3. A vector

$$\gamma = \begin{pmatrix} \gamma_1 \\ \vdots \\ \gamma_n \end{pmatrix}$$

will be called a *consistent ordering vector* of a matrix $A$ if
   a. $\gamma_i - \gamma_j = 1$ if $i$ and $j$ are associated and $i > j$;
   b. $\gamma_i - \gamma_j = -1$ if $i$ and $j$ are associated and $i < j$;

Note that every matrix has an ordering: we can just set $S_i = \{i\}$. It is *not* true that every matrix has a *consistent* ordering.

Consistent orderings, and consistent ordering vectors are closely related: the set $S_i$ in a consistent ordering is the set of $j$ such that $\gamma_j = i$, for a consistent ordering vector.

An ordering for a matrix is important because it allows us to *parallelize* the SOR method:

PROPOSITION 1.28. Suppose $A$ is an $n \times n$ matrix equipped with an ordering $\{S_1, \ldots, S_t\}$, and consider the following iteration procedure:

   **for** $j = 1$ **to** $t$ **do**
      **for** all $k$ such that $k \in S_j$
         **Compute** entries of $u_k^{(i+1)}$
         Set $u_k^{(i)} \leftarrow u_k^{(i+1)}$
      **endfor**
   **endfor**

This procedure is equivalent to the SOR method applied to a version of the linear system

$$Au = b$$

in which the coordinates have been re-numbered in some way. If the ordering of $A$ was *consistent*, then the iteration procedure above is *exactly* equivalent to the SOR algorithm[2].

In other words, instead of using computed values of $u^{(i+1)}$ *as soon as they are available*, we may compute all components for $u^{(i+1)}$ whose subscripts are in $S_1$, then use these values in computing other components whose subscripts are in $S_2$, etc. Each individual

---

[2]I.e., without re-numbering the coordinates.

"phase" of the computation can be done in parallel. In many applications, it is possible to use only *two* phases (i.e., the ordering only has sets $S_1$ and $S_2$).

Re-numbering the coordinates of the linear system

$$Au = b$$

does not change the *solution*. It is as if we solve the system

$$BAu = Bb$$

where $B$ is some permutation-matrix[3]. The solution is

$$u = A^{-1}B^{-1}Bb = A^{-1}b$$

— the same solution as before. Since the computations are being carried out in a different order than in equation (11) on page 173, the *rate* of convergence might be different.

PROOF. We will only give an intuitive argument. The definition of an ordering in 1.27 on page 174 implies that distinct elements $r, r$ in the same set $S_i$ are *independent*. This means that, in the formula (equation (11) on page 173) for $u^{(i+1)}$ in terms of $u^{(i)}$,

    1. the equation for $u_r^{(i+1)}$ does not contain $u_s^{(i+1)}$ on the right.
    2. the equation for $u_s^{(i+1)}$ does not contain $u_r^{(i+1)}$ on the right.

It follows that we can compute $u_r^{(i+1)}$ and $u_s^{(i+1)}$ *simultaneously*. The re-numbering of the coordinates comes about because we might do some computations in a different order than equation (11) would have done them. For instance, suppose we have an ordering with $S_1 = \{1, 2, 4\}$, and $S_2 = \{3, 5\}$, and suppose that component 4 is dependent upon component 3 — this does not violate our definition of an ordering. The original SOR algorithm would have computed component 3 before component 4 and may have gotten a different result than the algorithm based upon our ordering.

It is possible to show (see [165]) that if the ordering is *consistent*, the permutation $B$ that occurs in the re-numbering of the coordinates has the property that it doesn't map any element of the *lower triangular* half of $A$ into the *upper triangular* half, and vice-versa. Consequently, the phenomena described in the example above will *not* occur. □

The importance of a *consistent* ordering of a matrix[4] is that it is possible to give an explicit formula for the optimal *relaxation-coefficient* for such a matrix:

THEOREM 1.29. If the matrix $A$ is consistently-ordered, in the sense defined in 1.27 on page 174, then the SOR or the consistently-ordered iteration procedures for solving the system

$$Ax = b$$

---

[3]A matrix with exactly one 1 in each row and column, and all other entries equal to 0.

[4]Besides the fact that it produces computations *identical* to the SOR algorithm.

FIGURE V.3. Color-graph of a linear coloring

both converge if $\rho(Z(A)) < 1$. In both cases the optimum relaxation coefficient to use is

$$\omega = \frac{2}{1 + \sqrt{1 - \rho(Z(A))^2}}$$

where (as usual) $Z(A) = I - D(A)^{-1}A$, and $D(A)$ is the matrix of diagonal elements of $A$. If the relaxation coefficient has this value, then the spectral radius of the effective linear transformation used in the SOR (or consistently-ordered) iteration scheme is $\rho(\mathcal{L}_\omega) = \omega - 1$.

The last statement gives us a good measure of the rate of convergence of the SOR method (via 1.25 on page 173 and the fact that the 2-norm of a symmetric matrix is equal to the spectral radius).

We can expand $\omega$ into a Taylor series to get some idea of its size:

$$\omega = 1 + \frac{\mu^2}{4} + \frac{\mu^4}{8} + O(\mu^6)$$

and this results in a value of the effective spectral radius of the matrix of

$$\frac{\mu^2}{4} + \frac{\mu^4}{8} + O(\mu^6)$$

The proof of 1.29 is beyond the scope of this book — see chapter 6 of [165] for proofs. It is interesting that this formula does *not* hold for matrices that are not consistently ordered — [165] describes an extensive theory of what happens in such cases.

We will give a criterion for when a consistent ordering scheme exists for a given matrix. We have to make a definition first:

DEFINITION 1.30. Let $G$ be an undirected graph with $n$ vertices. A *coloring* of $G$ is an assignment of colors to the vertices of $G$ in such a way that no two adjacent vertices have the same color.

Given a coloring of $G$ with colors $\{c_1, \dots, c_k\}$, we can define the associated *coloring graph* to be a graph with vertices in a 1-1 correspondence with the colors $\{c_1, \dots, c_k\}$ and an edge between two vertices $c_1$ and $c_2$ if any vertex (of $G$) colored with color $c_1$ is adjacent to any vertex that is colored with color $c_2$.

A *linear coloring* of $G$ is a coloring whose associated coloring graph is a linear array of vertices (i.e., it consists of a single path, as in figure V.3).

PROPOSITION 1.31. The operation of finding a consistent-ordering of a matrix can be regarded as equivalent to solving a kind of *graph-coloring* problem:

Given a square matrix, $A$, construct a graph, $G$, with one vertex for each row (or column) of the matrix, and an edge connecting a vertex representing row $i$ to the vertex representing row $j$ if $j \neq i$ and $A_{ij} \neq 0$.

Then:

1. the ordering schemes of $A$ are in a $1 - 1$ correspondence with the colorings of $G$.
2. the consistent ordering schemes of $A$ are in a $1 - 1$ correspondence with the linear colorings of the graph $G$.

PROOF. Statement 1: Define the sets $\{S_j\}$ to be the sets of vertices of $G$ with the same color (i.e., set $S_1$ might be the set of "green" vertices, etc.). Now arrange these sets so as to satisfy the second condition of 1.27 on page 174. This essentially amounts to picking the smallest element of each set and arranging the sets so that these smallest elements are in *ascending order* as we go from 1 to the largest number.

Statement 2: Suppose $\{S_1, \ldots, S_k\}$ is a consistent ordering of the matrix. We will color vertex $i$ of $G$ with color $S_j$ where $S_j$ is the set containing $i$. The condition:

...for any pair of associated integers $0 \leq i, j \leq n$, such that $i \in S_t$,
- $j \in S_{t+1}$ if $j > i$;
- $j \in S_{t-1}$ if $j < i$;

implies that the associated coloring graph is linear. It implies that vertex $i$ in the coloring graph is *only* adjacent to vertices $i - 1$ and $i + 1$ (if they exist).

Conversely, given a linear coloring, we number the vertices of the coloring graph by assigning to a vertex (and its associated color) its *distance* from one end.

1. For each color $c_i$ arbitrarily order the rows with that color.
2. Associate with a row $i$ the pair $(c_{j_i}, o_i)$, where $c_{j_i}$ is the color of row $i$, and $o_i$ is the ordinal position of this row in the set of all rows with the same color.
3. Order the rows lexicographically by the pairs $(c_{j_i}, o_i)$.
4. Define the permutation $\pi$ to map row $i$ to its ordinal position in the ordering of all of the rows defined in the previous line.

It is not hard to verify that, after the permutation of the rows and columns of the matrix (or re-numbering of the coordinates in the original problem) that the matrix will be consistently-ordered with ordering vector whose value on a given coordinate is the number of the color of that coordinate. □

Here is an example of a consistently ordered matrix: Let

$$A = \begin{pmatrix} 4 & 0 & 0 & -1 \\ -1 & 4 & -1 & 0 \\ 0 & -1 & 4 & 0 \\ -1 & 0 & 0 & 4 \end{pmatrix}$$

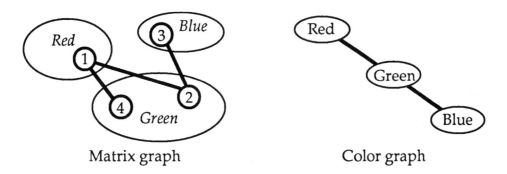

FIGURE V.4. Graph of the matrix $A$, and the associated color-graph

If we let $S_1 = \{1\}$, $S_2 = \{2,4\}$, and $S_3 = \{3\}$, we have a consistent ordering of $A$. The vector

$$\gamma = \begin{pmatrix} 1 \\ 2 \\ 3 \\ 2 \end{pmatrix}$$

is a consistent ordering vector for this matrix. The graph for this matrix is in figure V.4.

We conclude this section with an algorithm for determining whether a matrix *has* a consistent ordering (so that we can use the formula in 1.29 on page 176 to compute the optimum relaxation-factor). The algorithm is constructive in the sense that it actually finds an ordering if it exists. It is due to Young (see [165]):

ALGORITHM 1.32. Let $A$ be an $n \times n$ matrix. It is possible to determine whether $A$ has a consistent ordering via the following sequence of steps:

**Data:** vectors $\{\gamma_i\}$ and $\{\bar{\gamma}_i\}$
       sets $D$ initially $\{1\}$, $T = \{1\}$
       Boolean variable Cons $=$ TRUE

  $\gamma_1 \leftarrow 1, \bar{\gamma}_1 \leftarrow 1$
  **for** $j \leftarrow 2$ **to** $n$ **do**
    **if** $A_{1,j} \neq 0$ or $A_{j,1} \neq 0$ **then**
      $\gamma_j \leftarrow 2, \bar{\gamma}_j \leftarrow 2$
      $D \leftarrow D \cup \{j\}$
      $T \leftarrow T \cup \{j\}$
  **endfor**
    $D \leftarrow D \setminus \{1\}$
  **while** $T \neq \{1, \ldots, n\}$
    **if** $D \neq \emptyset$ **then**

$i \leftarrow$ Minimal element of $D$
**else**
$\quad i \leftarrow$ Minimal element of $\{1, \ldots, n\} \setminus T$
**endif**
**if** $j \notin T$ **then**
$\quad D \leftarrow D \cup \{j\}$
$\quad \bar{\gamma}_j \leftarrow 1 - \bar{\gamma}_i$
$\quad$**if** $j > i$ **then**
$\quad\quad \gamma_j \leftarrow \gamma_i + 1$
$\quad$**else**
$\quad\quad \gamma_j \leftarrow \gamma_i - 1$
$\quad$**endif**
**else** $\{ j \in T \}$
$\quad$**if** $\bar{\gamma}_j \neq 1 - \bar{\gamma}_i$ **then**
$\quad\quad$ Cons $\leftarrow$ FALSE
$\quad\quad$ **Exit**
$\quad$ **endif**
$\quad$**if** $j > i$ **then**
$\quad\quad$**if** $\gamma_j \neq \gamma_i + 1$ **then**
$\quad\quad\quad$ Cons $\leftarrow$ FALSE
$\quad\quad\quad$ **Exit**
$\quad\quad$ **endif**
$\quad$**else**
$\quad\quad$**if** $\gamma_j \neq \gamma_i - 1$ **then**
$\quad\quad\quad$ Cons $\leftarrow$ FALSE
$\quad\quad\quad$ **Exit**
$\quad\quad$ **endif**
$\quad$ **endif**
**endif**
**endwhile**

The variable Cons determines whether the matrix $A$ has a consistent ordering. If it is TRUE at the end of the algorithm, the vector $\gamma$ is a consistent ordering vector. This determines a consistent ordering, via the comment following 1.27, on page 175.

1.2.5. *Discussion.* We have only scratched the surface in our treatment of the theory of iterative algorithms for solving systems of linear equations. There are a number of important issues in finding parallel versions of these algorithms. For instance, the execution-time of such an algorithm depends upon:

1. The number of complete iterations required to achieve a desired degree of accuracy. This usually depends upon the norm or spectral radius of some matrix.

2. The number of parallel steps required to implement one complete iteration of an iteration-scheme. This depends upon the number of sets in an ordering (or consistent ordering) of a matrix.

In many cases the fastest parallel algorithm is the one that uses an ordering with the smallest number of sets in an ordering — even though that may lead to an iteration-matrix with a larger norm (than some alternative).

The reader may have noticed that we have required that the diagonal elements of the matrix $A$ be nonvanishing throughout this chapter. This is a very rigid condition, and we can eliminate it to some extent. The result is known at the *Richardson* Iteration scheme. It requires that we have some matrix $B$ that satisfies the conditions:

1. $B^{-1}$ is easy to compute exactly;
2. The spectral radius of $I - B^{-1}A$ is small (or at least $< 1$);

We develop this iteration-scheme via the following sequence of steps

$$Av = b$$
$$(A - B)v + Bv = b$$
$$B^{-1}(A - B)v + v = B^{-1}b$$
$$v = B^{-1}(B - A)v + B^{-1}b$$
$$v = (I - B^{-1}A)v + B^{-1}b$$

So our iteration-scheme is:

$$v^{(i+1)} = (I - B^{-1}A)v^{(i)} + B^{-1}b$$

Note that we really only need to know $B^{-1}$ in order to carry this out. The main result of the Pan-Reif matrix inversion algorithm (in the next section) gives an estimate for $B^{-1}$:

$$B^{-1} = A^{H}/(\|A\|_1 \cdot \|A\|_\infty)$$

(this is a slight re-phrasing of 1.40 on page 187). The results of that section show that $O(\lg n)$ parallel iterations of this algorithm are required to give a desired degree of accuracy (if $A$ is an invertible matrix). In general SOR algorithm is much faster than this scheme *if it can be used*.

Even the theory of consistent ordering schemes is well-developed. There are group-iteration schemes, and generalized consistent ordering schemes. In these cases it is also possible to compute the optimal relaxation coefficient. See [165] for more details.

**Exercises:**

1. Let
$$A = \begin{pmatrix} 4 & 2 & 3 \\ 2 & 4 & 1 \\ 3 & 0 & 4 \end{pmatrix}$$

Compute the spectral radius of this matrix. Will the SOR method converge in the problem

$$Ax = \begin{pmatrix} 2 \\ 3 \\ -5 \\ 1 \end{pmatrix}?$$

If so compute the optimum relaxation coefficient in the SOR method.

2. Show that a matrix that is consistently ordered has an ordering (non-consistent) that has only two sets. This means that the parallel version of the SOR algorithm (1.28 on page 175) has only *two* phases. (Hint: use 1.31 on page 177).

### 1.3. Power-series methods: the Pan-Reif Algorithm.

1.3.1. *Introduction.* The reader might have wondered what we do in the case where the diagonal elements of a matrix *vanish*, or the conditions on spectral radius are not satisfied.

In this section we present a variation on the iterative methods described above. The results in this section are of more theoretical than practical value since they require a larger number of processors than the iterative methods, and inversion of a matrix is generally not the most numerically stable method for solving a linear system.

In order to understand the algorithm for matrix inversion, temporarily forget that we are trying to invert a *matrix* — just assume we are inverting a number. Suppose we want to invert a number $u$ using a *power series*. Recall the well-known power-series for the inverse of $1 - x$:

$$1 + x + x^2 + x^3 \cdots$$

We can use this power series to compute $u^{-1}$ if we start with an estimate $a$ for $u^{-1}$ since $u^{-1} = u^{-1}a^{-1}a = (au)^{-1}a = (1 - (1 - au))^{-1}a$, where we use the power-series to calculate $(1 - (1 - au))^{-1}$. If $a$ is a good estimate for the inverse of $u$, then $au$ will be close to 1 and $1 - au$ will be close to 0 so that the power-series for $(1 - (1 - au))^{-1}$ will converge.

It turns out that all of this can be made to work for *matrices*. In order to reformulate it for matrices, we must first consider what it means for a power-series of matrices to converge. The simplest way for a power-series of matrices to converge is for all but a finite number of terms to vanish. For instance:

PROPOSITION 1.33. Let $M = (M_{ij})$ be an $n \times n$ *lower triangular* matrix. This is a matrix for which $M_{ik} = 0$ whenever $i \geq j$. Then $M^n = 0$, and

$$(I - M)^{-1} = I + M + M^2 + \cdots + M^{n-1}$$

PROOF. We prove that $M^n = 0$ by induction on $n$. It is easy to verify the result in the case where $n = 2$. Suppose the result is true for all $k \times k$, lower triangular matrices. Given a $k + 1 \times k + 1$ lower triangular matrix, $M$, we note that:

- Its first $k$ rows and columns form a $k \times k$ lower triangular matrix, $M'$.
- If we multiply $M$ by any other $k + 1 \times k + 1$ lower triangular matrix, $L$ with its $k \times k$ lower-triangular submatrix, $L'$, we note that the product is of the form:

$$\begin{pmatrix} M'L' & 0 \\ E & 0 \end{pmatrix}$$

where the $0$ in the upper right position represents a column of $k$ zeros, and $E$ represents a last row of $k$ elements.

It is not hard to see that $M^k$ will be of the form

$$\begin{pmatrix} 0 & 0 \\ E & 0 \end{pmatrix}$$

i.e., it will only have a single row with nonzero elements — the last. One further multiplication of this by $M$ will kill off all of the elements.

The remaining statement follows by direct computation:

$$(I - M)(I + M + M^2 + \cdots + M^{n-1}) = I - M^n = I$$

□

In somewhat greater generality:

COROLLARY 1.34. Let $A$ be an $n \times n$ matrix of the form

$$A = D - L$$

where $D$ is a diagonal matrix with all diagonal entries nonzero, and $L$ is a lower-triangular matrix. Then

$$D^{-1} = \begin{pmatrix} a_{11}^{-1} & 0 & \cdots & 0 \\ 0 & a_{22}^{-1} & \cdots & 0 \\ \vdots & \vdots & \ddots & \vdots \\ 0 & 0 & \cdots & a_{nn}^{-1} \end{pmatrix}$$

and

$$A^{-1} = D^{-1}(I + (D^{-1}L) + \cdots + (D^{-1}L)^{n-1})$$

Now we give a fast algorithm for *adding up* these power series:

PROPOSITION 1.35. Let $Z_k = \prod_{h=0}^{k-1}(I + R^{2^h})$. Then $Z_k = \sum_{i=0}^{2^k-1} R^i$.

PROOF. This follows by induction on $k$ — the result is clearly true in the case where $k = 1$. Now suppose the result is true for a given value of $k$ — we will prove it for the next higher value. Note that $Z_{k+1} = Z_k(I + R^{2^k})$. But this is equal to

$$(I + R^{2^k}) \sum_{i=0}^{2^k-1} R^i = \sum_{i=0}^{2^k-1} R^i + R^{2^k} \sum_{i=0}^{2^k-1} R^i$$

$$= \sum_{i=0}^{2^k-1} R^i + \sum_{i=0}^{2^k-1} R^{i+2^k}$$

$$= \sum_{i=0}^{2^{k+1}-1} R^i$$

□

COROLLARY 1.36. Let $M$ be an $n \times n$ matrix, all of whose entries vanish above the main diagonal, and such that $M$ has an inverse. Then there is a SIMD-parallel algorithm for computing $M^{-1}$ in $O(\lg^2 n)$ time using $n^{2.376}$ processors.

This algorithm first appeared in [120] and [65]. See [66] for related algorithms.

PROOF. We use 1.33, 1.34 and 1.35 in a straightforward way. The first two results imply that $M^{-1}$ is equal to a suitable power-series of matrices, with only $n$ nonzero terms, and the last result gives us a way to sum this power-series in $\lg n$ steps. The sum is a product of terms that each equal the identity matrix plus an iterated square of an $n \times n$ matrix. Computing this square requires $O(\lg n)$ time, using $n^{2.376}$ processors, by 1.2 on page 157. □

Now we will consider the question of how one finds an inverse of an arbitrary invertible matrix. In general, we cannot assume that some finite power of part of a matrix will vanish. We must be able to handle power series of matrices that converge more slowly. We need to define a measure of the *size* of a matrix (so we can easily quantify the statement that the terms of a power-series of matrices approach 0, for instance).

1.3.2. *The main algorithm.* Now we are in a position to discuss the Pan-Reif algorithm. Throughout this section, we will have a fixed $n \times n$ matrix, $A$. We will assume that $A$ is *invertible*.

DEFINITION 1.37. The following notation will be used throughout this section:
1. If $B$ is a matrix $R(B)$ is defined to be $I - BA$;
2. A matrix $B$ will be called a *sufficiently good estimate* for the inverse of $A$ if $\|R(B)\| = u(B) < 1$.

Now suppose $A$ is a nonsingular matrix, and $B$ is a sufficiently good estimate for the inverse (with respect to *some* matrix-norm) of $A$ (we will show how to get such an estimate later). Write $A^{-1} = A^{-1}B^{-1}B = (BA)^{-1}B = (I - R(B))^{-1}B$. We will use a *power series* expansion for $(I - R(B))^{-1}$ — the fact that $B$ was a sufficiently good estimate for the inverse of $A$ will imply that the series converges (with respect to the same matrix-norm).

The following results apply to *any* matrix-norms:

PROPOSITION 1.38. Suppose that $B$ is a sufficiently close estimate for the inverse of $A$. Then the power series $(I + R(B) + R(B)^2 + \cdots)B$ converges to the inverse of $A$. In fact, if $S_k$ is the $k^{\text{th}}$ partial sum of the series (i.e. $(I + R(B) + \cdots + R(B)^k)B$) then

$$\|A^{-1} - S_k\| \leq \frac{\|B\|u(B)^k}{(1 - u(B))}$$

which tends to 0 as $k$ goes to $\infty$.

PROOF. The inequalities in the remark above imply that $\|R(B)^i\| \leq \|R(B)\|^i$ and $\|S_k - S_{k'}\|$ (where $k > k'$) is $\leq \|B\|(u(B)^{k'} + \cdots + u(B)^k) \leq \|B\|u(B)^{k'}(1 + u(B) + u(B)^2 + \cdots) = \|B\|u(B)^{k'}/(1 - u(B))$ (we are making use of the fact that $u(B) < 1$ here). This tends to 0 as $k'$ and $k$ tend to 0 so the series *converges* (it isn't hard to see that the series will converge if the norms of the partial sums converge — this implies that the *result* of applying the matrix to *any vector* converges). Now $A^{-1} - S_k = (R(B)^k + \cdots)B = R(B)^k(I + R(B) + R(B)^2 + \cdots)B$. The second statement is clear. $\square$

We can use 1.35 on page 184 to add up this power series. We apply this result by setting $B_k^* = Z_k$ and $R(B) = R$.

It follows that

$$\|A^{-1} - B_k^*\| \leq \frac{\|B\|u(B)^{2^k}}{(1 - u(B))}$$

We will show that the series converges to an accurate estimate of $A^{-1}$ in $O(\lg n)$ steps. We need to know something about the value of $u(B)$. It turns out that this value will depend upon $n$ — in fact it will increase towards 1 as $n$ goes to 0 — we consequently need to know that it doesn't increase towards 1 too rapidly. It will be proved later in this section that:

**Fact:** $u(B) = 1 - 1/n^{O(1)}$ as $n \to \infty$, $|\lg \|B\|| \leq an^\beta$.

PROPOSITION 1.39. Let $\alpha$ be such that $u(B) \leq 1 - 1/n^\alpha$ and assume the two facts listed above and that $r$ bits of precision are desired in the computation of $A^{-1}$. Then $B \cdot B_k^*$ is a sufficiently close estimate of $A^{-1}$, where $an^\alpha \lg(n) + r + an^{\alpha+\beta} \leq 2^k$. It follows that $O(\lg n)$ terms in the computation of $\prod_{h=0}^{k-1} (I + R(B)^{2^h})$ suffice to compute $A^{-1}$ to the desired degree of accuracy.

PROOF. We want

$$\|A^{-1} - B_k^*\| \leq \frac{\|B\|u(B)^{2^k}}{(1 - u(B))} \leq 2^{-r}$$

Taking the logarithm gives

$$|\lg \|B\|| + 2^k \lg(u(B)) - \lg(1 - u(B)) \leq -r$$

The second fact implies that this will be satisfied if $2^k \lg(u(B)) - \lg(1 - u(B)) \leq -(r + an\beta)$. The first fact implies that this inequality will be satisfied if we have

$$2^k \lg\left(1 - \frac{1}{n^\alpha}\right) - \lg\left(\frac{1}{n^\alpha}\right) = 2^k \lg\left(1 - \frac{1}{n^\alpha}\right) + \alpha \lg(n) \leq -(r + an^\beta)$$

Now substituting the well-known power series $\lg(1-g) = -g - g^2/2 - g^3/3 \cdots$ gives $2^k(-1/n^\alpha - 1/2n^{2\alpha} - 1/3n^{3\alpha} \cdots) + \alpha \lg(n) \leq -(r + an^\beta)$. This inequality will be satisfied if the following one is satisfied (where we have replaced the power series by a strictly greater one): $2^k(-1/n^\alpha) + \alpha \lg(n) \leq -(r + an^\beta)$.

Rearranging terms gives $\alpha \lg(n) + r + an^\beta \leq 2^k/n^\alpha$ or $an^\alpha \lg(n) + r + an^{\alpha+\beta} \leq 2^k$, which proves the result. $\square$

Now that we have some idea of how the power series converges, we can state the most remarkable part of the work of Pan and Reif — their estimate for the inverse of $A$. This estimate is given by:

**1.40. Estimate for $A^{-1}$:**

$$B = A^{\mathrm{H}}/(\|A\|_1 \cdot \|A\|_\infty)$$

It is remarkable that this fairly simple formula gives an estimate for the inverse of an arbitrary nonsingular matrix. Basically, the matrix $R(B)$ will play the part of the matrix $Z(A)$ in the iteration-methods of the previous sections. One difference between the present results and the iteration methods, is that the present scheme converges *much more slowly* than the iteration-schemes discussed above. We must consequently, use many more processors in the computations — enough processors to be able to perform multiple iterations in parallel, as in 1.35.

Now we will work out an example to give some idea of how this algorithm works:
Suppose our initial matrix is

$$A = \begin{pmatrix} 1 & 2 & 3 & 5 \\ 3 & 0 & 1 & 5 \\ 0 & 1 & 3 & 1 \\ 5 & 0 & 0 & 3 \end{pmatrix}$$

Then our estimate for the inverse of $A$ is:

$$B = \frac{A^{\mathrm{H}}}{\|A\|_1 \cdot \|A\|_\infty} = \begin{pmatrix} 0.0065 & 0.0195 & 0 & 0.0325 \\ 0.0130 & 0 & 0.0065 & 0 \\ 0.0195 & 0.0065 & 0.0195 & 0 \\ 0.0325 & 0.0325 & 0.0065 & 0.0195 \end{pmatrix}$$

and

$$R(B) = I - BA = \begin{pmatrix} 0.7727 & -0.0130 & -0.0390 & -0.2273 \\ -0.0130 & 0.9675 & -0.0584 & -0.0714 \\ -0.0390 & -0.0584 & 0.8766 & -0.1494 \\ -0.2273 & -0.0714 & -0.1494 & 0.6104 \end{pmatrix}$$

The 2-norm of $R(B)$ turns out to be 0.9965 (this quantity is not easily computed, incidentally), so the power-series will converge. The easily computed norms are given by $\|R(B)\|_1 = 1.1234$ and $\|R(B)\|_\infty = 1.1234$, so they give no indication of whether the series will converge.

Now we compute the $B_k^*$ — essentially we use 1.1 and square $R(B)$ until we arrive at a power of $R(B)$ whose 1-norm is $< 0.00001$. This turns out to require 13 steps (which represents adding up $2^{13} - 1$ terms of the power series). We get:

$$B_{13}^* = \begin{pmatrix} 19.2500 & 24.3833 & 2.5667 & -16.6833 \\ 24.3833 & 239.9833 & -94.9667 & -21.8167 \\ 2.5667 & -94.9667 & 61.6000 & -7.7000 \\ -16.6833 & -21.8167 & -7.7000 & 19.2500 \end{pmatrix}$$

and

$$B_{13}^* B = \begin{pmatrix} -0.0500 & -0.1500 & 0.1000 & 0.3000 \\ 0.7167 & -0.8500 & -0.4333 & 0.3667 \\ -0.2667 & -0.2667 & 0.5333 & -0.0667 \\ 0.0833 & 0.2500 & -0.1667 & -0.1667 \end{pmatrix}$$

which agrees with $A^{-1}$ to the number of decimal places used here.

1.3.3. *Proof of the main result.* Recall that § 1.2.1 describes several different concepts of the size of a matrix: the 1-norm, the $\infty$-norm, and the 2-norm. The proof that the Pan-Reif algorithm works requires at  all three of these.  We will also need several inequalities that these norms satisfy.

LEMMA 1.41. The vector norms, the 1-norm, the $\infty$-norm, and the 2-norm defined in 1.9 on page 160 satisfy:

1. $\|v\|_\infty \le \|v\|_2 \le \|v\|_1$;
2. $\|v\|_1 \le n\|v\|_\infty$;
3. $\|v\|_2 \le (n^{1/2})\|v\|_\infty$;
4. $(n^{-1/2})\|v\|_1 \le \|v\|_2$;
5. $\|v\|_2^2 \le \|v\|_1 \cdot \|v\|_\infty$;

PROOF. $\|v\|_\infty \le \|v\|_2$ follows by *squaring both terms*; $\|v\|_2 \le \|v\|_1$ follows by squaring both terms and realizing that $\|v\|_1^2$ has *cross-terms* as well as the squares of single terms. $\|v\|_1 \le n\|v\|_\infty$ follows from the fact that each of the $n$ terms in $\|v\|_1$ is $\le$ the *maximum* such term. $\|v\|_2 \le (n^{1/2})\|v\|_\infty$ follows by a similar argument after first *squaring* both sides.

$(n^{-1/2})\|v\|_1 \le \|v\|_2$ follows from *Lagrange undetermined multipliers*. They are used to *minimize* $\sqrt{\sum_i |v_i|^2}$ subject to the *constraint* that $\sum_i |v_i| = 1$ (this is the same as minimizing $\sum_i |v_i|^2$) — it is not hard to see that the minimum occurs when all coordinates (i.e., all of the $|v_i|$) are *equal* and the value *taken on* by $\sqrt{\sum_i |v_i|^2}$ is $(n^{1/2})|v_0|$ in this case.

The last inequality follows by an argument like that used for ii — each term $|v_i|^2$ is $\le |v_i| \times$ the *maximum* such term.   $\square$

These relations imply the following relations among the corresponding *matrix norms*:

COROLLARY 1.42.      1. $\|M\|_2 \le (n^{1/2})\|M\|_\infty$;
2. $(n^{-1/2})\|M\|_1 \le \|M\|_2$;
3. $\|M\|_2^2 \le \|M\|_1 \cdot \|M\|_\infty$

Recall the formulas for the 1-norm and the $\infty$-norm in 1.18 on page 163 and 1.19 on page 164.

The rest of this section will be spent examining the theoretical basis for this algorithm. In particular, we will be interested in seeing why $B$, as given above, is a sufficiently good estimate for $A^{-1}$. The important property of $B$ is:

THEOREM 1.43.     1. $\|B\|_2 \leq 1/\|A\|_2 \leq 1/\max_{i,j} |a_{ij}|$;
2. $\|R(B)\|_2 \leq 1 - 1/((\text{cond } A)^2 n)$.

1. This result applies to the *2-norms* of the matrices in question.
2. The remainder of this section will be spent proving these statements. It isn't hard to see how these statements imply the main results (i.e. the facts cited above) — $\|B\|_2 \leq 1/\max_{i,j} |a_{ij}| \leq M$ and $\|R(B)\|_2 \leq 1 - 1/n^7 M^2$, by the remark above.
3. The proof of this theorem will make heavy use of the *eigenvalues* of matrices — 1.10 on page 160 for a definition of eigenvalues. The idea, in a nutshell, is that:
   a. the eigenvalues of $B$ and $R(B)$, as defined in 1.40, are closely related to those of $A$;
   b. the *eigenvalues* of any matrix are closely related to the *2-norm* of the matrix;

We can, consequently, prove an inequality for the 2-norm of $R(B)$ and use certain relations between norms (described in 1.41 on page 188) to draw conclusions about the values of the other norms of $R(B)$ and the convergence of the power-series.

Another way to think of this, is to suppose that the matrix $A$ is *diagonalizable* in a suitable way, i.e. there exists a nonsingular matrix $Z$ that preserves the 2-norm[5], and such that $D = Z^{-1}AZ$ is a *diagonal* matrix — its only nonzero entries lie on the main diagonal. For the time being forget about how to compute $Z$ — it turns out to be unnecessary to do so — we only need know that such a $Z$ *exists*. It turns out that it will be relatively easy to prove the main result in this case.

We make essential use of the fact that the 2-norm of $D$ is the same as that of $A$ (since $Z$ preserves 2-norms) — in other words, it is not affected by diagonalization. Suppose that $D$ is

$$\begin{pmatrix} \lambda_1 & 0 & \cdots & 0 \\ 0 & \lambda_2 & \cdots & 0 \\ \vdots & \vdots & \ddots & \vdots \\ 0 & 0 & \cdots & \lambda_n \end{pmatrix}$$

Then:

1. $\|D\|_2 = \|A\|_2 = \max_i |\lambda_i|$;
2. $A$ is nonsingular if and only if none of the $\lambda_i$ is 0.

---

[5]Such a matrix is called *unitary* – it is characterized by the fact that $Z^H = Z^{-1}$

Now, let $t = 1/(\|A\|_1 \cdot \|A\|_\infty)$. Then $\|A\|_2^2 \leq 1/t$, by 1.20 on page 165, and the 2-norm of $B$ is equal to $t \cdot \max_i |\lambda_i|$. The matrix $R(D)$ is equal to:

$$\begin{pmatrix} 1 - t\lambda_1^2 & 0 & \cdots & 0 \\ 0 & 1 - t\lambda_2^2 & \cdots & 0 \\ \vdots & \vdots & \ddots & \vdots \\ 0 & 0 & \cdots & 1 - t\lambda_n^2 \end{pmatrix}$$

and *its* 2-norm is equal to $\max_i (1 - t\lambda_i^2)$. If $A$ was nonsingular, then the $\lambda_i$ are all nonzero. The fact that $\|A\|_2^2 \leq 1/t$ and the fact that $\|A\|_2 = \max_i |\lambda_i|$ imply that $t\lambda_i^2 \leq 1$ for all $i$ and $\|D\|_2 = \max_i 1 - t\lambda_i^2 < 1$. But this is a 2-norm, so it is not sensitive to diagonalization. It is not hard to see that $R(D)$ is the diagonalization of $R(B)$ so we conclude that $\|R(B)\|_2 < 1$, if $A$ was nonsingular.

All of this depended on the assumption that $A$ could be diagonalized in this way. This is not true for all matrices — and the bulk of the remainder of this section involves some tricks that Reif and Pan use to reduce the general case to the case where $A$ can be diagonalized.

Recall the definition of *spectral radius* in 1.10 on page 160.

Applying 1.20 on page 165 to $W = A^H$ we get $\|A^H\|_2^2 \leq 1/t$ where $t = 1/(\|A\|_1 \cdot \|A\|_\infty)$ is defined in 1.39 on page 186. It follows, by 1.43 that $\|A^H\|_2 \leq 1/(t\|A\|_2)$.

The remainder of this section will be spent proving the inequality in line 2 of 1.43 on page 189.

COROLLARY 1.44. Let $\lambda$ be an eigenvalue of $A^H A$ and let $A$ be nonsingular. Then $1/\|A^{-1}\|_2^2 \leq \lambda \leq \|A\|_2^2$.

PROOF. $\lambda \leq \rho(A^H A) = \|A\|_2^2$ by definition and 1.15 on page 162. On the other hand $(A^H A)^{-1} = A^{-1}(A^{-1})^H$, so $(A^H A)^{-1}$ is a Hermitian positive definite matrix. Lemma 1.14 on page 162 implies that $1/\lambda$ is an eigenvalue of $(A^H A)^{-1}$. Consequently $1/\lambda \leq \rho((A^H A)^{-1}) = \rho(A^{-1}(A^{-1})^H) = \|A^{-1}\|_2^2$. $\square$

LEMMA 1.45. Let $B$ and $t = 1/(\|A\|_1 \cdot \|A\|_\infty)$ Let $\mu$ be an eigenvalue of $R(B) = I - BA$. Then $0 \leq \mu \leq 1 - 1/(\text{cond } A)^2 n$.

PROOF. Let $R(B)v = \mu v$ for $v \neq 0$. Then $(I - tA^H)v = v = tA^H v = \mu v$. Therefore $A^H Av = \lambda v$ for $\lambda = (1 - \mu)/t$ so $\lambda$ is an eigenvalue of $A^H A$. Corollary 1.44 implies that $1/\|A^{-1}\|_2^2 \leq \lambda = (1 - \mu)/t \leq \|A\|_2^2$. It immediately follows that $1 - t\|A\|_2^2 \leq \mu \leq 1 - t/\|A^{-1}\|_2^2$. It remains to use the definition of $t = 1/(\|A\|_1 \cdot \|A\|_\infty)$ and to apply statement 3 of 1.42 on page 188. $\square$

We are almost finished. We have bounded the eigenvalues of $R(B)$ and this implies a bound on the spectral radius. The bound on the spectral radius implies a bound on the 2-*norm* since $R(B)$ is Hermitian. Since $\mu$ is an *arbitrary* eigenvalue of $R(B)$,

$\rho(R(B)) \leq 1 - 1/((\text{cond } A)^2)$. On the other hand, $\|R(B)\|_2 = \rho(R(B))$ since $R(B) = I - tA^H A$ is Hermitian. This completes the proof of the second line of 1.43 on page 189.

**Exercises:**

1. Apply the algorithm given here to the development of an algorithm for determining whether a matrix is nonsingular. Is it possible for a matrix $A$ to be *singular* with $\|R(B)\| < 1$ (any norm)? If this happens, how can one decide whether $A$ was singular or not?

2. Write a C* program to implement this algorithm. Design the algorithm to run until a desired degree of accuracy is achieved — in other words do *not* make the program use the error-estimates given in this section.

3. From the sample computation done in the text, estimate the size of the constant of proportionality that appears in the '$O(\lg^2 n)$'.

4. Compute the inverse of the matrix

$$\begin{pmatrix} 2 & 0 & 0 & 0 \\ 1 & -1 & 0 & 0 \\ 3 & 0 & 3 & 0 \\ 1 & 2 & 3 & 4 \end{pmatrix}$$

using the algorithm of 1.34 on page 183.

5. If $A$ is a symmetric positive definite matrix (i.e., all of its eigenvalues are positive) show that

$$B = \frac{I}{\|A\|_\infty}$$

can be used as an approximate inverse of $A$ (in place of the estimate in 1.40 on page 187).

**1.4. Nonlinear Problems.** In this section we will give a very brief discussion of how the iterative methods developed in this chapter can be generalized to nonlinear problems. See [74] for a general survey of this type of problem and [101] and [15] for a survey of parallel algorithms

DEFINITION 1.46. Let $f: \mathbb{R}^n \to \mathbb{R}^n$ be a function that maps some region $M \subseteq \mathbb{R}^n$ to itself. This function will be called:

1. a *contracting map* on $M$ with respect to a (vector) norm $\| * \|$ if there exists a number $0 \leq \alpha < 1$ such that, for all pairs of points $x$ and $y$, in $M$

$$\|f(x) - f(y)\| \leq \alpha \|x - y\|$$

2. a *pseudo-contracting* map on $M$ (with respect to a vector norm) if there exists a point $x_0 \in M$ such that:
   a. $f(x_0) = x_0$ and there exists a number $\alpha$ between 0 and 1 such that for all $x \in M$

$$\|f(x) - x_0\| \leq \alpha \|x - x_0\|$$

Although these definition might seem a little abstract at first glance, it turns out that the property of being a contracting map is precisely a nonlinear version of the statement that the norm of a matrix must be $< 1$. The material in this section will be a direct generalization of the material in earlier section on the Jacobi method — see page 167.

Suppose $f$ is a pseudo-contracting map. Then, it is not hard to see that the iterating the application of $f$ to any point in space will result in a sequence of points that converge to $x_0$:

$$f(x), f(f(x)), f(f(f(x))), \cdots \to x_0$$

This is a direct consequence of the fact that the distance between any point and $x_0$ is reduced by a constant factor each time $f$ is applied to the parameter. This means that if we want to solve an equation like:

$$f(x) = x$$

we can easily get an *iterative* procedure for finding $x_0$. In fact the Jacobi (JOR, SOR) methods are just special cases of this procedure in which $f$ is a *linear function*.

As remarked above, the possibilities for exploiting parallelism in the nonlinear case are generally far less than in the linear case. In the linear case, if we have enough processors, we can parallelize multiple iterations of the iterative solution of a linear system — i.e. if we must compute

$$M^n x$$

where $M$ is some matrix, and $n$ is a large number, we can compute $M^n$ by repeated squaring — this technique is used in the Pan-Reif algorithm in § 1.3. In nonlinear problems, we generally do not have a compact and uniform way of representing $f^n$ (the $n$-fold composite of a function $f$), so we must perform the iterations one at a time. We can still exploit parallelism in computing $f$ — when there are a large number of variables in the problem under investigation.

We conclude this section with an example:

EXAMPLE 1.47. Let $M$ be an $k \times k$ matrix. If $v$ is an eigenvector of $M$ with a nonzero eigenvalue, $\lambda$, then

$$Mv = \lambda v$$

by definition. If $\| * \|$ is any norm, we get:

$$\|Mv\| = |\lambda| \|v\|$$

so we get

$$\frac{Mv}{\|Mv\|} = \frac{v}{\|v\|}$$

Consequently, we get the following nonlinear equation

$$f(w) = w$$

where $f(w) = Mw/\|Mw\|$. Its solutions are eigenvectors of $M$ of unit norm (all eigenvectors of $M$ are scalar multiples of these). This is certainly nonlinear since it involves dividing a linear function by $\|Mw\|$ which, depending upon the norm used may have square roots or the maximum function. This turns out to be a pseudo-contracting map to where $x_0$ (in the notation of definition 1.46) is the eigenvector of the eigenvalue of largest absolute value.

**1.5. A Parallel Algorithm for Computing Determinants.** In this section we will discuss an algorithm for computing determinants of matrices. The first is due to Csanky and appeared in [38]. In is of more theoretical than practical interest, since it uses $O(n^4)$ processors and is numerically unstable. The problem of computing the determinant of a matrix is an interesting one because there factors that make it seem as though there might not exist an **NC** algorithm.

- the definition of a determinant (1.5 on page 159) has an exponential number of terms in it.
- the only well-known methods for computing determinants involve variations on using the definition, or Gaussian Elimination. But this is known to be P-complete — see page 47.

Before Csanky published his paper [38], in 1976, the general belief was that no **NC** algorithm existed for computing determinants.

Throughout this section $A$ will denote an $n \times n$ matrix.

DEFINITION 1.48. If $A$ is an $n \times n$ matrix, the *trace* of $A$ is defined to be

$$\mathrm{tr}(A) = \sum_{i=1}^{n} A_{ii}$$

Recall the characteristic polynomial, defined in 1.10 on page 160. If the matrix is nonsingular

(13)
$$f(\lambda) = \det(\lambda \cdot I - A) = \prod_{i-1}^{n}(\lambda - \lambda_i)$$

(14)
$$= \lambda^n + c_1 \lambda^{n-1} + \cdots + c_{n-1}\lambda + c_n$$

where the $\{\lambda_i\}$ are the eigenvalues of the matrix. Direct computation shows that

(15)
$$\text{tr}(A) = \sum_{i=1}^{n} \lambda_i = c_1$$

It follows that the trace is an invariant quantity — in other words, transforming the matrix in a way that corresponds to a change in coordinate system in the vector-space upon which it acts, doesn't change the trace.

Setting $\lambda = 0$ into equation (14) implies that the determinant of $A$ is equal to $(-1)^n c_n$.

The first step in Csanky's algorithm for the determinant is to compute the powers of $A$: $A^2, A^3, \ldots, A^{n-1}$, and the trace of each. Set $s_k = \text{tr}(A^k)$.

PROPOSITION 1.49. If $A$ is a nonsingular matrix and $k \geq 1$ is an integer then

$$s_k = \text{tr}(A^k) = \sum_{i=1}^{n} \lambda_i^k$$

PROOF. Equation (15) shows that $\text{tr}(A^k) = \sum_{i=1}^{n} \mu(k)_i$, where the $\{\mu(k)_i\}$ are the eigenvalues of $A^k$, counted with their multiplicities. The result follows from the fact that the eigenvalues of $A^k$ are just $k^{\text{th}}$ powers of corresponding eigenvalues of $A$. This follows from definition of eigenvalue in 1.10 on page 160: it is a number with the property that there exists a vector $v$ such that $Av = \lambda v$. Clearly, if multiplying $v$ by $A$ has the effect of multiplying it by the scalar $\lambda$, then multiplying the same vector by $A^k$ will have the effect of multiplying it by the scalar $\lambda^k$. So $\mu(k)_i = \lambda_i^k$ and the result follows. $\square$

At this point we have the quantities $\sum_{i=1}^{n} \lambda_i, \sum_{i=1}^{n} \lambda_i^2, \ldots, \sum_{i=1}^{n} \lambda_i^{n-1}$. It turns out that we can compute $\prod_{i=1}^{n} \lambda_i$ from this information. It is easy to see how to do this in simple cases. For instance, suppose $n = 2$. Then

1. $s_1 = \lambda_1 + \lambda_2$
2. $s_2 = \lambda_1^2 + \lambda_2^2$

and, if we compute $s_1^2$ we get $\lambda_1^2 + 2\lambda_1\lambda_2 + \lambda_2^2$, so

$$\lambda_1 \lambda_2 = \frac{s_1^2 - s_2}{2}$$

There is a general method for computing the coefficients of a polynomial in terms sums of powers of its roots. This was developed by Le Verrier in 1840 (see [159] and [49] to compute certain elements of *orbits* of the first seven planets (that is all that were known at the time).

PROPOSITION 1.50. Let

$$p(x) = x^n + c_1 x^{n-1} + \cdots + c_{n-1} x + c_n$$

and let the roots of $p(x)$ be $x_1, \ldots, x_n$. Then

$$\begin{pmatrix} 1 & 0 & \cdots\cdots\cdots & 0 \\ s_1 & 2 & 0 & \cdots\cdots & 0 \\ \vdots & \ddots & \ddots & \ddots & \vdots \\ s_{k-1} & \cdots & s_1 & k & 0 & \vdots \\ \vdots & \ddots & \ddots & \ddots & \ddots & 0 \\ s_{n-1} & \cdots & s_{k-1} & \cdots & s_1 & n \end{pmatrix} \begin{pmatrix} c_1 \\ \vdots \\ c_n \end{pmatrix} = -\begin{pmatrix} s_1 \\ \vdots \\ s_n \end{pmatrix}$$

where $s_k = \sum_{i=1}^n x_i^k$.

PROOF. We will give an analytic proof of this result. If we take the derivative of $p(x)$, we get:

$$\frac{dp(x)}{dx} = nx^{n-1} + c_1(n-1)x^{n-2} + \cdots + c_{n-1}$$

We can also set

$$p(x) = \prod_{i=1}^n (x - x_i)$$

and differentiate this formula (using the product rule) to get:

$$\frac{dp(x)}{dx} = \sum_{i=1}^n \prod_{\substack{j=1 \\ j \neq i}}^n (x - x_j) = \sum_{i=1}^n \frac{p(x)}{x - x_i}$$

so we get

$$(16) \quad nx^{n-1} + c_1(n-1)x^{n-2} + \cdots + c_{n-1} = \sum_{i=1}^n \frac{p(x)}{x - x_i} = p(x) \sum_{i=1}^n \frac{1}{x - x_i}$$

Now we expand each of the terms $1/(x - x_i)$ into a power-series over $x_i/x$ to get

$$\frac{1}{x - x_i} = \frac{1}{x(1 - x_i/x)} = \frac{1}{x}\left(1 + \frac{x_i}{x} + \frac{x_i^2}{x^2} + \cdots\right)$$

Now we plug this into equation (16) to get

$$nx^{n-1} + c_1(n-1)x^{n-2} + \cdots + c_{n-1} = p(x)\sum_{i=1}^{n}\frac{1}{x-x_i}$$

$$= \frac{p(x)}{x}\sum_{i=1}^{n}\left(1 + \frac{x_i}{x} + \frac{x_i^2}{x^2} + \cdots\right)$$

$$= p(x)\left(\frac{n}{x} + \frac{s_1}{x^2} + \frac{s_2}{x^3} + \cdots\right)$$

Since the power-series converge for all sufficiently large values of $x$, the coefficients of $x$ must be the same in both sides of the equations. If we equate the coefficients of $x^{n-k-1}$ in both sides of this equation, we get the matrix equation in the statement.  □

Csanky's algorithm for the determinant is thus:

ALGORITHM 1.51. Given an $n \times n$ matrix $A$ we can compute the determinant by the following sequence of steps:

1. Compute $A^k$ in parallel for $k = 2, \ldots, n-1$. This can be done in $O(\lg^2 n)$ time using $O(nn^{2.376})$ processors;
2. Compute $s_k = \text{tr}(A^k)$ for $k = 1, \ldots, n-1$. This requires $O(\lg n)$ time using $O(n^2)$ processors;
3. Solve the matrix-equation in 1.50 for

$$\begin{pmatrix} c_1 \\ \vdots \\ c_n \end{pmatrix}$$

these are the coefficients of the characteristic polynomial of $A$. This equation can be solved in $O(\lg^2 n)$ time using $n^{2.376}$ processors. The only thing that has to be done is to invert the square matrix in the equation, and this can be done via the algorithm 1.36 on page 184.

Return $(-1)^n c_n$ as the determinant of $A$.

Note that we also get the values of $c_1$, $c_2$, etc., as an added bonus. There is no simple way to compute $c_n$ without also computing these other coefficients. The original paper of Csanky used these coefficients to compute $A^{-1}$ via the formula:

$$A^{-1} = -\frac{1}{c_n}(A^{n-1} + c_1 A^{n-2} + \cdots + c_{n-1}I)$$

Although this was the first published NC-algorithm for the inverse of an arbitrary invertible matrix, it is not currently used, since there are much better ones available[6].

---

[6]Better in the sense of using fewer processors, and being more numerically stable.

**1.6. Further reading.** Many matrix algorithms make use of the so-called $LU$ decomposition of a matrix (also called the Cholesky decomposition). Given a square matrix $M$, the Cholesky decomposition of $M$ is a formula

$$M = LU$$

where $L$ is a lower triangular matrix and $U$ is an upper triangular matrix. In [40], Datta gives an **NC** algorithm for finding the Cholesky decomposition of a matrix. It isn't entirely practical since it requires a parallel algorithm for the determinant (like that in § 1.5 above).

One topic we haven't touched upon here is that of *normal forms of matrices*. A normal form is a matrix-valued function of a matrix that determines (among other things) whether two matrices are *similar* (see the definition of similarity of matrices in 1.12 on page 162). If a given normal form of a matrix $M$ is denoted $F(M)$ (some other *matrix*), then two matrices $M_1$ and $M_2$ are similar if and only if $F(M_1) = F(M_2)$, exactly. There are a number of different normal forms of matrices including: Smith normal form and Jordan form.

Suppose $M$ is an $n \times n$ matrix with eigenvalues (in increasing order) $\lambda_1, \ldots, \lambda_k$, and suppose that $\lambda_i$ has multiplicity[7] $m_i$. Then the Jordan normal form of $M$ is the matrix

$$J(M) = \begin{pmatrix} Q_1 & & & & 0 \\ & Q_2 & & & \\ & & \cdot & & \\ & & & \cdot & \\ & & & & \cdot \\ 0 & & & & Q_k \end{pmatrix}$$

where $Q_i$ is an $m_i \times m_i$ matrix of the form

$$Q_i = \begin{pmatrix} \lambda_i & & & & 0 \\ 1 & \lambda_i & & & \\ & \cdot & \cdot & & \\ & & \cdot & \cdot & \\ & & & \cdot & \cdot \\ 0 & & & 1 & \lambda_i \end{pmatrix}$$

In [79], Kaltofen, Krishnamoorthy, and Saunders present parallel *randomized* algorithms for computing these normal forms. Note that, there algorithm must, among other things, compute the eigenvalues of a matrix.

---

[7]Recall that the multiplicity of the eigenvalue $\lambda_i$ is the highest power of $\lambda - \lambda_i$ that divides the characteristic polynomial of $M$. Equivalently, it is the dimension of the eigenspace of $\lambda_i$ — see 1.10 on page 160, and the exercises at the end of that section.

If we only want to know the *largest* eigenvalue of a matrix, we can use the *power method*, very briefly described in example 1.47 on page 193. If we only want the *eigenvalues* of a matrix, we can use the parallel algorithm developed by Kim and Chronopoulos in [86]. This algorithm is particularly adapted to finding the eigenvalues of *sparse* matrices. In [137], Sekiguchi, Sugihara, Hiraki, and Shimada give an implementation of an algorithm for eigenvalues of a matrix on a particular parallel computer (the Sigma-1).

## 2. The Discrete Fourier Transform

**2.1. Background.** Fourier Transforms are variations on the well-known Fourier Series. A Fourier Series was traditionally defined as an expansion of some function in a series of *sines* and *cosines* like:

$$f(x) = \sum_{k=0}^{\infty} a_k \sin(kx) + b_k \cos(kx)$$

Since sines and cosines are periodic functions[8] with period $2\pi$, the expansion will also have this property. Consequently, any expansion of this type will only be valid if $f(x)$ is a periodic function with the same period. It is easy to transform this series (by a simple scale-factor) to make the period equal to any desired value — we will stick to the basic case shown above. *If they exist,* such expansions have many applications

- If $f(x)$ is equal to the sum given above, it will periodic — a *wave* of some sort — and we can regard the terms $\{a_k \sin(kx), b_k \cos(kx)\}$ as the *components* of $f(x)$ of various *frequencies*[9]. We can regard the expansion of $f(x)$ into a Fourier series as a decomposition of it into its components of various frequencies. This has many applications to signal-processing, time-series analysis, etc.
- Fourier series are very useful in finding solutions of certain types of partial differential equations.

Suppose $f(x)$ is the function equal to $x$ when $-\pi < x \leq \pi$ and periodic with period $2\pi$ (these two statements define $f(x)$ completely). Then its Fourier series is:

$$(17) \qquad\qquad f(x) = 2\sum_{k=1}^{\infty}(-1)^{k+1}\frac{\sin(kx)}{k}$$

Figures V.5 through V.8 illustrate the convergence of this series when $-\pi/2 < x \leq \pi/2$. In each case a partial sum of the series is plotted alongside $f(x)$ to show how the partial sums get successively closer.

- Figure V.5 compares $f(x)$ to $2\sin(x)$,
- figure V.6 plots it against $2\sin(x) - \sin(2x)$,

---

[8]In other words $\sin(x + 2\pi) = \sin(x)$ and $\cos(x + 2\pi) = \cos(x)$ for all values of $x$
[9]The frequency of $a_k \sin(kx)$ is $k/2\pi$.

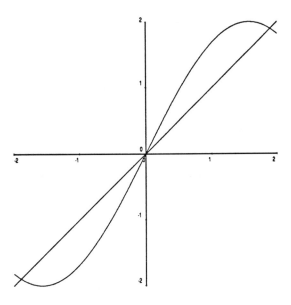

FIGURE V.5. First term

- figure V.7 plots it against $2\sin(x) - \sin(2x) + 2*\sin(3x)/3$, and
- figure V.8 plots it against $2\sin(x) - \sin(2x) + 2*\sin(3x)/3 - \sin(4x)/2$.

This series is only valid over a small range of values of $x$ — the interval $[-\pi, \pi]$. This series is often re-written in terms of exponentials, via the formula:

$$e^{ix} = \cos(x) + i\sin(x)$$

where $i = \sqrt{-1}$. We get a series like:

$$f(x) = \sum_{k > \infty}^{\infty} A_k e^{ikx}$$

We can compute the coefficients $A_k$ using the *orthogonality property* of the function $e^{ikx}$:

(18)

$$\int_{-\pi}^{\pi} e^{ikx} e^{i\ell x}\, dx = \int_{-\pi}^{\pi} e^{ix(k+\ell)}\, dx = \begin{cases} 2\pi & \text{if } k = -\ell \text{ (because } e^{ix(k+\ell)} = e^0 = 1) \\ 0 & \text{otherwise, because } e^{2\pi i(k+\ell)} = e^0 = 1 \end{cases}$$

so

$$A_k = \frac{1}{2\pi} \int_{-\pi}^{\pi} f(x) e^{-ikx}\, dx$$

We would like to *discretize* this construction — to define it in such a way that $x$ only takes on a finite number of values. The crucial observation is that if $x$ is of the form

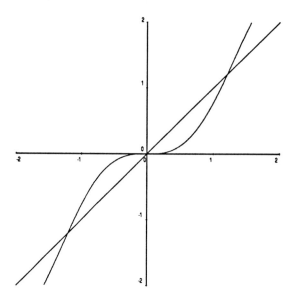

FIGURE V.6. First two terms

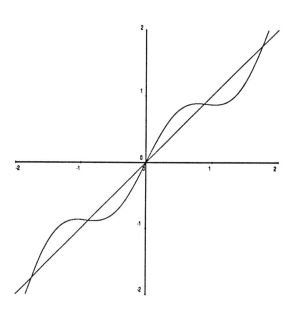

FIGURE V.7. First three terms

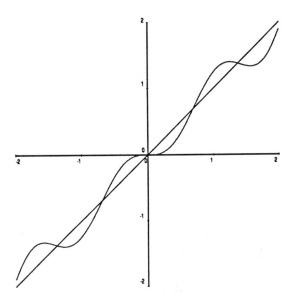

FIGURE V.8. First four terms

$2\pi j/n$ for some integers $j$ and $n$, then

$$e^{ikx} = e^{2\pi ijk/n} = e^{2\pi ikj/n+2\pi i} = e^{2\pi i(k+n)j/n}$$

(because $e^{it}$ is periodic with period $2\pi$. It follows that the exponential that occurs in the $k^{th}$ term is the same as the exponential that occurs in the $k + n^{th}$ term. Our series only really has $n$ terms in it:

(19)
$$f(2\pi j/n) = \sum_{k=0}^{n-1} S_k e^{2\pi ijk/n}$$

where

(20)
$$S_k = \sum_{m>-\infty}^{\infty} A_{k+mn}$$

where $k$ runs from 0 to $n - 1$.

This serves to motivate the material on the Discrete Fourier Transform in the rest of this section.

**2.2. Definition and basic properties.** There are many variations on the Discrete Fourier Transform, but the basic algorithm for handling the computations are the same in all cases.

Suppose we are given a sequence of numbers $\{a_0, \ldots, a_{n-1}\}$ — these represent the *values* of $f(x)$ at $n$ points — and a principal $n^{th}$ root of 1. This principal $n^{th}$ root of 1 plays the part of the functions $\sin(x)$ and $\cos(x)$ (both). The *powers* of $\omega$ will play the part of $\sin(kx)$ and $\cos(kx)$. This is a complex number $\omega$ with the property that $\omega^n = 1$ and $\omega^k \neq 1$ for all $1 \leq k < n$ or $\omega^k = 1$ only if $k$ is an exact multiple of $n$. For instance, $e^{2\pi i/n}$ is a principal $n^{\text{th}}$ root of 1.

Let $A = \{a_0, \ldots, a_{n-1}\}$ be a sequence of (real or complex) numbers. The Fourier Transform, $\mathcal{F}_\omega(A)$, of this sequence with respect to $\omega$ is defined to be the sequence $\{b_0, \ldots, b_{n-1}\}$ given by

$$(21) \qquad\qquad b_i = \sum_{j=0}^{n-1} a_j \omega^{ij}$$

Discrete Fourier Transforms have many applications:

1. Equations (19) and (20) relate the discrete Fourier Transform with Fourier series. It follows that all of the applications of Fourier series discussed on page 198 also apply here.
2. They give rise to many fast algorithms for taking *convolutions* of sequences. Given two sequences $\{a_0, \ldots, a_{n-1}\}$ and $\{b_0, \ldots, b_{n-1}\}$, their (length-$n$) *convolution* is a sequence $\{c_0, \ldots, c_{n-1}\}$ given by:

$$(22) \qquad\qquad c_k = \sum_{i=0}^{n-1} a_i b_{k-i \bmod n}$$

It turns out that the Fourier Transform of the convolution of two sequences is just the *termwise* product of the Fourier Transforms of the sequences. The way to see this is as follows:

a. Pretend that the sequences $\{a_0, \ldots, a_{n-1}\}$ and $\{b_0, \ldots, b_{n-1}\}$ are sequences of coefficients of two polynomials:

$$A(x) = \sum_{i=0}^{n-1} a_i x^i$$

$$B(x) = \sum_{j=0}^{n-1} b_j x^j$$

b. Note that the coefficients of the product of the polynomials are the convolution of the $a$- and the $b$-sequences:

$$(A \cdot B)(x) = \sum_{k=0}^{n-1} \left( \sum_{i+j=k} a_i b_j \right) x^k$$

and

$$(A \cdot B)(\omega) = \sum_{k=0}^{n-1} \left( \sum_{i+j=k \bmod n} a_i b_j \right) \omega^k$$

where $\omega$ is an $n^{\text{th}}$ root of 1. The $\mod n$ appears here because the product-polynomial above will be of degree $2n - 2$, but the powers of an $n^{\text{th}}$ root of 1 will "wrap around" back[10] to 0.

c. Also note that the formulas for the Fourier Transform simply correspond to plugging the powers of the principal root of 1 into the polynomials $A(x)$ and $B(x)$:

If $\{\mathcal{F}_\omega(a)(0), \ldots, \mathcal{F}_\omega(a)(n-1)\}$ and $\{\mathcal{F}_\omega(b)(0), \ldots, \mathcal{F}_\omega(b)(n-1)\}$ are the Fourier Transforms of the two sequences, then equation 21 shows that $\mathcal{F}_\omega(a)(i) = A(\omega^i)$ and $\mathcal{F}_\omega(b)(i) = B(\omega^i)$, so the Fourier Transformed sequences are just values taken on by the polynomials when evaluated at certain points. The conclusion follows from the fact that, when you multiply two polynomials, their values at corresponding points get multiplied. The alert reader might think that this implies that Discrete Fourier Transforms have applications to computer algebra. This is very much the case — see §5.1 of chapter VI for more details. Page 429 has a sample program that does symbolic computation by using a form of the Fast Fourier Transform.

It follows that, if we could compute *inverse* Fourier transforms, we could compute the convolution of the sequences by:

- taking the Fourier Transforms of the sequences;
- multiplying the Fourier Transforms together, term by term.
- Taking the Inverse Fourier Transform of this termwise product.

Convolutions have a number of important applications:

a. the applications to symbolic computation mentioned above (and discussed in more detail in a later chapter.

b. if $\{a_0, \ldots, a_{n-1}\}$ and $\{b_0, \ldots, b_{n-1}\}$ are *bits* of two $n$-bit binary numbers then the *product* of the two numbers is the sum

$$\sum_{i=0}^{n} c_i 2^i$$

where $\{c_0, \ldots, c_{n-1}\}$ is the convolution of the $a$, and $b$-sequences. It is, of course, very easy to multiply these terms by powers of 2. It follows that convolutions have applications to binary multiplication. These algorithms can be incorporated into VLSI designs.

3. A two-dimensional version of the Fourier Transform is used in *image-processing*. In the two-dimensional case the summation over $j$ in formula 21 is replaced by a double sum over two subscripts, and the sequences are all double-indexed.

---

[10]Notice that this "wrapping around" will *not* occur if the sum of the degrees of the two polynomials being multiplied is $< n$.

The Fourier Transform of a bitmap can be used to extract graphic *features* from the original image.

The idea here is that Fourier Transforms express image-functions[11] in terms of periodic functions. They, consequently, extract *periodic* information from the image or recognize repeating patterns.

We will, consequently be very interested in finding fast algorithms for computing Fourier Transforms and their *inverses*. It turns out that the *inverse* of a Fourier Transform is essentially nothing but another type of *Fourier Transform*. In order to *prove* this, we need

PROPOSITION 2.1. Suppose $\omega$ is a principal $n^{\text{th}}$ root of 1, and $j$ is an integer $0 \leq j \leq n - 1$. Then:

(23)
$$\sum_{i=0}^{n-1} \omega^{ij} = \begin{cases} n & \text{if } j = 0 \\ 0 & \text{if } 1 \leq j \leq n - 1 \end{cases}$$

PROOF. In order to see this suppose

$$S = \sum_{i=0}^{n-1} \omega^{ij}$$

Now, if we multiply $S$ by $\omega^j$, the sum *isn't changed* — in other words

(24)
$$\omega^j S = \sum_{i=0}^{n-1} \omega^{(i+1)j}$$
$$\text{because } \omega^n = 1$$
$$= \sum_{i=0}^{n-1} \omega^{ij}$$
$$= S$$

This implies that $(\omega^j - 1)S = 0$. Since $\omega$ is a *principal* $n^{\text{th}}$ root of 1, $\omega^j \neq 1$ which implies that $S = 0$, since the *only number* that isn't changed by being multiplied by a nonzero number is *zero*. □

The upshot of all of this is that the Inverse Fourier Transform is essentially the same as another Fourier Transform:

THEOREM 2.2. Suppose $A = \{a_0, \ldots, a_{n-1}\}$ is a sequence of $n$ numbers, $\omega$ is a principal $n^{\text{th}}$ root of 1 and the sequence $B = \{b_0, \ldots, b_{n-1}\}$ is the Fourier Transform of $A$ with respect to $\omega$. Let the sequence $C = \{c_0, \ldots, c_{n-1}\}$ be the Fourier Transform

---

[11]I. e., the functions whose value is the intensity of light at a point of the image

of $B$ with respect to $\omega^{-1}$ (which is also a principal $n^{\text{th}}$ root of 1. Then $c_i = na_i$ for all $0 \le i < n$.

It follows that we can invert the Fourier Transform with respect to $\omega$ by taking the Fourier Transform with respect to $\omega^{-1}$ and dividing by $n$.

We can prove this statement by straight computation — the Fourier transform of $B = \{b_0, \dots, b_{n-1}\}$ with respect to $\omega^{-1}$ is

$$\sum_{i=0}^{n-1} b_i \omega^{-1 ij} = \sum_{i=0}^{n-1} b_i \omega^{-ij}$$

$$= \sum_{i=0}^{n-1} \left( \sum_{k=0}^{n-1} a_k \omega^{ik} \right) \omega^{-ij}$$

$$= \sum_{i=0}^{n-1} \sum_{k=0}^{n-1} a_k \omega^{i(k-j)}$$

$$= \sum_{k=0}^{n-1} a_k \sum_{i=0}^{n-1} \omega^{i(k-j)}$$

and now we use formula 23 to conclude that

$$\sum_{i=0}^{n-1} \omega^{i(k-j)} = \begin{cases} n & \text{if } k = j \\ 0 & \text{if } k \ne j \end{cases}$$

and the last sum must be $na_j$.

We conclude this section by analyzing the time and space complexity of implementing equation (21).

1. On a *sequential computer* we clearly need $O(n^2)$-time and space.
2. On a PRAM we can compute the Fourier Transform in $O(\lg n)$ time using $O(n^2)$ processors. This algorithm stores $a_i \omega^{ij}$ in processor $P_{i,j}$, and uses the sum-algorithm described on page 69 to compute the summations.

**2.3. The Fast Fourier Transform Algorithm.** In 1942 Danielson and Lanczos developed an algorithm for computing the Fourier Transform that executed in $O(n \lg n)$ time — see [39]. The title of this paper was "Some improvements in practical Fourier analysis and their application to X-ray scattering from liquids" — it gives some idea of how one can apply Fourier Transforms. They attribute their method to ideas of König and Runge that were published in 1924 in [90].

In spite of these results, the Fast Fourier Transform is generally attributed to Cooley and Tuckey[12] who published [32] in 1965. It is an algorithm for computing Fourier

---

[12]This is a curious example of the "failure to communicate" that sometimes manifests itself in academia — publication of results doesn't necessary cause them to become "well-known."

Transforms that, in the sequential case, is considerably faster than the straightforward algorithm suggested in equation (21). In the sequential case, the Fast Fourier Transform algorithm executes in $O(n \lg n)$-time. On a PRAM computer it executes in $O(\lg n)$ time like the straightforward implementation of formula (21), but only requires $O(n)$ processors. This algorithm has a simple and ingenious idea behind it.

Suppose $n = 2^k$ and we have a sequence of $n$ numbers $\{a_0, \ldots, a_{n-1}\}$, and we want to take its Fourier Transform. As mentioned above, this is exactly the same as evaluating the polynomial $a_0 + a_1 x + \cdots + a_n x^{n-1}$ at the points $\{\omega, \omega^2, \ldots, \omega^{n-1}\}$, where $\omega$ is a principal $n^{\text{th}}$ root of 1. The secret behind the Fast Fourier Transform is to notice a few facts:

- $\omega^2$ is a principal $n/2^{\text{th}}$ root of 1.
- We can write the polynomial $p(x) = a_0 + a_1 x + \cdots + a_n x^{n-1}$ as

(25) $$p(x) = r(x) + x s(x)$$

where

$$r(x) = a_0 + a_2 x^2 + a_4 x^4 + \cdots + a_{n-2} x^{n-2}$$
$$s(x) = a_1 + a_3 x^2 + a_5 x^4 + \cdots + a_{n-1} x^{n-2}$$

This means that we can evaluate $p(x)$ at the powers of $\omega$ by

1. Splitting the sequence of coefficients into the even and the odd subsequences — forming the polynomials $r(x)$ and $s(x)$.
2. Evaluating $r(x)$ and $s(x)$ at the powers of $\omega^2$.
3. Plugging the results into equation (25).

This observations gives rise to the following algorithm for the Discrete Fourier Transform:

ALGORITHM 2.3. Let $n$ a positive even integer. We can compute the Fourier Transform of the sequence $A_0, \ldots, A_{n-1}$, with respect to a primitive $n^{\text{th}}$ root of 1, $\omega$, by the following sequence of operations:

1. "unshuffle" the sequence into odd and even subsequences: $A_{\text{odd}}$ and $A_{\text{even}}$.
2. Compute $\mathcal{F}_{\omega^2}(A_{\text{odd}})$ and $\mathcal{F}_{\omega^2}(A_{\text{even}})$.
3. Combine the results together using (25). We get:

(26) $$\mathcal{F}_\omega(A)(i) = \mathcal{F}_{\omega^2}(A_{\text{even}})(\lfloor i/2 \rfloor \bmod n/2) + \omega^i \mathcal{F}_{\omega^2}(A_{\text{odd}})(\lfloor i/2 \rfloor \bmod n/2)$$

Now we analyze the time-complexity of this algorithm. Suppose that $T(n)$ is the time required to compute the Fourier Transform for a sequence of size $n$ (using this algorithm). The algorithm above shows that:

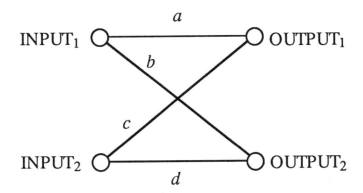

FIGURE V.9

1. In parallel: $T(n) = T(n/2) + 1$, whenever $n$ is an even number. It is not hard to see that this implies that $T(2^k) = k$, so the parallel execution-time of the algorithm is $O(\lg n)$.
2. Sequentially $T(n) = 2T(n/2) + n/2$, whenever $n$ is an even number. Here, the additional term of $n/2$ represents the step in which the Fourier Transforms of the odd and even sub-sequences are combined together. If we write $T(2^k) = a_k 2^k$ we get the following formula for the $a_k$:

$$a_k 2^k = 2 \cdot a_{k-1} 2^{k-1} + 2^{k-1}$$

and, when we divide by $2^k$, we get:

$$a_k = a_{k-1} + 1/2$$

which means that $a_k$ is proportional to $k$, and the sequential execution-time of this algorithm is $O(n \lg n)$. This is still faster than the original version of the Fourier Transform algorithm in equation (21) on page 202.

Although it is possible to program this algorithm directly, most practical programs for the FFT use an *iterative* version of this algorithm. It turns out that the simplest way to describe the iterative form of the FFT algorithm involves representing it *graphically*. Suppose the circuit depicted in figure V.9 represents the process of forming linear combinations:

$$\text{OUTPUT}_1 \leftarrow a \cdot \text{INPUT}_1 + c \cdot \text{INPUT}_2$$
$$\text{OUTPUT}_2 \leftarrow b \cdot \text{INPUT}_1 + d \cdot \text{INPUT}_2$$

We will refer to figure V.9 as the *butterfly diagram* of the equations above.

Then we can represent equation (26) by the diagram in figure V.10. In this figure, the shaded patches of the graphs represent "black boxes" that compute the Fourier transforms of the odd and even subsequences, with respect to $\omega^2$. All lines without

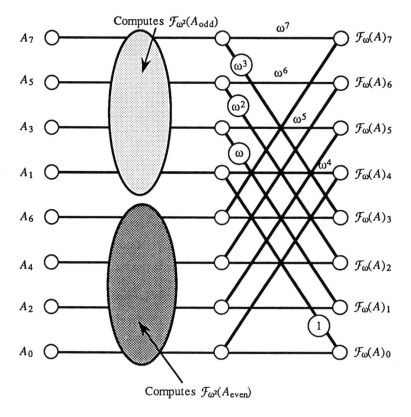

FIGURE V.10. Graphic representation of the FFT algorithm

any values indicated for them are assumed to have values of 1. Diagonal lines with values not equal to 1 have their values enclosed in a bubble.

In order to understand the recursive behavior of this algorithm, we will need the following definition:

DEFINITION 2.4. Define the *unshuffle* operation of size $2m$ to be the permutation:

$$U_m = \begin{pmatrix} 0 & 2 & 4 & 6 & \ldots & 2m-2 & 1 & \ldots & 2m-1 \\ 0 & 1 & 2 & 3 & \ldots & m-1 & m & \ldots & 2m-1 \end{pmatrix}$$

Here, we use the usual notation for a permutation — we map elements of the upper row into corresponding elements of the lower row.

Let $n = 2^k$ and define the *complete unshuffle* operation on a sequence, $V$ of $n$ items to consist of the result of:

- Performing $U_n$ on $V$;
- Subdividing $V$ into its first and last $n/2$ elements, and performing $U_{n/2}$ independently on these subsequences.
- Subdividing $V$ into 4 disjoint intervals of size $n/4$ and performing $U_{n/4}$ on each of these.
- This procedure is continues until the intervals are of size 2.

We use the notation $\mathcal{C}_n$ to denote the complete unshuffle.

For instance, a complete unshuffle of size 8 is:

$$\mathcal{C}_8 = \begin{pmatrix} 0 & 1 & 2 & 3 & 4 & 5 & 6 & 7 \\ 0 & 4 & 2 & 6 & 1 & 5 & 3 & 7 \end{pmatrix}$$

PROPOSITION 2.5. Let $n = 2^k$, where $k$ is a positive integer. Then the complete unshuffle, $\mathcal{C}_n$, is given by the permutation:

$$\mathcal{C}_n = \begin{pmatrix} 0 & 1 & 2 & \ldots & n-1 \\ e(0,k) & e(1,k) & e(2,k) & \ldots & e(n-1,k) \end{pmatrix}$$

where the function $e(*, k)$ is defined by:

If the $k$-bit binary representation of $j$ is $b_{k-1} \ldots b_0$, define $e(k, j)$ to be the binary number given by $b_0 \ldots b_{k-1}$.

In other words, we just *reverse* the $k$-bit binary representation of a number in order to calculate the e-function. Note that this permutation is *idempotent*, i.e., $\mathcal{C}_n^2 = 1$, since the operation of reversing the order of the bits in a binary representation of a number is also idempotent. This means that $\mathcal{C}_n^{-1} = \mathcal{C}_n$.

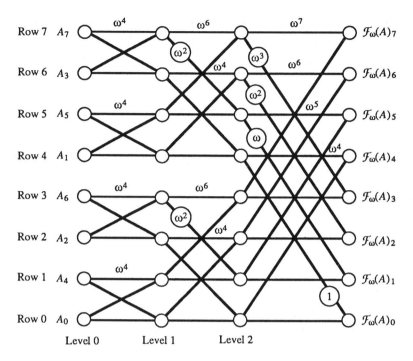

FIGURE V.11. Complete butterfly diagram for the FFT algorithm on 8 inputs

PROOF. We prove this by induction on $k$ (where $n = 2^k$). If $k = 1$ then the result is trivial. We assume the result is true for some value of $k$, and we prove this for $k + 1$. When $n = 2^{k+1}$, the first unshuffle operation moves all even-indexed elements to the lower half of the interval and all odd-indexed elements to the upper half, and then performs the unshuffle-operations for $n = 2^k$ on each of the sub-intervals. Suppose we have a number, $m$ with the binary representation $b_{k+1} b_k \ldots b_1 b_0$.

Claim: The process of unshuffling the numbers at the top level has the effect of performing a *cyclic right shift* of the binary-representations.

This is due to the fact that *even* numbers get moved to a position equal to half their original value — this is a right-shift of 1 bit. Odd numbers, $s$, get sent to $2^k + \lfloor s \rfloor$ — so the 1-bit in the $0^{\text{th}}$ position gets shifted out of the number, but is added in to the left end of the number.

It follows that our number $m$ gets shifted into position $b_0 b_{k+1} b_k \ldots b_1$. Now we perform the complete shuffle permutations on the upper and lower halves of the whole interval. This reverses the remaining bits of the binary-representation of numbers, by the inductive hypothesis. Our number $m$ gets shifted to position $b_0 b_1 \ldots b_k b_{k+1}$. $\square$

We can plug diagrams for the Fourier transforms of the odd and even subsequences into this diagram, and for the odd and even subsequences of *these* sequences. We ultimately get the diagram in figure V.11.

2.6. Close examination of figure V.11 reveals several features:

1. If a diagonal line from level $i$ connects rows $j$ and $k'$, then the binary representations of $j$ and $j'$ are identical, except for the $i^{\text{th}}$ bit. (Here we are counting the bits from right to left — the number of a bit corresponds to the power of 2 that the bit represents.)

2. The Fourier Transforms from level $i$ to level $i+1$ are performed with respect to a primitive $2^{i\text{th}}$ root of 1, $\eta_i = \omega^{2^{k-i-1}}$. This is due to the fact that, as we move to the left in the diagram, the root of 1 being used is squared.

3. A line from level $i$ to level $i+1$ has a coefficient attached to it that is $\neq 1$ if and only if the left end of this line is on a row $r$ whose binary-representation has a 1 in position $i$. This is because in level $i$, the whole set of rows is subdivided into subranges of size $2^i$. Rows whose number has a 1 in bit-position $i$ represent the top-halves of these subranges. The top half of each subrange gets the Fourier Transform of the odd-subsequence, and this is multiplied by a suitable power of $\eta_i$, defined above.

4. A line from level $i$ to level $i+1$ whose left end is in row $r$, and with the property that the binary representation of $r$ has a 1 in bit-position $i$ (so it has a nonzero power of $\eta_i$) has a coefficient of $\eta_i^{r'}$, where $r'$ is the row number of the right end of the line. This is a direct consequence of equation (26) on page 206, where the power of $\omega$ was equal to the number of the subscript on the output. We also get:

   • $r'$ may be taken modulo $2^i$, because we use it as the exponent of a $2^{i\text{th}}$ root of 1.
   • If the line in question is horizontal, $r' \equiv r \mod 2^i$.
   • If the line was diagonal, $r' \equiv \hat{r} \mod 2^i$, where $\hat{r}$ has a binary-representation that is the same as that of $r$, except that the $i^{\text{th}}$bit position has a 0 — see line 1, above.
   • This power of $\eta_i$ is equal to $\omega^{2^{k-i-1}r'}$, by line 2.

The remarks above imply that none of the observations are *accidents* of the fact that we chose a Fourier Transform of 8 data-items. It is normal practice to "unscramble" this diagram so that the input-data is in ascending order and the output is permuted. We get the diagram in figure V.12.

In order to analyze the effect of this "unscrambling" operation we use the description of the complete unshuffle operation in 2.5 on page 209.

Now we can describe the Fast Fourier Transform algorithm. We simply modify the general rule computing the power of $\omega$ in 2.6 on page 210. Suppose $n = 2^k$ is the size of the sequences in question. We define functions $e(r,j)$, $c_0(r,j)$ and $c_1(r,j)$ for all integers $0 \leq r \leq k-1$ and $0 \leq j < n$,

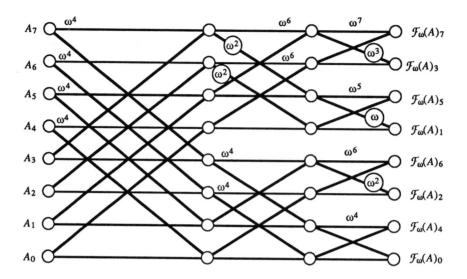

FIGURE V.12. "Unscrambled" FFT circuit.

1. If the $k$-bit binary representation of $j$ is $b_{k-1} \cdots b_0$, then $e(r,j)$ is the number whose binary representation is $b_{k-r-1} b_{k-r} \cdots b_{k-1} 0 \cdots 0$;

2. $c_0(r,j)$ is a number whose $k$-bit binary representation is the same as that of $j$, except that the $k-r+1^{\text{th}}$ bit is 0. In the scrambled diagram, we used the $r^{\text{th}}$ bit, but now the bits in the binary representation have been reversed.

3. $c_1(r,j)$ is a number whose binary representation is the same as that of $j$, except that the $k-r+1^{\text{th}}$ bit is 1. See the remark in the previous line.

Note that, for every value of $r$, every number, $i$, between 0 and $n-1$ is equal to either $c_0(r,i)$ or $c_1(r,i)$.

Given this definition, the Fast Fourier Transform Algorithm is:

ALGORITHM 2.7. Under the assumptions above, let $A = \{a_0, \ldots, a_{n-1}\}$ be a sequence of numbers. Define sequences $\{F_{i,j}\}$, $0 \le r \le k-1$, $0 \le j \le n-1$ via:

1. $F_{0,*} = A$;
2. For all $0 \le j \le n-1$,

$$F_{t+1,c_0(t,j)} = F_{t,c_0(t,j)} + \omega^{e(t,c_0(t,j))} F_{t,c_1(t,j)}$$

$$F_{t+1,c_1(t,j)} = F_{t,c_0(t,j)} + \omega^{e(t,c_1(t,j))} F_{t,c_1(t,j)}$$

Then the sequence $F_{k,*}$ is equal to the shuffled Fourier Transform of $A$:

(27)                          $\mathcal{F}_\omega(A)(e(k,i)) = F_{k,i}$

| $j$ | 0 | 1 | 2 | 3 | 4 | 5 | 6 | 7 |
|---|---|---|---|---|---|---|---|---|
| $r = 3$ | 0 | 4 | 2 | 6 | 1 | 5 | 3 | 7 |
| $r = 2$ | 0 | 0 | 4 | 4 | 2 | 2 | 6 | 6 |
| $r = 1$ | 0 | 0 | 0 | 0 | 4 | 4 | 4 | 4 |

TABLE V.1. Sample computations of $e(r, j)$

If we unshuffle this sequence, we get the Fourier Transform, $\mathcal{F}_\omega(A)$ of $A$.

We need to find an efficient way to compute the $e(r, j)$-functions. First, note that $e(k, j)$ is just the result of *reversing* the binary representation of $j$. Now consider $e(k-1, j)$: this is the result of taking the binary representation of $j$, reversing it, deleting the first bit (the *leftmost* bit, in our numbering scheme) and adding a 0 on the right. But this is just the result of doing a left-shift of $e(k, j)$ and truncating the result to $k$ bits by deleting the high-order bit. It is not hard to see that for all $i$, $e(i, j)$ is the result of a similar operation on $e(i + 1, j)$. In the C language, we could write this as (e(i+1,j)<<1)%n, where % is the **mod**-operation.

Since we actually only need $\omega^{e(r,j)}$, for various values of $r$ and $j$, we usually can avoid calculating the $e(r, j)$ entirely. We will only calculate the $e(k, *)$ and $\omega^{e(k,*)}$. We will then calculate the remaining $\omega^{e(r,j)}$ by setting $\omega^{e(r,j)} = \left(\omega^{e(r-1,j)}\right)^2$. Taking these squares is the same as multiplying the exponents by 2 and reducing the result modulo $n$.

The remaining functions that appear in this algorithm (namely $c_0(*, *)$ and $c_1(*, *)$) are trivial to compute.

Table V.1 gives some sample computations in the case where $k = 3$ and $n = 8$.

Here is a sample program for implementing the Fast Fourier Transform. In spite of the comment above about not needing to calculate the $e(r, j)$ for all values of $r$, we do so in this program. The reason is that it is easier to compute the powers of $\omega$ directly (using sines and cosines) than to carry out the squaring-operation described above. See page 429 for an example of an FFT program that doesn't compute all of the $e(r, j)$.

```c
#include <stdio.h>
#include <math.h>
shape [8192]linear;
/* Basic structure to hold the data—items. */
struct compl
{
    double re;
    double im;
};
typedef struct compl complex;
```

```
/* Basic structure to hold the data—items. */

complex:linear in_seq;      /* Input data. */
complex:linear work_seq;    /* Temporary variables, and
                             * output data */

int:linear e_vals[13];      /* Parallel array to hold
                             * the values of the e(k,j) */

complex:linear omega_powers[13];   /* Parallel array to
                                    * hold the values of
                                    * omega^e(r,j). */

void main()
{
    int i, j;
    int k = 13;        /* log of number of
                        * data—points. */
    int n = 8192;      /* Number of data—points. */

    /*
     * This block of code sets up the e_vals and the
     * omega_powers arrays.
     */

    with (linear)
    {
        int i;
        int:linear p = pcoord(0);
        int:linear temp;
            e_vals[k−1]= 0;
        for (i = 0; i < n; i++)
        {
            [i]in_seq.re = (double) i) / ((double) n;
            [i]in_seq.im = 0.;
        }
        for  (i = 0; i < k; i++)
        {
            e_vals[k−1]<<= 1;
```

```
   e_vals[k−1]+= p % 2;
   p >>= 1;
}
omega_powers[k − 1].re
 = cos(2.0 * M_PI * (double:linear) e_vals[k−1]/ (double) n);

omega_powers[k − 1].im
 = sin(2.0 * M_PI * (double:linear) e_vals[k−1]/ (double) n);
for (i = 1; i < k; i++)
{
 e_vals[k − 1 − i]= (e_vals[k − i]<< 1) % n;
 omega_powers[k − 1 − i].re
   =
   cos(2.0 * M_PI * ((double:linear) e_vals[k − 1 − i]/ (double) n));
 omega_powers[k − 1 − i].im
   =
   sin(2.0 * M_PI * ((double:linear) e_vals[k − 1 − i]/ (double) n));
}
work_seq.re = in_seq.re;
work_seq.im = in_seq.im;
p = pcoord(0);
for (i = 0; i < k; i++)
{
   complex:linear save;

   save.re = work_seq.re;
   save.im = work_seq.im;
   temp = p & (~(1 << (k−i−1)));  /* Compute c0(r,i).  The
                     '(k−i−1' is due to the fact
                     that the number of the bits
                     in the definition of c0(r,i)
                     is from 1 to k, rather than
                     k−1 to 0. */
   where (p == temp)
   {
      int:linear t = temp | (1 << (k−i−1));
                     /* Compute c1(r,i).  The
                      '(k−i−1' is due to the fact
                      that the number of the bits
                      in the definition of c0(r,i)
```

```
                        is from 1 to k, rather than
                        k−1 to 0. */
            [temp]work_seq.re = [temp]save.re
                + [temp]omega_powers[i].re *[t]save.re
                − [temp]omega_powers[i].im *[t]save.im;
            [temp]work_seq.im = [temp]work_seq.im
                + [temp]omega_powers[i].re *[t]save.im
                + [temp]omega_powers[i].im *[t]save.re;
            [t]work_seq.re = [temp]save.re +
                [t]omega_powers[i].re *[t]save.re
                − [t]omega_powers[i].im *[t]save.im;
            [t]work_seq.im = [temp]save.im +
                [t]omega_powers[i].re *[t]save.im
                + [t]omega_powers[i].im *[t]save.re;
        }
    }
    with(linear)
     where(pcoord(0)<n)
        {
          save_seq.re=work_seq.re;
          save_seq.im=work_seq.im;
          [e_vals[k−1]]work_seq.re=save_seq.re;
          [e_vals[k−1]]work_seq.im=save_seq.im;
        }
 }
 for  (i = 0; i < n; i++)
 {
    printf("Value %d, real part=%g\n",i,[i]work_seq.re);
    printf("Value %d, imaginary part=%g\n",i,[i]work_seq.im);
 }
}
```

Now we will analyze the *cost* of doing these computations. The original Discrete Fourier Transform executes sequentially in $O(n^2)$ time, and on a SIMD, PRAM it clearly executes in $O(\lg n)$-time, using $O(n^2)$ processors. The main advantage in using the Fast Fourier Transform algorithm is that it reduces the number of processors required. On a SIMD, PRAM machine it clearly executes in $O(\lg n)$-time using $n$ processors.

**Exercises:**

1. Express the $\mathcal{C}_n$ permutation defined in 2.5 on page 209 in terms of the generic ASCEND or DESCEND algorithms on page 68.

2. The alert reader may have noticed that the FFT algorithm naturally fits into the framework of the generic DESCEND algorithm on page 68. This implies that we can find very efficient implementations of the FFT algorithm on the network architectures of chapter III on page 67. Find efficient implementations of the FFT on the hypercube and the cube-connected cycles architectures.

3. Equation (20) on page 201 implies a relationship between the Discrete Fourier Transform and the coefficients in a Fourier Series. Suppose that it is known (for some reason) that $A_k = 0$ for $|k| > n$. Also assume that $f(x)$ is an *even function*: $f(x) = f(-x)$[13]. Show how to compute the nonzero coefficients of the Fourier series for $f(x)$ from the Discrete Fourier Transform, performed upon some finite set of values of $f(x)$.

4. Consider the number $n$ that has been used throughout this section as the size of the sequence being transformed (or the order of the polynomial being evaluated at principal roots of unity, or the order of the principal root of unity being used). The Fast Fourier Transform algorithm gives a fast procedure for computing Fourier Transforms when $n = 2^k$. Suppose, instead, that $n = 3^k$. Is there a fast algorithm for computing the discrete Fourier Transform in this case? Hint: Given a polynomial

$$p(x) = \sum_{i=0}^{3^k-1} a_i x^i$$

we can re-write it in the form

$$p(x) = u(x) + xv(x) + x^2 w(x)$$

where

$$u(x) = \sum_{i=0}^{3^{k-1}-1} a_{3i} x^{3i}$$

$$u(x) = \sum_{i=0}^{3^{k-1}-1} a_{3i+1} x^{3i}$$

$$u(x) = \sum_{i=0}^{3^{k-1}-1} a_{3i+2} x^{3i}$$

If such a modified Fast Fourier Transform algorithm exists, how does its execution time compare with that of the standard Fast Fourier Transform.

---

[13]This implies that $A_k = A_{-k}$.

**2.4. Eigenvalues of cyclic matrices.** In this section we will give a simple application of Discrete Fourier Transforms that will be used in later sections of the book. In certain cases we can use Discrete Fourier Transforms to easily compute the eigenvalues of matrices.

DEFINITION 2.8. An $n \times n$ matrix $A$ is called a *cyclic matrix* or *circulant* if there exists a function $f$ such that
$$A_{i,j} = f(i - j \bmod n)$$
Here, we assume the indices of $A$ run from 0 to $n - 1$[14].

Note that cyclic matrices have a great deal of symmetry. Here is an example:
$$\begin{pmatrix} 1 & 2 & 3 \\ 3 & 1 & 2 \\ 2 & 3 & 1 \end{pmatrix}$$

It is not hard to see that the function $f$ has the values: $f(0) = 1$, $f(1) = 3$, $f(2) = 2$. In order to see the relationship between cyclic matrices and the Fourier Transform, we multiply a vector, $v$ by the matrix $A$ in the definition above:
$$(Av)_i = \sum_{j=0}^{n-1} A_{i,j} v_j = \sum_{j=0}^{n-1} f(i - j \bmod n) v_j$$

so the act of taking the product of the vector by the matrix is the same a taking the *convolution* of the vector by the function $f$ (see the definition of convolution on page 202).

It follows that we can compute the product of the vector by $A$ via the following sequence of operations:

1. Select some primitive $n^{\text{th}}$ root of 1, $\omega$
2. Form the FFT of $f$ with respect to $\omega$:
$$\{\mathcal{F}_\omega(f)(0), \ldots, \mathcal{F}_\omega(f)(n-1)\}$$
3. Form the FFT of $v$ with respect to $\omega$:
$$\{\mathcal{F}_\omega(v)(0), \ldots, \mathcal{F}_\omega(v)(n-1)\}$$
4. Form the elementwise product of these two sequences:
$$\{\mathcal{F}_\omega(f)(0) \cdot \mathcal{F}_\omega(v)(0), \ldots, \mathcal{F}_\omega(f)(n-1) \cdot \mathcal{F}_\omega(v)(n-1)\}$$
5. This resulting sequence is the Fourier Transform of $Av$:
$$\{\mathcal{F}_\omega(Av)(0), \ldots, \mathcal{F}_\omega(Av)(n-1)\}$$

---

[14]It is not hard to modify this definition to accommodate the case where they run from 1 to $n$.

While this may seem to be a convoluted way to multiply a vector by a matrix, it is interesting to see what effect this has on the basic equation for the eigenvalues and eigenvectors of $A$ —

$$Av = \lambda v$$

becomes

$$\mathcal{F}_\omega(f)(i) \cdot \mathcal{F}_\omega(v)(i) = \lambda \mathcal{F}_\omega(v)(i)$$

for all $i = 0, \ldots, n-1$ (since $\lambda$ is a *scalar*). Now we will try to solve these equations for values of $\lambda$ and nonzero vectors $\mathcal{F}_\omega(v)(*)$. It is easy to see what the solution is if we re-write the equation in the form:

$$(\mathcal{F}_\omega(f)(i) - \lambda) \cdot \mathcal{F}_\omega(v)(i) = 0$$

for all $i$. Since there is no summation here, there can be no cancellation of terms, and the only way these equations can be satisfied is for

1. $\lambda = \mathcal{F}_\omega(f)(i)$ for some value of $i$;
2. $\mathcal{F}_\omega(v)(i) = 1$ and $\mathcal{F}_\omega(v)(j) = 0$ for $i \neq j$.

This *determines* the possible values of $\lambda$, and we can also solve for the eigenvectors associated with these values of $\lambda$ by taking an inverse Fourier Transform of the $\mathcal{F}_\omega(v)(*)$ computed above.

THEOREM 2.9. Let $A$ be an $n \times n$ cyclic matrix with $A_{i,j} = f(i - j \bmod n)$. Then the eigenvalues of $A$ are given by

$$\lambda_i = \mathcal{F}_\omega(f)(i)$$

and the eigenvector corresponding to the eigenvalue $\lambda_i$ is

$$\mathcal{F}_{\omega^{-1}}(\delta_{i,*})/n$$

where $\delta_{i,*}$ is the sequence $\{\delta_{i,0}, \ldots, \delta_{i,n-1}\}$ and

$$\delta_{i,j} = \begin{cases} 1 & \text{if } i = j \\ 0 & \text{otherwise} \end{cases}$$

Recall that $\mathcal{F}_{\omega^{-1}}(*)/n$ is just the inverse Fourier Transform — see 2.2 on page 204.

We will conclude this section with an example that will be used in succeeding material.

DEFINITION 2.10. For all $n > 1$ define the $n \times n$ matrix $\mathcal{Z}(n)$ via:

$$\mathcal{Z}(n)_{i,j} = \begin{cases} 1 & \text{if } i - j = \pm 1 \bmod n \\ 0 & \text{otherwise} \end{cases}$$

This is clearly a cyclic matrix with $f$ given by

$$f(i) = \begin{cases} 1 & \text{if } i = 1 \text{ or } i = n - 1 \\ 0 & \text{otherwise} \end{cases}$$

We will compute the eigenvalues and eigenvectors of this $\mathcal{Z}(n)$. Let $\omega = e^{2\pi i/n}$ and compute the Fourier Transform of $f$ with respect to $\omega$:

$$\lambda_i = \mathcal{F}_\omega(f)(j) = e^{2\pi i j/n} + e^{2\pi i j (n-1)/n}$$
$$= e^{2\pi i j/n} + e^{-2\pi i j/n}$$

Since $n - 1 \equiv -1 \mod n$

$$= 2\cos(2\pi j/n)$$

since $\cos(x) = e^{ix} + e^{-ix}/2$. Note that these eigenvalues are not all distinct — the symmetry of the cosine function implies that

(28) $$\lambda_i = \lambda_{n-i} = 2\cos(2\pi i/n)$$

so there are really only $\left\lceil \frac{n}{2} \right\rceil$ *distinct* eigenvalues. Now we compute the eigenvectors associated with these eigenvalues. The eigenvector associated with the eigenvalue $\lambda_j$ is the inverse Fourier Transform of $\delta_{j,*}$. This is

$$v(j) = \{e^{-2\pi i j \cdot 0/n}/n, \ldots, e^{-2\pi i j \cdot (n-1)/n}/n\}$$
$$= \{1, \ldots, e^{-2\pi i j \cdot (n-1)/n}/n\}$$

Since $\lambda_j = \lambda_{n-j}$, so these are not really different eigenvalues, any *linear combination* of $v(j)$ and $v(n-j)$ will *also* be a valid eigenvector[15] associated with $\lambda_j$ — the resulting eigenspace is 2-dimensional. If we don't like to deal with complex-valued eigenvectors we can form linear combinations that cancel out the imaginary parts:

$$n(v(j)/2 + v(n-j)/2)_k = \left(e^{-2\pi i j \cdot k/n} + e^{-2\pi i j \cdot (n-k)/n}\right)/2$$
$$= \cos(2\pi j k/n)$$

and

$$n(-v(j)/2 + v(n-j)/2i)_k = \left(-e^{-2\pi i j \cdot k/n} + e^{-2\pi i j \cdot (n-k)/n}\right)/2$$
$$= \sin(2\pi j k/n)$$

---

[15]See exercise 1 on page 166 and its solution in the back of the book.

so we may use the two vectors

(29)        $w(j) = \{1, \cos(2\pi j/n), \cos(4\pi j/n), \ldots, \cos(2\pi j(n-1)/n)\}$

(30)        $w'(j) = \{0, \sin(2\pi j/n), \sin(4\pi j/n), \ldots, \sin(2\pi j(n-1)/n)\}$

as the basic eigenvectors associated with the eigenvalue $\lambda_j$.

Notice that the formula for the eigenvectors of $A$ in 2.9 contains no explicit references to the function $f$. This means that all cyclic matrices have the *same eigenvectors*[16], namely

$$v(j) = \{1, \ldots, e^{-2\pi i j \cdot (n-1)/n}/n\}$$

(our conversion of this expression into that in (30) made explicit use of the fact that many of the eigenvalues of $\mathcal{Z}(n)$ were equal to each other).

**Exercises:**

1. Compute the eigenvalues and eigenvectors of the matrix

$$\begin{pmatrix} 1 & 2 & 3 \\ 3 & 1 & 2 \\ 2 & 3 & 1 \end{pmatrix}$$

2. Compute the eigenvalues and eigenvectors of the matrix

$$\begin{pmatrix} 0 & 1 & 1 & 1 \\ 1 & 0 & 1 & 1 \\ 1 & 1 & 0 & 1 \\ 1 & 1 & 1 & 0 \end{pmatrix}$$

3. Recall that the Discrete Fourier Transform can be computed with respect to an arbitrary primitive $n^{\text{th}}$ root of 1. What effect does *varying* this primitive $n^{\text{th}}$ root of 1 have on the computation of the eigenvalues and eigenvectors of an $n \times n$ cyclic matrix?

4. Give a formula for the *determinant* of a cyclic matrix.

5. Give a formula for the spectral radius of the matrices $\mathcal{Z}(n)$.

---

[16]They don't, however, have the same eigenvalues.

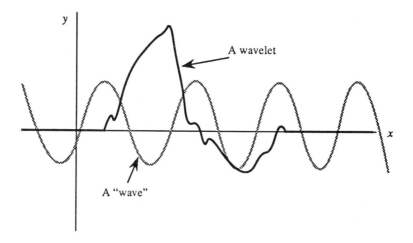

FIGURE V.13. An example of a wavelet

## 3. Wavelets

**3.1. Background.** In this section, we will discuss a variation on Fourier Expansions that has gained a great deal of attention in recent years. It has many applications to image analysis and data-compression. We will only give a very abbreviated overview of this subject — see [107] and [145] for a more complete description.

Recall the Fourier Expansions

$$f(x) = \sum_{k=0}^{\infty} a_k \sin(kx) + b_k \cos(kx)$$

The development of wavelets was originally motivated by efforts to find a kind of Fourier Series expansion of *transient* phenomena[17]. If the function $f(x)$ has large *spikes* in its graph, the Fourier series expansion converges very slowly. This makes some intuitive sense — it is not surprising that it is difficult to express a function with sharp transitions or discontinuities in terms of smooth functions like sines and cosines. Furthermore, if $f(x)$ is localized in space (i.e., vanishes outside an interval) it may be hard to expand $f(x)$ in terms of sines and cosines, since these functions take on nonzero values over the entire $x$-axis.

We solve this problem by trying to expand $f(x)$ in series involving functions that *themselves* may have such spikes and vanish outside of a small interval. These functions are called *wavelets*. The term "wavelet" comes from the fact that these functions are vanish outside an interval. If a periodic function like $\sin(x)$ has a sequence of peaks and valleys over the entire $x$-axis, we think of this as a "wave", we think of a function with, say only small number of peaks or valleys, as a wavelet — see figure V.13. Incidentally,

---

[17]Wavelet expansions grew out of problems related to *seismic analysis* — see [60].

the depiction of a wavelet in figure V.13 is accurate in that the wavelet is "rough" — in many cases, wavelets are *fractal functions*, for reasons we will discuss a little later.

If we want to expand arbitrary functions like $f(x)$ in terms of wavelets, $w(x)$, like the one in figure V.13, several problems are immediately apparent:

1. How do we handle functions with spikes "sharper" than that of the main wavelet? This problem is solved in conventional Fourier series by *multiplying* the variable $x$ by integers in the expansion. For instance, $\sin(nx)$ has peaks that are $1/n^{\text{th}}$ as wide as the peaks of $\sin(x)$. For various reasons, the solution that is used in wavelet-expansions, is to multiply $x$ by a *power of 2* — i.e., we expand in terms of $w(2^j x)$, for different values of $j$. This procedure of changing the scale of the $x$-axis is called *dilation*.

2. Since $w(x)$ is only nonzero over a small finite interval, how do we handle functions that are nonzero over a much larger range of $x$-values? This is a problem that doesn't arise in conventional Fourier series because they involve expansions in functions that are nonzero over the whole $x$-axis. The solution use in wavelet-expansions is to *shift* the function $w(2^j x)$ by an integral distance, and to form linear combinations of these functions: $\sum_k w(2^j x - k)$. This is somewhat akin to taking the individual wavelets $w(2^j x - k)$ and assemble them together to form a *wave*. The reader may wonder what has been gained by all of this — we "chopped up" a wave to form a wavelet, and we are re-assembling these wavelets back into a wave. The difference is that we have direct control over how far this wave extends — we may, for instance, only use a *finite number* of displaced wavelets like $w(2^j x - k)$.

The upshot of this discussion is that a general wavelet expansion of of a function is *doubly indexed* series like:

$$(31) \qquad f(x) = \sum_{\substack{-1 \leq j < \infty \\ -\infty < k < \infty}} A_{jk} w_{jk}(x)$$

where

$$w_{jk}(x) = \begin{cases} w(2^j x - k) & \text{if } j \geq 0 \\ \phi(x - k) & \text{if } j = -1 \end{cases}$$

The function $w(x)$ is called the *basic wavelet* of the expansion and $\phi(x)$ is called the *scaling function* associated with $w(x)$.

We will begin by describing methods for computing suitable functions $w(x)$ and $\phi(x)$. We will usually want conditions like the following to be satisfied:

$$(32) \qquad \int_{-\infty}^{\infty} w_{j_1 k_1}(x) w_{j_2 k_2}(x) \, dx = \begin{cases} 2^{-j_1} & \text{if } j_1 = j_2 \text{ and } k_1 = k_2 \\ 0 & \text{otherwise} \end{cases}$$

—these are called *orthogonality conditions*. Compare these to equation (18) on page 199.

The reason for these conditions is that they make it very easy (at least in principle) to compute the yoefficients in the basic wavelet expansion in equation (31): we simply multiply the entire series by $w(2^j x - k)$ or $\phi(x - i)$ and integrate. All but one of the terms of the result vanish due to the orthogonality conditions (equation (32)) and we get:

$$(33) \qquad A_{jk} = \frac{\int_{-\infty}^{\infty} f(x) w_{jk}(x)\, dx}{\int_{-\infty}^{\infty} w_{jk}^2(x)\, dx} = 2^j \int_{-\infty}^{\infty} f(x) w_{jk}(x)\, dx$$

In order to construct functions that are only nonzero over a finite interval of the $x$-axis, and satisfy the basic orthogonality condition, we carry out a sequence of steps. We begin by computing the *scaling function* associated with the wavelet $w(x)$.

A scaling function for a wavelet must satisfy the conditions[18]:

1.  Its support (i.e., the region of the $x$-axis over which it takes on nonzero values) is some finite interval. This is the same kind of condition that wavelets themselves must satisfy. This condition is simply a consequence of the basic concept of a wavelet-series.

2.  It satisfies the basic *dilation equation*:

$$(34) \qquad \phi(x) = \sum_{i > -\infty}^{\infty} \xi_i \phi(2x - i)$$

Note that this sum is not as imposing as it appears at first glance — the previous condition implies that only a finite number of the $\{\xi_i\}$ can be nonzero. We write the sum in this form because we don't want to specify any *fixed* ranges of subscripts over which the $\{\xi_i\}$ may be nonzero.

This condition is due to Daubechies — see [41]. It is the heart of her theory of wavelet-series. It turns out to imply that the wavelet-expansions are orthogonal and easy to compute.

3.  Note that any multiple of a solution of equation (34) is also a solution. We select a preferred solution by imposing the condition

$$(35) \qquad \int_{-\infty}^{\infty} \phi(x)\, dx = 1$$

One points come to mind when we consider these conditions from a *computer science* point of view:

Equation (34), the finite set of nonzero values of $\phi(x)$ at integral points, and the finite number of nonzero $\{\xi_i\}$ completely determine $\phi(x)$. They determine it at all *dyadic points* (i.e., values of $x$ of the form $p/q$, where

---

[18]Incidentally, the term scaling function, like the term wavelet refers to a whole *class* of functions that have certain properties.

$q$ is a power of 2). For virtually all modern computers, such points are the only ones that exist, so $\phi(x)$ is completely determined.

Of course, from a *function theoretic* point of view $\phi(x)$ is far from being determined by its dyadic values. What is generally done is to perform a iterative procedure: we begin by setting $\phi_0(x)$ equal to some simple function like the box function equal to 1 for $0 \leq x < 1$ and 0 otherwise. We then define

$$(36) \qquad \phi_{i+1}(x) = \sum_{k>-\infty}^{\infty} \xi_k \phi_i(2x - k)$$

It turns out that this procedure converges to a limit $\phi(x) = \phi_\infty(x)$, that satisfied equation (34) exactly. Given a suitable scaling-function $\phi(x)$, we define the associated wavelet $w(x)$ by the formula

$$(37) \qquad w(x) = \sum_{i>-\infty}^{\infty} (-1)^i \xi_{1-i} \phi(2x - i)$$

We will want to impose some conditions upon the coefficients $\{\xi_i\}$.

DEFINITION 3.1. The defining coefficients of a system of wavelets will be assumed to satisfy the following two conditions:

1. **Condition O:** This condition implies the orthogonality condition of the wavelet function (equation (32) on page 223):

$$(38) \qquad \sum_{k>-\infty}^{\infty} \xi_k \xi_{k-2m} = \begin{cases} 2 & \text{if } m = 0 \\ 0 & \text{otherwise} \end{cases}$$

   The orthogonality relations mean that if a function can be expressed in terms of the wavelets, we can easily *calculate* the coefficients involved, via equation (33) on page 224.

2. **Condition A:** There exists a number $p > 1$, called the *degree of smoothness* of $\phi(x)$ and the associated wavelet $w(x)$, such that

$$\sum_{k>-\infty}^{\infty} (-1)^k k^m \xi_k = 0, \text{ for all } 0 \leq m \leq p - 1$$

   It turns out that wavelets are generally *fractal functions* — they are not differentiable unless their degree of smoothness is $> 2$.

   This condition guarantees that the functions that interest us[19] *can* be expanded in terms of wavelets. If $\phi(x)$ is a scaling function with a degree of smoothness

---

[19]This term is deliberately vague.

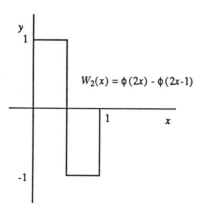

FIGURE V.14. The Haar Wavelet

equal to $p$, it is possible to expand the functions $1, x, \ldots, x^{p-1}$ in terms of series like

$$\sum_{j>-\infty}^{\infty} A_n \phi(x-n)$$

In order for wavelets to be significant to us, they (and their scaling functions) must be derived from a sequence of coefficients $\{\xi_i\}$ with a degree of smoothness $> 0$.

Daubechies discovered a family of wavelets $W_2$, $W_4$, $W_6, \cdots$ whose defining coefficients (the $\{\xi_i\}$) satisfy these conditions in [41]. All of these wavelets are based upon scaling functions that result from iterating the box function

$$\phi_0(x) = \begin{cases} 1 & \text{if } 0 \le x < 1 \\ 0 & \text{otherwise} \end{cases}$$

in the dilation-equation (36) on page 225. The different elements of this sequence of wavelets are only distinguished by the sets of coefficients used in the iterative procedure for computing $\phi(x)$ and the corresponding wavelets.

(Note: this function, *must vanish* at one of the endpoints of the interval $[0,1]$.) This procedure for computing wavelets (i.e., plugging the box function into equation (34) and repeating this with the result, etc.) is not very practical. It is computationally expensive, and only computes approximations to the desired result[20].

Fortunately, there is a simple, fast, and *exact* algorithm for computing wavelets at all *dyadic points* using equation (34) and the values of $\phi(x)$ at integral points. Furthermore, from the perspective of computers, the dyadic points are the only ones that exist. We

[20]This slow procedure has theoretical applications — the proof that the wavelets are orthogonal (i.e., satisfy equation (32) on page 223) is based on this construction.

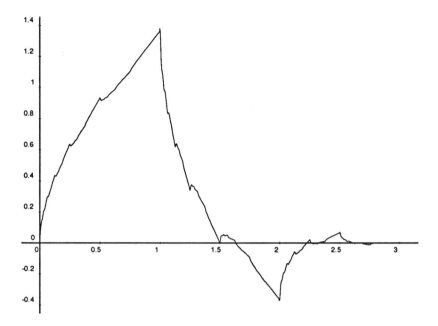

FIGURE V.15. Daubechies degree-4 scaling function

just perform a recursive computation of $\phi(x)$ at points of the form $i/2^{k+1}$ using the values at points of the form $i/2^k$ and the formula

$$\phi(i/2^{k+1}) = \sum_{-\infty < m < \infty} \xi_m \phi(\frac{i}{2^k} - m)$$

It is often possible for the dilation-equation to imply relations between the values of a scaling function at distinct integral points. We must choose the value at these points in such a way as to satisfy the dilation-equation.

EXAMPLE 3.2. **Daubechies' $W_2$ Wavelet.** This is the simplest element of the Daubechies sequence of wavelets. This family is defined by the fact that the coefficients of the dilation-equation are $\xi_0 = \xi_1 = 1$, and all other $\xi_i = 0$.

In this case $\phi(x) = \phi_0(x)$, the box function. The corresponding *wavelet*, $W_2(x)$ has been described long before the development of wavelets — it is called the *Haar function*. It is depicted in figure V.14.

EXAMPLE 3.3. **Daubechies' $W_4$ Wavelet.** Here we use the coefficients $\xi_0 = (1 + \sqrt{3})/4$, $\xi_1 = (3 + \sqrt{3})/4$, $\xi_2 = (3 - \sqrt{3})/4$, and $\xi_3 = (1 - \sqrt{3})/4$ in the dilation-equation. This wavelet has smoothness equal to 2, and its scaling function $\phi(x)$ is called $D_4(x)$. We can compute the scaling function, $D_4(x)$ at the dyadic points by the recursive procedure described above. We cannot pick the values of $\phi(1)$ and

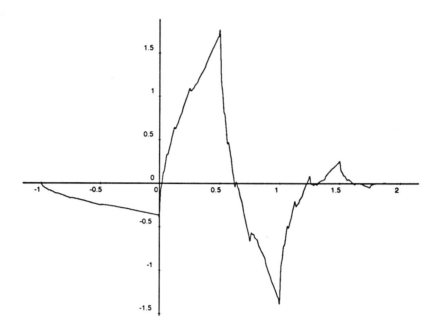

FIGURE V.16. Daubechies degree-4 wavelet

$\phi(2)$ arbitrarily because they are not independent of each other in the equation for $\phi(x) = D_4(x)$.

They satisfy the equations

$$\phi(1) = \frac{3 + \sqrt{3}}{4}\phi(1) + \frac{1 + \sqrt{3}}{4}\phi(2)$$

$$\phi(2) = \frac{1 - \sqrt{3}}{4}\phi(1) + \frac{3 - \sqrt{3}}{4}\phi(2)$$

This is an eigenvalue problem[21] like

$$\Xi x = \lambda x$$

where $x$ is the vector composed of $\phi(1)$ and $\phi(2)$, and $\Xi$ is the matrix

$$\begin{pmatrix} \frac{3+\sqrt{3}}{4} & \frac{1+\sqrt{3}}{4} \\ \frac{1-\sqrt{3}}{4} & \frac{3-\sqrt{3}}{4} \end{pmatrix}$$

The problem only has a solution if $\lambda = 1$ is a valid eigenvalue of $\Xi$ — in this case the correct values of $\phi(1)$ and $\phi(2)$ are given by some scalar multiple of the eigenvector associated with the eigenvalue 1.

---

[21]See 1.10 on page 160 for the definition of an eigenvalue.

This matrix ($\Xi$) does have an eigenvalue of $1$, and its associated eigenvector is

$$\begin{pmatrix} \frac{1+\sqrt{3}}{2} \\ \frac{1-\sqrt{3}}{2} \end{pmatrix}$$

and these become, from top to bottom, our values of $\phi(1)$ and $\phi(2)$, respectively. The scaling function, $\phi(x)$, is called $D_4(x)$ in this case[22] and is plotted in figure V.15.

Notice the *fractal* nature of the function. It is actually much more irregular than it appears in this graph. The associated wavelet is called $W_4$ and is depicted in figure V.16.

**Exercises:**

1. Write a program to compute $D_4(x)$ and $W_4(x)$ at dyadic points, using the recursive algorithm described above. What is the running-time of the algorithm? Generally we measure the extent to which $D_4(x)$ has been computed by measuring the "fineness of the mesh" upon which we know the values of this function — in other words, $1/2^n$.
2. What is the execution-time of a parallel algorithm for computing $\phi(x)$ in general?

**3.2. Discrete Wavelet Transforms.** Now we are in a position to discuss how one does a discrete version of the wavelet transform. We will give an algorithm for computing the wavelet-series in equation (31) on page 223 up to some finite value of $j$ — we will compute the $A_{j,k}$ for $j \leq p$. We will call the parameter $p$ the fineness of mesh of the expansion, and $2^{-p}$ the mesh-size. the algorithm that we will discuss, are closely related to the Mallat pyramid algorithm, but not entirely identical to it.

DEFINITION 3.4. Define $B_{r,j}$ by

$$B_{r,j} = 2^r \int_{-\infty}^{\infty} \phi(2^r x - j) f(x)\, dx$$

where $0 \leq r \leq p$ and $-\infty < j < \infty$.

---

[22] In honor of Daubechies.

These quantities are important because they allow us to compute the coefficients of the wavelet-series.

In general, the $B_{r,j}$ are nonzero for at most a *finite* number of values of $j$:

PROPOSITION 3.5. In the notation of 3.4, above, suppose that $f(x)$ is only nonzero on the interval $a \leq x \leq b$ and $\phi(x)$ is only nonzero on the interval $0 \leq x \leq R$. Then $B_{r,j} = 0$ unless $\mathcal{L}(r) \leq j \leq \mathcal{U}(r)$, where $\mathcal{L}(r) = \lfloor 2^r a - R \rfloor$ and $\mathcal{U}(r) = \lceil 2^r b \rceil$.

We will follow the convention that $\mathcal{L}(-1) = \lfloor a - R \rfloor$, and $\mathcal{U}(-1) = \lceil b \rceil$

PROOF. In order for the integral 3.4 to be nonzero, it is at least necessary for the domains in which $f(x)$ and $\phi(2^r x - j)$ are nonzero to intersect. This means that

$$0 \leq 2^r x - j \leq R$$
$$a \leq x \leq b$$

If we add $j$ to the first inequality, we get:

$$j \leq 2^r x \leq j + R$$

or

$$2^r x - R \leq j \leq 2^r x$$

The second inequality implies the result.  □

The first thing to note is that the quantities $B_{p,j}$ determine the $B_{r,j}$ for all values of $r$ such that $0 \leq r < p$:

PROPOSITION 3.6. For all values of $r \leq p$

$$B_{r,j} = \sum_{m > -\infty}^{\infty} \frac{\xi_{m-2j} B_{r+1,m}}{2}$$

PROOF. This is a direct consequence of the basic dilation equation (34) on page 224:

$$B_{r,j} = 2^r \int_{-\infty}^{\infty} \phi(2^r x - j) f(x)\, dx$$

$$= 2^r \int_{-\infty}^{\infty} \sum_{s > -\infty}^{\infty} \xi_s \phi(2(2^r x - j) - s) f(x)\, dx$$

$$= 2^r \int_{-\infty}^{\infty} \sum_{s > -\infty}^{\infty} \xi_s \phi(2^{r+1} x - 2j - s) f(x)\, dx$$

setting $m = 2j + s$

$$= 2^r \int_{-\infty}^{\infty} \sum_{m > -\infty}^{\infty} \xi_{m-2j} \phi(2^{r+1} x - m) f(x)\, dx$$

$$= 2^r \sum_{m > -\infty}^{\infty} \xi_{m-2j} \int_{-\infty}^{\infty} \phi(2^{r+1} x - m) f(x)\, dx$$

□

The definition of $w(x)$ in terms of $\phi(x)$ implies that

PROPOSITION 3.7. Let $A_{r,k}$ denote the coefficients of the wavelet-series, as defined in equation (33) on page 224. Then

$$A_{r,k} = \begin{cases} \displaystyle\sum_{m > -\infty}^{\infty} (-1)^m \frac{\xi_{1-m+2k} B_{r+1,m}}{2} & \text{if } r \geq 0 \\[2ex] B_{-1,k} & \text{if } r = -1 \end{cases}$$

PROOF. This is a direct consequence of equation (37) on page 225. We take the definition of the $A_{r,k}$ and plug in equation (37):

$$A_{r,k} = 2^r \int_{-\infty}^{\infty} f(x) w(2^r x - k)\, dx$$

$$= 2^r \int_{-\infty}^{\infty} f(x) \sum_{s > -\infty}^{\infty} (-1)^s \xi_{1-s} \phi(2(2^r x - k) - s)\, dx$$

$$= 2^r \int_{-\infty}^{\infty} f(x) \sum_{s > -\infty}^{\infty} (-1)^s \xi_{1-s} \phi(2^{r+1} x - 2k - s)\, dx$$

now we set $m = 2k + s$

$$= 2^r \int_{-\infty}^{\infty} f(x) \sum_{m > -\infty}^{\infty} (-1)^m \xi_{1-m+2k} \phi(2^{r+1} x - m)\, dx$$

$$= 2^r \sum_{m > -\infty}^{\infty} (-1)^m \xi_{1-m+2k} \int_{-\infty}^{\infty} f(x) \phi(2^{r+1} x - m)\, dx$$

□

These results motivate the following definitions:

**DEFINITION 3.8.** Define the following two matrices:

1. $L(r)_{i,j} = \xi_{j-2i}/2.$
2. $H(r)_{i,j} = (-1)^j \xi_{1-j+2i}/2.$
   Here $i$ runs from $\mathcal{L}(r-1)$ to $\mathcal{U}(r-1)$ and $j$ runs from $\mathcal{L}(r)$ to $\mathcal{U}(r)$, in the notation of 3.5.

For instance, if we are working with the Daubechies wavelet, $W_4(x)$, and $p = 3$, $a = 0$, $b = 1$, $R = 2$, then $\mathcal{L}(3) = -2$, $\mathcal{U}(3) = 8$ and

$$
L(3) =
\begin{pmatrix}
\frac{3-\sqrt{3}}{2} & \frac{1-\sqrt{3}}{2} & 0 & 0 & 0 & 0 & 0 & 0 & 0 & 0 \\
\frac{1+\sqrt{3}}{2} & \frac{3+\sqrt{3}}{2} & \frac{3-\sqrt{3}}{2} & \frac{1-\sqrt{3}}{2} & 0 & 0 & 0 & 0 & 0 & 0 \\
0 & 0 & \frac{1+\sqrt{3}}{2} & \frac{3+\sqrt{3}}{2} & \frac{3-\sqrt{3}}{2} & \frac{1-\sqrt{3}}{2} & 0 & 0 & 0 & 0 \\
0 & 0 & 0 & 0 & \frac{1+\sqrt{3}}{2} & \frac{3+\sqrt{3}}{2} & \frac{3-\sqrt{3}}{2} & \frac{1-\sqrt{3}}{2} & 0 & 0 \\
0 & 0 & 0 & 0 & 0 & 0 & \frac{1+\sqrt{3}}{2} & \frac{3+\sqrt{3}}{2} & \frac{3-\sqrt{3}}{2} & \frac{1-\sqrt{3}}{2} \\
0 & 0 & 0 & 0 & 0 & 0 & 0 & 0 & \frac{1+\sqrt{3}}{2} & \frac{3+\sqrt{3}}{2}
\end{pmatrix}
$$

and

$$
H(3) =
$$

$$
\begin{pmatrix}
0 & 0 & 0 & 0 & 0 & 0 & 0 & 0 & 0 & 0 \\
\frac{3+\sqrt{3}}{2} & -\frac{1+\sqrt{3}}{2} & 0 & 0 & 0 & 0 & 0 & 0 & 0 & 0 \\
\frac{1-\sqrt{3}}{2} & -\frac{3-\sqrt{3}}{2} & \frac{3+\sqrt{3}}{2} & -\frac{1+\sqrt{3}}{2} & 0 & 0 & 0 & 0 & 0 & 0 \\
0 & 0 & \frac{1-\sqrt{3}}{2} & -\frac{3-\sqrt{3}}{2} & \frac{3+\sqrt{3}}{2} & -\frac{1+\sqrt{3}}{2} & 0 & 0 & 0 & 0 \\
0 & 0 & 0 & 0 & \frac{1-\sqrt{3}}{2} & -\frac{3-\sqrt{3}}{2} & \frac{3+\sqrt{3}}{2} & -\frac{1+\sqrt{3}}{2} & 0 & 0 \\
0 & 0 & 0 & 0 & 0 & 0 & \frac{1-\sqrt{3}}{2} & -\frac{3-\sqrt{3}}{2} & \frac{3+\sqrt{3}}{2} & -\frac{1+\sqrt{3}}{2}
\end{pmatrix}
$$

Our algorithm computes all of the $A_{r,k}$, given the values of $B_{p+1,k}$:

**ALGORITHM 3.9.** This algorithm computes a wavelet-expansion of the function $f(x)$ with mesh-size $2^{-p}$.

- **Input:** The quantities $\{B_{p+1,j}\}$, where $j$ runs from $\mathcal{L}(p+1)$ to $\mathcal{U}(p+1)$. There are $n = \lceil 2^{p+1}(b-a) + R \rceil$ such inputs (in the notation of 3.5 on page 230);
- **Output:** The values of the $A_{k,j}$ for $-1 \le k \le p$, and, for each value of $k$, $j$ runs from $\mathcal{L}(k)$ to $\mathcal{U}(k)$. There are approximately $\lceil 2^{p+1}(b-a) + R \rceil$ such outputs.

  for $k \leftarrow p$ down to 1 do
     Compute $B_{k,*} \leftarrow L(k+1)B_{k+1,*}$
     Compute $A_{k,*} \leftarrow H(k+1)B_{k+1,*}$
  endfor

Here, the arrays $L$ and $H$ are defined in 3.8 above. The array $B_{k,*}$ has half as many nonzero entries as $B_{k+1,*}$.

Since the number of nonzero entries in each row of the $L(*)$ and $H(*)$ arrays is so small, we generally incorporate this number in the constant of proportionality in estimating the execution-time of the algorithm.

With this in mind, the *sequential* execution-time of an iteration of this algorithm is $O(2^k)$. If we have $2^{p+1}$ processors available, the parallel execution-time (on a CREW-SIMD computer) is *constant*.

The total sequential execution-time of this algorithm is $O(n)$ and the total parallel execution-time (on a CREW-SIMD computer with $n$ processors) is $O(\lg n)$.

The $A_{k,*}$ are, of course, the coefficients of the wavelet-expansion. The only elements of this algorithm that look a little mysterious are the quantities

$$B_{p+1,j} = 2^{p+1} \int_{-\infty}^{\infty} \phi(2^{p+1}x - j)f(x)\,dx$$

First we note that $\int_{-\infty}^{\infty} \phi(u)\,du = 1$ (by equation (35) on page 224), so $\int_{-\infty}^{\infty} \phi(2^{p+1}x - j)\,dx = 2^{-(p+1)}$ (set $u = 2^{p+1} - j$, and $dx = 2^{-(p+1)}du$) and

$$B_{p+1,j} = \frac{\int_{-\infty}^{\infty} \phi(2^{p+1}x - j)f(x)\,dx}{\int_{-\infty}^{\infty} \phi(2^{p+1}x - j)\,dx}$$

so that $B_{p+1,j}$ is nothing but a *weighted average* of $f(x)$ weighted by the function $\phi(2^{p+1}x - j)$. Now note that this weighted average is really being taken over a small interval $0 \le 2^{p+1}x - j \le R$, where $[0, R]$ is the range of values over which $\phi(x) \ne 0$. This is always some finite interval — for instance if $\phi(x) = D_4(x)$ (see figure V.15 on page 227), this interval is $[0, 3]$. This means that $x$ runs from $j2^{-(p+1)}$ to $(j + R)2^{-(p+1)}$.

At this point we make the assumption:

> **The width of the interval $[j2^{-(p+1)}, (j + R)2^{-(p+1)}]$, is small enough that $f(x)$ doesn't vary in any appreciable way over this interval. Consequently, the weighted average is equal to $f(j2^{-(p+1)})$.**

So we begin the inductive computation of the $A_{k,j}$ in 3.9 by setting $B_{p+1,j} = f(j/2^{p+1})$. We regard the set of values $\{f(j/2^{p+1})\}$ with $0 \le j < 2^{p+1}$ as the *inputs* to the discrete wavelet transform algorithm.

The *output* of the algorithm is the set of wavelet-coefficients $\{A_{k,j}\}$, with $-1 \le k \le p$, $-\infty < j < \infty$. Note that $j$ actually only takes on a finite set of values — this set is usually small and depends upon the type of wavelet under consideration. In the case of the Haar wavelet, for instance $0 \le j \le 2^k - 1$, if $k \le 0$, and $j = 0$ if $k = -1$. In the case of the Daubechies $W_4$ wavelet this set is a little larger, due to the fact that there are more nonzero defining coefficients $\{\xi_i\}$.

Now we will give a fairly detailed example of this algorithm. Let $f(x)$ be the function defined by:

$$f(x) = \begin{cases} 0 & \text{if } x \leq 0 \\ x & \text{if } 0 < x \leq 1 \\ 0 & \text{if } x > 1 \end{cases}$$

We will expand this into a wavelet-series using the degree-4 Daubechies wavelet defined in 3.3 on page 227. We start with mesh-size equal to $2^{-5}$, so $p = 4$, and we define $B_{5,*}$ by

$$B_{5,i} = \begin{cases} 0 & \text{if } i \leq 0 \\ i/32 & \text{if } 1 \leq i \leq 32 \\ 0 & \text{if } i > 32 \end{cases}$$

In the present case, the looping phase of algorithm 3.9 involves the computation:

$$B_{k,i} = \frac{1+\sqrt{3}}{8}B_{k+1,2i} + \frac{3+\sqrt{3}}{8}B_{k+1,2i+1} + \frac{3-\sqrt{3}}{8}B_{k+1,2i+2} + \frac{1-\sqrt{3}}{8}B_{k+1,2i+3}$$

$$A_{k,i} = \frac{1-\sqrt{3}}{8}B_{k+1,2i-2} - \frac{3-\sqrt{3}}{8}B_{k+1,2i-1} + \frac{3+\sqrt{3}}{8}B_{k+1,2i} - \frac{1+\sqrt{3}}{8}B_{k+1,2i+1}$$

This can be done in constant parallel time (i.e., the parallel execution-time is independent of the number of data-points).

- **Iteration 1:** The $B_{4,*}$ and the wavelet-coefficients, $A_{4,*}$ are all zero except for the following cases:

$$\begin{cases} B_{4,-1} = \dfrac{1}{256} - \dfrac{\sqrt{3}}{256} \\ B_{4,j} = \dfrac{4j+3-\sqrt{3}}{64} \text{ for } 0 \leq j \leq 14 \\ B_{4,15} = \dfrac{219}{256} + \dfrac{29\sqrt{3}}{256} \\ B_{4,16} = 1/8 + \dfrac{\sqrt{3}}{8} \end{cases} \qquad \begin{cases} A_{4,0} = -\dfrac{1}{256} - \dfrac{\sqrt{3}}{256} \\ A_{4,16} = \dfrac{33}{256} + \dfrac{33\sqrt{3}}{256} \\ A_{4,17} = 1/8 - \dfrac{\sqrt{3}}{8} \end{cases}$$

Now we can calculate $B_{3,*}$ and $A_{3,*}$:

- **Iteration 2:**

$$\begin{cases} B_{3,-2} = \dfrac{1}{512} - \dfrac{\sqrt{3}}{1024} \\ B_{3,-1} = \dfrac{11}{256} - \dfrac{29\sqrt{3}}{1024} \\ B_{3,j} = \dfrac{8j+9-3\sqrt{3}}{64} \text{ for } 0 \leq j \leq 5 \\ B_{3,6} = \dfrac{423}{512} - \dfrac{15\sqrt{3}}{1024} \\ B_{3,7} = \dfrac{121}{256} + \dfrac{301\sqrt{3}}{1024} \\ B_{3,8} = 1/16 + \dfrac{\sqrt{3}}{32} \end{cases} \qquad \begin{cases} A_{3,-1} = \dfrac{1}{1024} \\ A_{3,0} = \dfrac{1}{1024} - \dfrac{5\sqrt{3}}{512} \\ A_{3,7} = -\dfrac{33}{1024} \\ A_{3,8} = \dfrac{5\sqrt{3}}{512} - \dfrac{65}{1024} \\ A_{3,9} = -1/32 \end{cases}$$

- **Iteration 3:**

$$\left\{\begin{array}{l} B_{2,-2} = \dfrac{35}{2048} - \dfrac{39\sqrt{3}}{4096} \\[2mm] B_{2,-1} = \dfrac{259}{2048} - \dfrac{325\sqrt{3}}{4096} \\[2mm] B_{2,0} = \dfrac{21}{64} - \dfrac{7\sqrt{3}}{64} \\[2mm] B_{2,1} = \dfrac{37}{64} - \dfrac{7\sqrt{3}}{64} \\[2mm] B_{2,2} = \dfrac{1221}{2048} + \dfrac{87\sqrt{3}}{4096} \\[2mm] B_{2,3} = \dfrac{813}{2048} + \dfrac{1125\sqrt{3}}{4096} \\[2mm] B_{2,4} = \dfrac{5}{256} + \dfrac{3\sqrt{3}}{256} \end{array}\right\} \qquad \left\{\begin{array}{l} A_{2,-1} = \dfrac{23}{4096} - \dfrac{\sqrt{3}}{512} \\[2mm] A_{2,0} = -\dfrac{27}{4096} - \dfrac{3\sqrt{3}}{256} \\[2mm] A_{2,3} = \dfrac{15\sqrt{3}}{512} - \dfrac{295}{4096} \\[2mm] A_{2,4} = \dfrac{315}{4096} - \dfrac{35\sqrt{3}}{256} \\[2mm] A_{2,5} = -\dfrac{1}{256} - \dfrac{\sqrt{3}}{256} \end{array}\right\}$$

- **Iteration 4:**

$$\left\{\begin{array}{l} B_{1,-2} = \dfrac{455}{8192} - \dfrac{515\sqrt{3}}{16384} \\[2mm] B_{1,-1} = \dfrac{2405}{8192} - \dfrac{2965\sqrt{3}}{16384} \\[2mm] B_{1,0} = \dfrac{2769}{8192} - \dfrac{381\sqrt{3}}{16384} \\[2mm] B_{1,1} = \dfrac{2763}{8192} + \dfrac{3797\sqrt{3}}{16384} \\[2mm] B_{1,2} = \dfrac{7}{1024} + \dfrac{\sqrt{3}}{256} \end{array}\right\} \qquad \left\{\begin{array}{l} A_{1,-1} = \dfrac{275}{16384} - \dfrac{15\sqrt{3}}{2048} \\[2mm] A_{1,0} = -\dfrac{339}{16384} - \dfrac{67\sqrt{3}}{4096} \\[2mm] A_{1,2} = \dfrac{531}{16384} - \dfrac{485\sqrt{3}}{4096} \\[2mm] A_{1,3} = -\dfrac{1}{512} - \dfrac{\sqrt{3}}{1024} \end{array}\right\}$$

- **Iteration 5:** In this phase we complete the computation of the wavelet-coefficients: these are the $A_{0,*}$ and the $B_{0,*} = A_{-1,*}$.

$$\left\{\begin{array}{l} B_{0,-2} = \dfrac{4495}{32768} - \dfrac{5115\sqrt{3}}{65536} \\[2mm] B_{0,-1} = \dfrac{2099}{16384} - \dfrac{3025\sqrt{3}}{32768} \\[2mm] B_{0,1} = \dfrac{19}{8192} + \dfrac{11\sqrt{3}}{8192} \end{array}\right\} \qquad \left\{\begin{array}{l} A_{0,-1} = \dfrac{2635}{65536} - \dfrac{155\sqrt{3}}{8192} \\[2mm] A_{0,0} = \dfrac{919\sqrt{3}}{16384} - \dfrac{5579}{32768} \\[2mm] A_{0,2} = -\dfrac{5}{8192} - \dfrac{3\sqrt{3}}{8192} \end{array}\right\}$$

We will examine the convergence of this wavelet-series. The $A_{-1,*}$ terms are:

$$S_{-1} = \left(\frac{4495}{32768} - \frac{5115\sqrt{3}}{65536}\right) D_4(x+2) + \left(\frac{2099}{16384} - \frac{3025\sqrt{3}}{32768}\right) D_4(x+1)$$

$$+ \left(\frac{19}{8192} + \frac{11\sqrt{3}}{8192}\right) D_4(x)$$

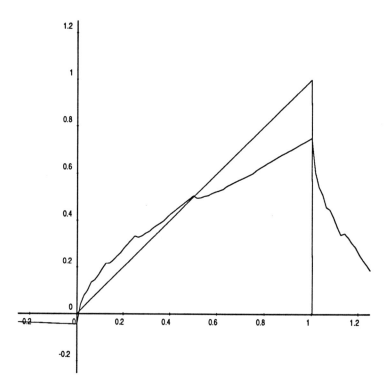

FIGURE V.17

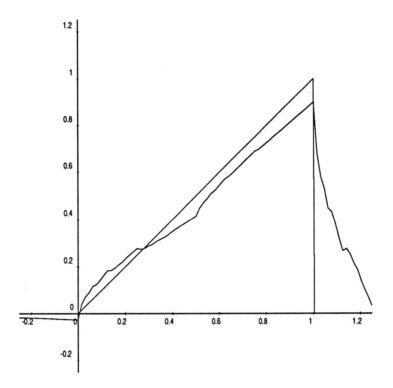

FIGURE V.18

This expression is analogous to the *constant term* in a Fourier series.

It is plotted against $f(x)$ in figure V.17 — compare this (and the following plots with the partial-sums of the Fourier series in figures V.5 to V.8 on page 199. If we add in the $A_{0,*}$-terms we get:

$$S_0(x) = S_{-1}(x) + \left( \frac{2635}{65536} - \frac{155\sqrt{3}}{8192} \right) W_4(x+1) + \left( \frac{919\sqrt{3}}{16384} - \frac{5579}{32768} \right) W_4(x)$$
$$- \left( \frac{5}{8192} + \frac{3\sqrt{3}}{8192} \right) W_4(x-2)$$

It is plotted against the original function $f(x)$ in figure V.18. The next step involves adding in the $A_{1,*}$-terms

$$S_1(x) = S_0(x) + \left( \frac{275}{16384} - \frac{15\sqrt{3}}{2048} \right) W_4(2x+1) - \left( \frac{339}{16384} + \frac{67\sqrt{3}}{4096} \right) W_4(2x)$$
$$- \left( \frac{531}{16384} - \frac{485\sqrt{3}}{4096} \right) W_4(2x-2) - \left( \frac{1}{512} + \frac{\sqrt{3}}{1024} \right) W_4(2x-3)$$

Figure V.19 shows how the wavelet-series begins to approximate $f(x)$.

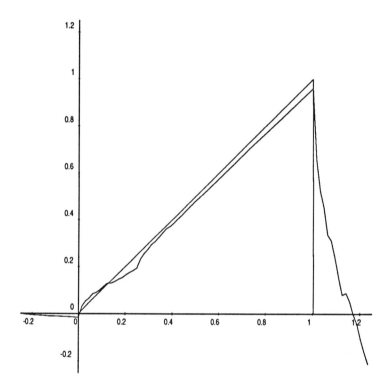

FIGURE V.19

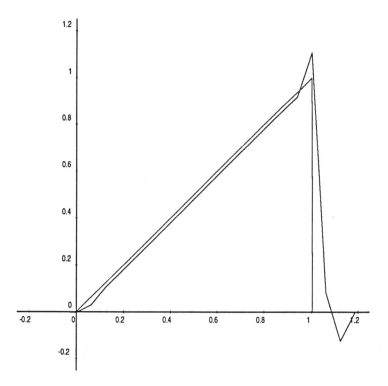

FIGURE V.20

The $A_{3,*}$ contribute:

$$S_2(x) = S_1(x) + \left(\frac{23}{4096} - \frac{\sqrt{3}}{512}\right) W_4(4x+1) - \left(\frac{27}{4096} + \frac{3\sqrt{3}}{256}\right) W_4(4x)$$

$$+ \left(\frac{15\sqrt{3}}{512} - \frac{295}{4096}\right) W_4(4x-3)$$

$$+ \left(\frac{315}{4096} - \frac{35\sqrt{3}}{256}\right) W_4(4x-4) - \left(\frac{1}{256} + \frac{\sqrt{3}}{256}\right) W_4(4x-5)$$

Incidentally, this series (and, in fact, Fourier series) converges in following sense (we will not prove this)

(39)
$$\lim_{n\to\infty} \int_{-\infty}^{\infty} |f(x) - S_n(x)|^2 \, dx \to 0$$

This, roughly speaking, means that the *area* of the space *between* the graphs of $f(x)$ and $S_n(x)$ approaches $0$ as $n$ approaches $\infty$. This does not necessarily mean that

$S_n(x) \rightarrow f(x)$ for all values of $x$. It is interesting that there are points $x_0$ where $\lim_{n \to \infty} S_n(x) \neq f(x) - x = 1$ is such a point[23]. Equation (39) implies that the total area of this set of points is zero. Luckily, most of the applications of wavelets only require the kind of convergence described in equation (39).

We will conclude this section with a discussion of the *converse* of algorithm 3.9 — it is an algorithm that computes partial sums of a wavelet-series, given the coefficients $\{A_{j,k}\}$. Although there is a straightforward algorithm for doing this that involves simply plugging values into the functions $w(2^j x - k)$ and plugging these values into the wavelet-series, there is also a faster algorithm for this. This algorithm is very similar to the algorithm for computing the $\{A_{j,k}\}$ in the first place. We make use of the recursive definition of the scaling and wavelet-functions in equation (34) on page 224.

PROPOSITION 3.10. In the notation of 3.8 on page 232 define the following two sets of matrices:

1. $L(r)_{i,j}^* = \xi_{i-2j}$. Here $i$ runs from $\mathcal{L}(r)$ to $\mathcal{U}(r)$ and $j$ runs from $\mathcal{L}(r-1)$ to $\mathcal{U}(r-1)$.

2. $H(r)_{i,j}^* = (-1)^j \xi_{1-i+2j}$. Here $i$ runs from $\mathcal{L}(r) + 1$ to $\mathcal{U}(r)$ and $j$ runs from $\mathcal{L}(r-1)$ to $\mathcal{U}(r-1)$.

Also, let $\mathbb{R}(u,v)$ denote the $v - u$-dimensional subspace of $\mathbb{R}^\infty$ spanned by coordinates $u$ through $v$.

Then

$$L(r)L(r)^* = I \text{ on } \mathbb{R}(\mathcal{L}(r-1), \mathcal{U}(r-1))$$
$$H(r)H(r)^* = I \text{ on } \mathbb{R}(\mathcal{L}(r-1) + 1, \mathcal{U}(r-1))$$

and

$$H(r)L(r)^* = 0$$
$$L(r)H(r)^* = 0$$

PROOF. This follows directly from the orthogonality conditions (Condition O) on the $\{\xi_i\}$ on page 225. We get

$$(L(r)L(r)^*)_{i,k} = \sum_j \xi_{j-2i} \xi_{j-2k}/2$$

$$= \sum_\ell \xi_\ell \xi_{\ell - 2(k-i)}/2 \text{ (Setting } \ell = j - 2i)$$

$$= \begin{cases} 1 & \text{if } i = k \\ 0 & \text{otherwise} \end{cases} \text{ (by equation (38))}$$

---

[23]This is a well-known phenomena in Fourier series — it is called Gibbs phenomena.

and

$$(H(r)H(r)^*)_{i,k} = (-1)^{j+j} \sum_j \xi_{1-j+2i}\xi_{1-j+2k}/2$$

$$= \sum_\ell \xi_\ell \xi_{\ell+2(k-i)}/2 \text{ (Setting } \ell = 1 - j + 2i)$$

$$= \begin{cases} 1 & \text{if } i = k \\ 0 & \text{otherwise} \end{cases} \text{ (by equation (38))}$$

We also have

$$(H(r)L(r)^*)_{i,k} = (-1)^j \sum_j \xi_{1-j+2i}\xi_{j-2k}/2$$

Setting $\ell = 1 - j + 2i$

$$= \sum_\ell (-1)^{1-\ell}\xi_\ell \xi_{1-\ell+2(i-k)}/2$$

Now we can pair up each term $(-1)^\ell \xi_\ell \xi_{1-\ell+2(i-k)}/2$ with a term $(-1)^{1-\ell}\xi_{1-\ell+2(i-k)}\xi_\ell/2$, so the total is 0.

The remaining identity follows by a similar argument. □

This implies:

COROLLARY 3.11. The maps $L(r)^*L(r)$ and $H(r)^*H(r)$ satisfy the equations:

$$L(r)^*L(r)L(r)^*L(r) = L(r)^*L(r)$$
$$H(r)^*H(r)H(r)^*H(r) = H(r)^*H(r)$$
$$H(r)^*H(r)L(r)^*L(r) = 0$$
$$L(r)^*L(r)H(r)^*H(r)) = 0$$
$$L(r)^*L(r) + H(r)^*H(r) = I$$

The last equation applies to the space $\mathbb{R}(\mathcal{L}(r) + 1, \mathcal{U}(r))$ (in the notation of 3.10).

PROOF. The first four equations follow immediately from the statements of 3.10. The last statement follows from the fact that

$$L(r)(L(r)^*L(r) + H(r)^*H(r)) = L$$
$$H(r)(L(r)^*L(r) + H(r)^*H(r)) = H$$

so that $L(r)^*L(r) + H(r)^*H(r)$ is the identity on the images of $L(r)$ and $H(r)$. The span of $L(r)$ and $H(r)$ is the entire space of dimension $\lceil 2^r(b-a) + R - 1 \rceil$ because the rank of $L(r)$ is $\lceil 2^{r-1}(b-a) + R \rceil$ and that of $\lceil 2^{r-1}(b-a) + R - 1 \rceil$ and the ranks add up to this. $\square$

This leads to the reconstruction-algorithm for wavelets:

ALGORITHM 3.12. Given the output of algorithm 3.9 on page 232, there exists an algorithm for reconstructing the inputs to that algorithm that $O(n)$ sequential time, and $O(\lg n)$ parallel time with a CREW-PRAM computer with $O(n)$ processors.

- Input: The values of the $A_{k,j}$ for $-1 \leq k \leq p$, and, for each value of $k$, $j$ runs from $\mathcal{L}(k)$ to $\mathcal{U}(k)$. There are approximately $\lceil 2^{p+1}(b-a) + R \rceil$ such inputs.
- Output:The quantities $\{B_{p+1,j}\}$, where $j$ runs from $\lfloor 2^{p+1}a \rfloor$ to $\mathcal{U}(p+1)$. There are $n = \lceil 2^{p+1}(b-a) + R \rceil$ such inputs (in the notation of 3.5 on page 230).

The algorithm amounts to a loop:

**for** $k \leftarrow -1$ **to** $p$ **do**
  **Compute** $B_{k+1,*} \leftarrow L(k+1)^* B_{k,*} + H(k+1)^* A_{k,*}$
**endfor**

PROOF. This is a straightforward consequence of 3.10 above, and the main formulas of algorithm 3.9 on page 232. $\square$

**3.3. Discussion and Further reading.** The defining coefficients for the Daubechies wavelets $W_{2n}$ for $n > 2$ are somewhat complex — see [41] for a general procedure for finding them. For instance, the defining coefficients for $W_6$ are

$$c_0 = \frac{\sqrt{5 + 2\sqrt{10}}}{16} + 1/16 + \frac{\sqrt{10}}{16}$$

$$c_1 = \frac{\sqrt{10}}{16} + \frac{3\sqrt{5 + 2\sqrt{10}}}{16} + \frac{5}{16}$$

$$c_2 = 5/8 - \frac{\sqrt{10}}{8} + \frac{\sqrt{5 + 2\sqrt{10}}}{8}$$

$$c_3 = 5/8 - \frac{\sqrt{10}}{8} - \frac{\sqrt{5 + 2\sqrt{10}}}{8}$$

$$c_4 = \frac{5}{16} - \frac{3\sqrt{5 + 2\sqrt{10}}}{16} + \frac{\sqrt{10}}{16}$$

$$c_5 = 1/16 - \frac{\sqrt{5 + 2\sqrt{10}}}{16} + \frac{\sqrt{10}}{16}$$

In [146], Sweldens and Piessens give formulas for approximately computing coefficients of wavelet-expansions:

$$B_{r,j} = 2^r \int_{-\infty}^{\infty} \phi(2^r x - j) f(x) \, dx$$

(defined in 3.4 on page 229). For the Daubechies wavelet $W_4(x)$ the simplest case of their algorithm gives:

$$B_{r,j} \approx f\left(\frac{2j + 3 - \sqrt{3}}{2^{r+1}}\right)$$

(the accuracy of this formula increases with increasing $r$). This is more accurate than the estimates we used in the example that appeared in pages 233 to 239 (for instance, it is *exact* if $f(x) = x$). We didn't go into this approximation in detail because it would have taken us too far afield.

The general continuous wavelet-transform of a function $f(x)$ with respect to a wavelet $w(x)$, is given by

$$\mathcal{T}_f(a, b) = \frac{1}{\sqrt{a}} \int_{-\infty}^{\infty} \bar{w}\left(\frac{x - b}{a}\right) f(x) \, dx$$

where $\bar{*}$ denotes complex conjugation. The two variables in this function correspond to the two indices in the wavelet-series that we have been discussing in this section. This definition was proposed by Morlet, Arens, Fourgeau and Giard in [9]. It turns out that we can recover the function $f(x)$ from its wavelet-transform via the formula

$$f(x) = \frac{1}{2\pi C_h} \int_{-\infty}^{\infty} \int_{-\infty}^{\infty} \frac{\mathcal{T}_f(a, b)}{\sqrt{|a|}} w\left(\frac{x - b}{a}\right) da \, db$$

where $C_h$ is a suitable constant (the explicit formula for $C_h$ is somewhat complicated, and not essential for the present discussion).

The two algorithms 3.9 and 3.12 are, together, a kind of wavelet-analogue to the FFT algorithm. In many respects, the fast wavelet transform and its corresponding reconstruction algorithm are simpler and more straightforward than the FFT algorithm.

Wavelets that are used in image processing are two-dimensional. It is possible to get such wavelets from one-dimensional wavelets via the process of taking the *tensor-product*. This amounts to making definitions like:

$$W(x, y) = w(x)w(y)$$

The concept of wavelets predate their "official" definition in [60].

Discrete Wavelet-transforms of images that are based upon tensor-products of the Haar wavelet were known to researchers in image-processing — such transforms are known as *quadtree* representations of images. See [72] for a parallel algorithm for image analysis that makes use of wavelets. In [88], Knowles describes a specialized VLSI design for performing wavelet transforms — this is for performing image-compression "on the fly". Wavelets were also used in edge-detection in images in [108].

Many authors have defined systems of wavelets that remain nonvanishing over the entire $x$-axis. In every case, these wavelets decay to 0 in such a way that conditions like equation 32 on page 223 are still satisfied. The wavelets of Meyer decay like $1/x^k$ for a suitable exponent $k$ — see [111].

See [17] for an interesting application of wavelets to *astronomy* — in this case, the determination of large-scale structure of the universe.

**Exercises:**

1. Write a C* program to compute wavelet-coefficients using algorithm 3.9 on page 232.

2. Improve the program above by modifying it to minimize roundoff-error in the computations, taking into account the following two facts:
   - We only compute wavelets at dyadic points.
   - The coefficients $\{\xi_i\}$ used in most wavelet-expansions involve rational numbers, and perhaps, a few irrational numbers with easily-described properties — like $\sqrt{3}$.

3. Find a wavelet-series for the function

$$f(x) = \begin{cases} 1 & \text{if } 0 \leq x \leq 1/3 \\ 0 & \text{otherwise} \end{cases}$$

4. Suppose that

$$S_n(x) = \sum_{i=-1}^{n} \sum_{j>-\infty}^{\infty} A_{k,j} w(2^k x - j)$$

is a partial sum of a wavelet-series, as in 3.12 on page 242. Show that this partial sum is equal to

$$S_n(x) = \sum_{j>-\infty}^{\infty} B_{n,j} \phi(2^n x - j)$$

so that wavelet-series correspond to series of *scaling functions*.

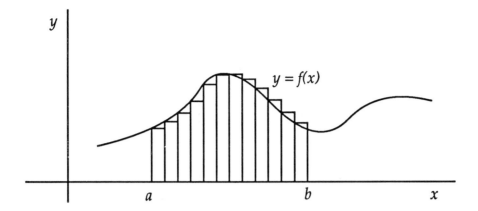

FIGURE V.21. Rectangular approximation to a definite integral

## 4. Numerical Evaluation of Definite Integrals

**4.1. The one-dimensional case.** In this section we will discuss a fairly straightforward application of parallel computing. See § 1.1.1 on page 442 for related material.

Suppose we have a definite integral

$$A = \int_a^b f(x)\, dx$$

where $f(x)$ is some function and $a$ and $b$ are numbers. It often happens that there is no simple closed-form expression for the indefinite integral

$$\int f(x)\, dx$$

and we are forced to resort to numerical methods. Formulas for numerically calculating the approximate value of a definite integral are sometimes called *quadrature formulas*. The simplest of these is based upon the definition of an integral. Consider the diagram in figure V.21. The integral is defined to be the limit of total area of the rectangular strips, as their width approaches 0. Our numerical approximation consists in measuring this area with strips whose width is *finite*. It is easy to compute the area of this union of rectangular strips: the area of a rectangle is the product of the width and the height. If the strips are aligned so that the upper right corner of each strip intersects the graph of the function $y = f(x)$, we get the formula:

(40)
$$A \approx \sum_{i=0}^{n-1} \frac{b-a}{n} f(a + i(b-a)/n)$$

where $n$ is the number of rectangular strips. The accuracy of the approximation increases as $n \to \infty$.

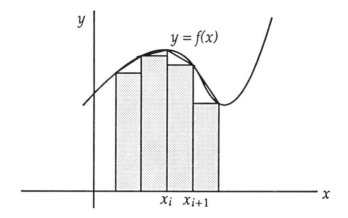

FIGURE V.22. Trapezoidal approximation to a definite integral

In general, it is better to align the strips so that top center point intersects the graph $y = f(x)$. This results in the formula:

(41)
$$A \approx \sum_{i=0}^{n-1} \frac{1}{n} f(a + i(b - a)/n + 1/2n)$$

We can get a better approximation by using geometric figures that are rectangles that lie entirely below the curve $y = f(x)$ and are capped with triangles — see figure V.22, where $x_i = a + (b - a)i/n$. This is called the *trapezoidal approximation* to the definite integral. That this is better than the original rectangular approximation is visually apparent — there is less space between the curve $y = f(x)$ and the top boundary of the approximating area. We will not give any other proof of its superiority here. The geometric figure that looks like a rectangle capped with a triangle is called a trapezoid, and the area of a trapezoid whose base is $r$ and whose sides are $s_1$ and $s_2$ is

$$\frac{r(s_1 + s_2)}{2}$$

The area of the trapezoid from $a + i(b - a)/n$ to $a + (i + 1)(b - a)/n$ is, thus:

$$(x_{i+1} - x_i) \left( \frac{1}{2} f(x_i) + \frac{1}{2} f(x_{i+1}) \right)$$

If we add up the areas of all of the trapezoidal strips under the graph from $a$ to $b$, we get

$$
\begin{aligned}
A \approx & \sum_{i=0}^{n-1} \left( (a + \frac{(i+1)(b-a)}{n}) - (a + \frac{i(b-a)}{n}) \right) \\
& \cdot \left( \frac{1}{2} f(a + \frac{i(b-a)}{n}) + \frac{1}{2} f(a + \frac{(i+1)(b-a)}{n}) \right) \\
= & \frac{b-a}{n} \sum_{i=0}^{n-1} \left( \frac{1}{2} f(a + \frac{i(b-a)}{n}) + \frac{1}{2} f(a + \frac{(i+1)(b-a)}{n}) \right) \\
= & \frac{b-a}{2n} f(a) + \frac{b-a}{2n} f(b) + \frac{b-a}{n} \sum_{i=1}^{n-1} f \left( a + \frac{i(b-a)}{n} \right)
\end{aligned}
$$

(42)

Notice that this formula looks very much like equation (40) on page 245, except for the fact that the values at the endpoints are divided by 2. This equation is called the *Trapezoidal Rule* for computing the definite integral. It is very easy to program on a SIMD computer. Here is a C* program:

```
/* This is a C* program for Trapezoidal Integration.  We assume given
 * a function 'f' that computes the function to be integrated.
 * The number 'n' will be equal to 8192 here. */

#include <stdio.h>
#include <math.h>

shape [8192]linear;
int N;


double:current f(double:current);  /* Header for a function to be integrated */

double:linear addval; /* The values that are actually added to the integral. */
double:linear xval;   /* The values of the x-coordinate.                     */
double a,b,intval;    /* The end-points of the interval of integration    * /

void main()
{
  intval=0;
  N=8192;
  with(linear)
    xval=a+(b−a)*((double)pcoord(0))/(((double)N);
```

```
with(linear)
  where((pcoord(0)==0)|(pcoord(0)==N−1))
    {
     intval+=f(xval);
     intval/=2.;
    }
  else
       intval+=f(xval)/((double)N);
  /* 'intval' now contains the approximate value of the
   * int(f(x),x=a..b). */
}
```

The execution-time of this algorithm is $O(\lg n)$.

In order to improve accuracy further, we can approximate the function being integrated by *polynomials*. Given $k$ points $(x_1, y_1), \ldots, (x_n, y_n)$, we can find a unique degree-$k$ polynomial that passes through these $k$ points. Consider the polynomials

$$p_i(x) = \frac{(x - x_1)(x - x_2) \cdots (x - x_{i-1})(x - x_{i+1}) \cdots (x - x_k)}{(x_i - x_1)(x_i - x_2) \cdots (x_i - x_{i-1})(x_i - x_{i+1}) \cdots (x_i - x_k)}$$

This is a degree-$k$ polynomial with the property that:

$$p_i(x_j) = \begin{cases} 0 & \text{if } i \neq j \\ 1 & \text{if } i = j \end{cases}$$

It is not hard to see that

(43)                          $$p(x) = y_1 p_1(x) + \cdots + y_k p_k(x)$$

will pass though the $k$ points $(x_1, y_1), \ldots, (x_k, y_k)$. This technique of finding a function that fits a given set of data-points is known as Lagrange's Interpolation formula. We can develop integration algorithms that are based upon the idea of every sequence of $k$ successive points $(x_i, f(x_i)), \ldots, (x_{i+k-1}, f(x_{i+k-1}))$ by a polynomial of degree-$k - 1$ and then integrating this polynomial *exactly*.

In order to simplify this discussion somewhat, we will assume that $a = 0$ and $b = k - 1$. This transformation is *reasonable* because:

$$\int_a^b f(x)\, dx = \frac{(b - a)}{k - 1} \int_0^{k-1} f\left(a + \frac{(b - a)}{k - 1}u\right) du = \frac{(b - a)}{k - 1} \int_0^{k-1} \hat{f}(u)\, du$$

where $\hat{f}(u) = f\left(a + \dfrac{(b - a)}{k - 1}u\right)$.

(44)    $$\int_0^{k-1} \hat{f}(u)\, du = \hat{f}(0) \int_0^{k-1} Q_{0,k}(u)\, du + \cdots + \hat{f}(k - 1) \int_0^{k-1} Q_{k-1,k}(u)\, du$$

where

$$Q_{i,k}(x) = \frac{x(x-1)\cdots(x-(i-1))(x-(i+1))\cdots(x-(k-1))}{i(i-1)\cdots(1)(-1)\cdots(i-(k-1))}$$

We can compute the values of the integrals

$$A_j = \int_0^{k-1} Q_{j,k}(u)\,du$$

once and for all, and use the same values for integrations that have different values of $n$.

If we want to integrate from $a$ to $b$, we can transform this back to that range to get:

(45)
$$\int_a^b f(x)\,dx \approx \frac{b-a}{k-1} \cdot \sum_{i=0}^{k-1} f\left(a + \frac{(b-a)i}{n}\right) A_i$$

with the same values of the $\{A_j\}$ as before. This is known as the Newton-Cotes Integration algorithm with *degree-parameter* $k$ and $k$ *data-points*. It is interesting to note that, since Newton-Cotes Algorithm for a given value of $k$ approximates $f(x)$ by a polynomial of degree $k - 1$ and *exactly* integrates *these*, it follows that

> **If $f(x)$ is a polynomial of degree $\leq k - 1$, Newton-Cotes Algorithm (with degree-parameter $k$) computes the integral of $f(x)$ exactly.**

Now we will consider the case where we use $n$ data-points, where $n$ is some multiple of $k - 1$. We subdivide the range $[a, b]$ into $t = n/(k-1)$ subranges: $[a, a + (b - a)/t], [a + (b - a)/t, a + 2(b - a)/t], \ldots, [a + (t - 1)(b - a)/t, b]$:

$$\int_a^b f(x)\,dx = \int_a^{a+(b-a)/t} f(x)\,dx + \cdots + \int_{a+(t-1)(b-a)/t}^b f(x)\,dx$$

Our integration formula for Newton-Cotes integration with degree-parameter $k$ and $n$ data-points is:

$$\int_a^b f(x)\,dx \approx \frac{(b-a)}{n} \cdot \left\{ \sum_{i=0}^{k-1} f\left(a + \frac{(b-a)i}{n}\right) \cdot A_i \right.$$

$$+ \sum_{i=k-1}^{2(k-1)} f\left(a + \frac{(b-a)i}{n}\right) \cdot A_{i-(k-1)}$$

$$+ \cdots$$

$$\left. + \sum_{i=n-(k-1)}^{n} f\left(a + \frac{(b-a)i}{n}\right) \cdot A_{i-n+(k-1)} \right\}$$

We will conclude this section by discussing several integration algorithms that are based upon this general approach.

CLAIM 4.1. The Trapezoidal Rule is a special case of Newton-Cotes algorithm with $k = 2$.

If $k = 2$, we have the two coefficients $A_0$ and $A_1$ to compute:

$$A_0 = \int_0^1 Q_{0,2}(x)\,dx = \int_0^1 \frac{x-1}{-1}\,dx = \frac{1}{2}$$

and

$$A_1 = \int_0^1 Q_{1,2}(x)\,dx = \int_0^1 \frac{x}{+1}\,dx = \frac{1}{2}$$

and it is not hard to see that equation (45) coincides with that in the Trapezoidal Rule.

If $k = 3$ we get an algorithm called Simpson's Rule:

Here

$$A_0 = \int_0^2 Q_{0,3}(x)\,dx = \int_0^2 \frac{(x-1)(x-2)}{(-1)(-2)}\,dx = \frac{1}{3}$$

$$A_1 = \int_0^2 Q_{1,3}(x)\,dx = \int_0^2 \frac{(x)(x-2)}{(1)(-1)}\,dx = \frac{4}{3}$$

$$A_2 = \int_0^2 Q_{2,3}(x)\,dx = \int_0^2 \frac{(x)(x-1)}{(2)(1)}\,dx = \frac{1}{3}$$

and our algorithm for the integral is:

ALGORITHM 4.2. Let $f(x)$ be a function, let $a < b$ be numbers, and let $n$ be an even integer $> 2$. Then Simpson's Rule for approximately computing the integral of $f(x)$ over the range $[a, b]$ with $n$ data-points, is

(47)

$$\int_a^b f(x)\,dx \approx \frac{b-a}{3n} \cdot \left\{ f(a) + f(b) + \sum_{i=1}^{n-1} f(a + (b-a)i/n) \cdot \left\{ \begin{array}{ll} 2 & \text{if } i \text{ is even} \\ 4 & \text{if } i \text{ is odd} \end{array} \right\} \right\}$$

If we let $k = 4$ we get another integration algorithm, called Simpson's 3/8 Rule. Here

$$A_0 = \frac{3}{8}$$

$$A_1 = \frac{9}{8}$$

$$A_2 = \frac{9}{8}$$

$$A_3 = \frac{3}{8}$$

and the algorithm is:

ALGORITHM 4.3. Let $f(x)$ be a function, let $a < b$ be numbers, and let $n$ be a multiple of 3. Then Simpson's 3/8-Rule for approximately computing the integral of $f(x)$ over the range $[a, b]$ with $n$ data-points, is

$$\int_a^b f(x)\, dx \approx \frac{3(b-a)}{8n} \cdot \left\{ f(a) + f(b) \right.$$

$$\left. + \sum_{i=1}^{n-1} f(a + (b-a)i/n) \cdot \begin{cases} 2 & \text{if } i \bmod 3 = 0 \\ 3 & \text{otherwise} \end{cases} \right\}$$

The reader can find many other instances of the Newton-Cotes Integration Algorithm in [1].

**4.2. Higher-dimensional integrals.** The methods of the previous section can easily be extended to multi-dimensional integrals. We will only discuss the simplest form of such an extension — the case where we simply integrate over each coordinate via a one-dimensional method. Suppose we have two one-dimensional integration formulas of the form

$$\int_a^b g(x)\, dx \approx \frac{b-a}{n} \sum_{i=0}^n B_i g\left(a + \frac{i(b-a)}{n}\right)$$

$$\int_c^d g(y)\, dy \approx \frac{d-c}{m} \sum_{j=0}^m B_i' g\left(c + \frac{i(d-c)}{m}\right)$$

where $n$ is some fixed large number — for instance, in Simpson's Rule $B_0 = B_n = 1/3$, $B_{2k} = 4$, $B_{2k+1} = 2$, where $1 \leq k < n/2$. Given this integration algorithm, we get the following two dimensional integration algorithm:

ALGORITHM 4.4. Let $f(x, y)$ be a function of two variables and let $R$ be a rectangular region with $a \leq x \leq b, c \leq y \leq d$. Then the following equation gives an approximation to the integral of $f(x, y)$ over $R$:

(49)

$$\iint_R f(x, y)\, dx\, dy \approx \frac{(b-a)(d-c)}{nm} \sum_{i=0}^m \sum_{j=0}^n B_i B_j' f\left(a + \frac{i(b-a)}{n}, c + \frac{j(d-c)}{m}\right)$$

Here, we are assuming that the two one dimensional-integrations have different numbers of data-points, $n$, and $m$.

Here is a sample C* program for computing the integral

$$\iint_R \frac{1}{\sqrt{1 + x^2 + y^2}}\, dx\, dy$$

where $R$ is the rectangle with $1 \leq x \leq 2$ and $3 \leq y \leq 4$. We use Simpson's 3/8-rule:

```
#include <stdio.h>
#include <math.h>
int NX=63;        /* This is equal to n in the formula for
                   * a two—dimensional integral. */
int NY=126;       /* This is equal to m in the formula for
                   * a two—dimensional integral. */
shape [64][128]twodim;
shape [8192]linear;

double:linear B,Bpr;
    /* B corresponds to the coefficients B_i */
    /* Bpr corresponds to the coefficients B'_i */

double a,b,c,d;

/* This function computes the function to be integrated. */
double:twodim funct(double:twodim x,
        double:twodim y)
{
    return (1./sqrt(1.+x*x+y*y));
}



double:twodim coef,x,y;
double intval;
void main()
{
    a=1.;
    b=2.;
    c=3.;
    d=4.;

    with(twodim) coef=0.;

/* Compute the coefficients in the boundary cases.
 This is necessary because the formula for
 Simpson's 3/8—Rule has special values for the
 coefficients in the high and low end of the range */

    [0]B=3.*(b-a)/(8.*(double)(NX));
```

```
[0]Bpr=3.*(d−c)/(8.*(double)(NY));
[NX−1]B=3.*(b−a)/(8.*(double)(NX));
[NY−1]Bpr=3.*(d−c)/(8.*(double)(NY));
```

/* Compute the coefficients in the remaining cases. */
```
with(linear)
  where(pcoord(0)>0)
    {
      where((pcoord(0) % 3)==0)
        {
          B=6.*(b−a)/(8.*(double)(NX));
          Bpr=6.*(d−c)/(8.*(double)(NY));
        }
      else
        {
          B=9.*(b−a)/(8.*(double)(NX));
          Bpr=9.*(d−c)/(8.*(double)(NY));
        };
    };
  with(twodim)
    where((pcoord(0)<NX)|(pcoord(1)<NY))
      {

        /* Compute the x and y coordinates. */
        x=(b−a)*((double)pcoord(0))/((double)NX);
        y=(d−c)*((double)pcoord(1))/((double)NY);

        /* Evaluate the integral. */
        intval+=[pcoord(0)]B*[pcoord(1)]Bpr*funct(x,y);
      };
    printf("The integral is %g",intval);
}
```

**4.3. Discussion and Further reading.** There are several issues we haven't addressed. In the one-dimensional case there are the *Gaussian*-type of integration formulas in which we form a sum like

$$\int_a^b f(x)\,dx \approx \frac{b-a}{n} \sum_{i=0}^{n} C_i f(x_i)$$

where the $\{x_i\}$ are *not* equally-spaced numbers. These algorithms have the property that, like the Newton-Cotes algorithm, the Gauss algorithm with parameter $k$ computes the integral exactly when $f(x)$ is a polynomial of degree $\leq k - 1$. They have the *additional* property that

The Gauss algorithm with degree-parameter $k$ computes the integral of $f(x)$ with *minimal* error, when $f(x)$ is a polynomial of degree $\leq 2k - 1$.

There are several variations of the Gauss integration algorithms:

1. Gauss-Laguerre integration formulas. These compute integrals of the form

$$\int_0^\infty e^{-x} f(x)\, dx$$

   and are optimal for $f(x)$ a polynomial.

2. Gauss-Hermite integration formulas. They compute integrals of the form

$$\int_{-\infty}^\infty e^{-x^2} f(x)\, dx$$

3. Gauss-Chebyshev (Gauss-Чебышев) integration formulas. These are for integrals of the form

$$\int_{-1}^1 \frac{f(x)}{\sqrt{1 - x^2}}\, dx$$

   See [42] for more information on these algorithms. The book of tables, [1], has formulas for all of these integration algorithms for many values of the degree-parameter.

4. Wavelet integration formulas. In [146], Sweldens and Piessens give formulas for approximately computing integrals like

$$B_{r,j} = 2^r \int_{-\infty}^\infty \phi(2^r x - j) f(x)\, dx$$

   which occur in wavelet expansions of functions — see 3.4 on page 229. For the Daubechies wavelet $W_4(x)$ the simplest form of their algorithm gives:

$$B_{r,j} \approx f\left( \frac{2j + 3 - \sqrt{3}}{2^{r+1}} \right)$$

   (the accuracy of this formula increases as $r$ increases.

In multi-dimensional integration we have only considered the case where the region of integration is rectangular. In many cases it is possible to *transform* non-rectangular regions of integration into the rectangular case by a suitable change of coordinates. For instance, if we express the coordinates $x_i$ in terms of another coordinate system $x_i'$ we can transforms integrals over the $x_i$ into integrals over the $x_i'$ via:

$$\int_D f(x_1, \ldots, x_t)\, dx_1 \ldots dx_t = \int_{D'} f(x_1, \ldots, x_t) \det(J)\, dx_1' \ldots dx_t'$$

where $\det(J)$ is the determinant (defined in 1.5 on page 159) of the $t \times t$ matrix $J$ defined by

$$J_{i,j} = \frac{\partial x_i}{\partial x'_j}$$

For example, if we want to integrate over a disk, $D$, of radius 5 centered at the origin, we transform to polar coordinates — $x = r\cos(\theta)$, $y = r\sin(\theta)$, and

$$J = \begin{pmatrix} \dfrac{\partial x}{\partial r} & \dfrac{\partial x}{\partial \theta} \\ \dfrac{\partial y}{\partial r} & \dfrac{\partial y}{\partial \theta} \end{pmatrix} = \begin{pmatrix} \cos(\theta) & -r\sin(\theta) \\ \sin(\theta) & r\cos(\theta) \end{pmatrix}$$

and $\det(J) = r\left(\sin^2(\theta) + \cos^2(\theta)\right) = r$. We get

$$\iint_D f(x,y)\,dx\,dy = \int_0^5 \int_0^{2\pi} f(r\cos(\theta), r\sin(\theta)) \cdot r\,dr\,d\theta$$

We can easily integrate this by numerical methods because the new region of integration is rectangular: $0 \le r \le 5$, $0 \le \theta \le 2\pi$. Volume I of [116] lists 11 coordinate systems that can be used for these types of transformations. High-dimensional numerical integration is often better done using *Monte Carlo* methods — see § 1.1.1 on page 442 for a description of Monte Carlo methods.

**Exercises:**

1. Write a C* program that implements Simpson's 3/8-Rule (using equation (48) on page 251). How many processors can be used effectively? Use this program to compute the integral:

$$\int_0^4 \frac{1}{\sqrt{1 + |2 - x^3|}}\,dx$$

2. Derive the formula for Newton-Cotes Algorithm in the case that $k = 5$.
3. Numerically evaluate the integral

$$\iint_D \sin^2(x) + \sin^2(y)\,dx\,dy$$

where $D$ is the disk of radius 2 centered at the origin.

4. The following integral is called the *elliptic integral of the first kind*

$$\int_0^1 \frac{dx}{\sqrt{(1-x^2)(1-k^2x^2)}}$$

Write a C* program to evaluate this integral for $k = 1/2$. The correct value is 1.6857503548 . . . .

## 5. Partial Differential Equations

We conclude this chapter with algorithms for solving partial differential equations. We will focus on second-order partial differential equations for several reasons:

1. Most of the very basic partial differential equations that arise in physics are of the second-order;
2. These equations are sufficiently complex that they illustrate many important concepts;

We will consider three broad categories of these equations:

- Elliptic equations.
- Parabolic equations.
- Hyperbolic equations.

Each of these categories has certain distinctive properties that can be used to solve it numerically.

In every case we will replace the partial derivatives in the differential equations by *finite differences*. Suppose $f$ is a function of several variables $x_1, \ldots, x_n$. Recall the definition of $\partial f / \partial x_i$:

$$\frac{\partial f(x_1, \ldots, x_n)}{\partial x_i} = \lim_{\delta \to 0} \frac{f(x_1, \ldots, x_i + \delta, \ldots, x_n) - f(x_1, \ldots, x_n)}{\delta}$$

The *simplest* finite-difference approximation to a partial derivative involves picking a small nonzero value of $\delta$ and replacing all of the partial derivatives in the equations by finite differences. We can solve the finite-difference equations on a computer and hope that the error in replacing differential equations by difference equations is not too great.

We initially get

$$\frac{\partial^2 f(x_1, \ldots, x_n)}{\partial x_i^2}$$
$$\approx \frac{f(x_1, \ldots, x_i + 2\delta, \ldots, x_n) - 2f(x_1, \ldots, x_i + \delta, \ldots, x_n) + f(x_1, \ldots, x_n)}{\delta^2}$$

In general we will want to use the formula

$$\frac{\partial^2 f(x_1, \ldots, x_n)}{\partial x_i^2}$$
$$\approx \frac{f(x_1, \ldots, x_i + \delta, \ldots, x_n) - 2f(x_1, \ldots, x_n) + f(x_1, \ldots, x_i - \delta, \ldots, x_n)}{\delta^2}$$

—this is also an approximate formula for the second-partial derivative in the sense that it approaches $\partial^2 f(x_1, \ldots, x_n)/\partial x_i^2$ as $\delta \to 0$. We prefer it because it is symmetric about $x$.

All of the differential equations we will study have an expression of the form

$$\nabla^2 \psi = \frac{\partial^2 \psi}{\partial x_1^2} + \frac{\partial^2 \psi}{\partial x_2^2} + \cdots + \frac{\partial^2 \psi}{\partial x_n^2}$$

in the $n$-dimensional case. We will express this in terms of finite differences. Plugging the finite-difference expression for $\partial^2 f(x_1, \ldots, x_n)/\partial x_i^2$ into this gives:

(51) $\nabla^2 \psi =$
$$\dfrac{\begin{array}{c}\psi(x_1 + \delta, \ldots, x_n) + \psi(x_1 - \delta, x_2, \ldots, x_n) + \cdots + \psi(x_1, \ldots, x_n + \delta) \\ + \psi(x_1, \ldots, x_n - \delta) - 2n\psi(x_1, \ldots, x_n)\end{array}}{\delta^2}$$

This has an interesting interpretation: notice that

(52) $\psi_{\text{average}} =$
$$\dfrac{\begin{array}{c}\psi(x_1 + \delta, \ldots, x_n) + \psi(x_1 - \delta, x_2, \ldots, x_n) + \cdots \\ + \psi(x_1, \ldots, x_n + \delta) + \psi(x_1, \ldots, x_n - \delta)\end{array}}{2n}$$

can be regarded as the *average* of the function-values $\psi(x_1 + \delta, \ldots, x_n)$, $\psi(x_1 - \delta, x_2, \ldots, x_n)$, .... Consequently, $\nabla^2 \psi/2n$ can be regarded as the difference between the *value* of $\psi$ at a point and the *average* of the values of $\psi$ at *neighboring* points.

## 5.1. Elliptic Differential Equations.

### 5.1.1. *Basic concepts.* 
Elliptic partial differential equations are equations of the form

(53) $$A_1 \frac{\partial^2 \psi}{\partial x_1^2} + \cdots + A_n \frac{\partial^2 \psi}{\partial x_2^2} + B_1 \frac{\partial \psi}{\partial x_1} + \cdots + B_n \frac{\partial \psi}{\partial x_n} + C\psi = 0$$

where $A_1, \ldots, A_n, B_1, \ldots, B_n$, and $C$ are functions of the $x_1, \ldots, x_n$ that all satisfy the conditions $A_i \geq m$, $B_i \geq M$, for some positive constants $m$ and $M$, and $C \leq 0$. It turns out that, when we attempt the numerical solution of these equations, it will be advantageous for them to be in their so-called *self-adjoint* form. The general *self-adjoint* elliptic partial differential equation is

(54) $$\frac{\partial}{\partial x_1}\left(A_1' \frac{\partial \psi}{\partial x_1}\right) + \cdots + \frac{\partial}{\partial x_n}\left(A_n' \frac{\partial \psi}{\partial x_n}\right) + C'\psi = 0$$

where $A_1', \ldots, A_n'$, and $C'$ are functions of the $x_1, \ldots, x_n$ that all satisfy the conditions $A_i' \geq M$, for some positive constant $M$ and $C' \leq 0$. The numerical methods presented here are guaranteed to converge if the equation is self-adjoint. See § 5.1.2 on page 272 for a more detailed discussion of this issue.

Here is an elliptic differential equation called the Poisson Equation — it is fairly typical of such equations:

$$(55) \qquad \nabla^2 \psi(x_1, \ldots, x_n) = \sigma(x_1, \ldots, x_n)$$

where $\psi(x_1, \ldots, x_n)$ is the unknown function for which we are solving the equation, and $\sigma(x_1, \ldots, x_n)$ is some given function. This equation is clearly self-adjoint.

We will focus upon one elliptic partial differential equation that occurs in physics — it is also the *simplest* such equation that is possible. The following partial differential equation is called Laplace's equation for gravitational potential in empty space[24]:

$$(56) \qquad \nabla^2 \psi(x_1, \ldots, x_n) = 0$$

In this case $\psi(x, y, z)$ is gravitational potential. This is a quantity that can be used to compute the *force* of gravity by taking its single partial derivatives:

$$(57) \qquad F_x = -Gm\frac{\partial \psi}{\partial x}$$

$$(58) \qquad F_y = -Gm\frac{\partial \psi}{\partial y}$$

$$(59) \qquad F_z = -Gm\frac{\partial \psi}{\partial z}$$

where $G$ is Newton's Gravitational Constant ($= 6.673 \times 10^{-8} \mathrm{cm}^3/\mathrm{g}\ \mathrm{sec}^2$) and $m$ is the mass of the object being acted upon by the gravitational field (in grams).

Partial differential equations have an infinite number of solutions — we must select a solution that is relevant to the problem at hand by imposing *boundary conditions*. We assume that our unknown function $\psi$ satisfies the partial differential equation in some domain, and we *specify* the values it must take on at the *boundary* of that domain. Boundary conditions may take many forms:

1. If the domain of solution of the problem is finite (for our purposes, this means it is contained within some large but finite cube of the appropriate dimension), the boundary conditions might state that $\psi$(boundary) takes on specified values.
2. If the domain is infinite, we might require that $\psi(x, y, z) \to 0$ as $x^2 + y^2 + z^2 \to \infty$.

Here is an example:

---

[24]It can also be used for electrostatic potential of a stationary electric field.

EXAMPLE 5.1. Suppose that we have a infinitely flexible and elastic rubber sheet stretched over a rectangle in the $x$-$y$ plane, where the rectangle is given by

$$0 \leq x \leq 10$$
$$0 \leq y \leq 10$$

Also suppose that we push the rubber sheet up to a height of 5 units over the point $(1, 2)$ and push it down 3 units over the point $(7, 8)$. It turns out that the height of the rubber sheet over any point $(x, y)$ is a function $\psi(x, y)$ that satisfies Laplace's equation, except at the boundary points. We could compute this height-function by solving for $\psi$ where:

- $\psi$ satisfies the two-dimensional form of Laplace's equation

$$\frac{\partial^2 \psi}{\partial x^2} + \frac{\partial^2 \psi}{\partial y^2} = 0$$

  for

$$0 < x < 10$$
$$0 < y < 10$$
$$(x, y) \neq (1, 2)$$
$$(x, y) \neq (7, 8)$$

- (Boundary conditions). $\psi(0, y) = \psi(x, 0) = \psi(10, y) = \psi(x, 10) = 0$, and $\psi(1, 2) = 5$, $\psi(7, 8) = -3$.

There is an extensive theory of how to solve partial differential equations *analytically* when the domain has a geometrically simple shape (like a square or circle, for instance). This theory essentially breaks down for the example above because the domain is very irregular — it is a square *minus two points*.

If we plug the numeric approximation to $\nabla^2 \psi$ into this we get:

(60)
$$\frac{\psi(x + \delta, y, z) + \psi(x - \delta, y, z) + \psi(x, y + \delta, z) + \psi(x, y - \delta, z)}{\delta^2} \atop {+\psi(x, y, z + \delta) + \psi(x, y, z - \delta) - 6\psi(x, y, z)} = 0$$

We can easily rewrite this as:

(61)    $$\psi(x, y, z) = \frac{\begin{array}{c}\psi(x + \delta, y, z) + \psi(x - \delta, y, z) + \psi(x, y + \delta, z) \\ +\psi(x, y - \delta, z) + \psi(x, y, z + \delta) + \psi(x, y, z - \delta)\end{array}}{6}$$

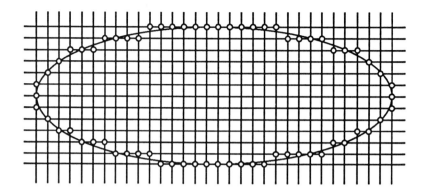

FIGURE V.23. Digitized form of oval-shaped region

This essentially states that the value of $\psi$ at any point is equal to the average of its values at certain neighboring points[25]. This implies that a solution to this equation cannot have a maximum or minimum in the region where the equation is satisfied (we usually assume that it *isn't* satisfied on the boundary of the region being considered, or at some selected points of that region). In any case we will use this form of the numerical equation to derive an algorithm for solving it. The algorithm amounts to:

1. Overlay the region in question with a rectangular grid whose *mesh size* is equal to the number $\delta$. This also requires "digitizing" the boundary of the original region to accommodate the grid. We will try to compute the values of the function $\psi$ at the grid points.
2. Set $\psi$ to zero on all grid points in the *interior* of the region in question, and to the assigned boundary values on the grid points at the boundary (and, possibly at some interior points). See figure V.23.
   Here the "digitized" boundary points have been marked with small circles.
3. Solve the resulting finite-difference equations for the values of $\psi$ on the grid-points.

This procedure *works* because:

The finite-difference equation is nothing but a system of *linear equations* for $\psi(x, y, z)$ (for values of $x$, $y$, and $z$ that are integral multiples of $\delta$). This system is very large — if our original grid was $n \times m$, we now have $nm - b$ equations in $nm - b$ unknowns, where $b$ is the number of boundary points (they are not unknowns). For instance, if our digitized region is a $4 \times 5$ rectangular region, with a total of 14 boundary points, we number the 6 interior points in an arbitrary way, as depicted in figure V.24. This

---

[25]This also gives a vague intuitive justification for the term "elliptic equation" — the value of $\psi$ at any point is determined by its values on an "infinitesimal sphere" (or ellipsoid) around it.

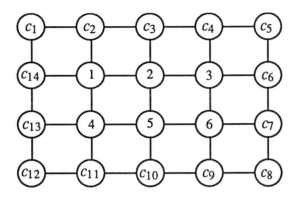

FIGURE V.24. A digitized region with interior points numbered

results in the system of linear equations:

$$
(62)\quad
\begin{pmatrix}
1 & -1/4 & 0 & -1/4 & 0 & 0 \\
-1/4 & 1 & 0 & 0 & -1/4 & 0 \\
0 & -1/4 & 1 & 0 & 0 & -1/4 \\
-1/4 & 0 & 0 & 1 & -1/4 & 0 \\
0 & -1/4 & 0 & -1/4 & 1 & -1/4 \\
0 & 0 & -1/4 & 0 & -1/4 & 1
\end{pmatrix}
\cdot
\begin{pmatrix}
\psi_1 \\ \psi_2 \\ \psi_3 \\ \psi_4 \\ \psi_5 \\ \psi_6
\end{pmatrix}
=
\begin{pmatrix}
c_2/4 + c_{14}/4 \\
c_3/4 \\
c_4/4 + c_6/4 \\
c_{11}/4 + c_{13}/4 \\
c_{10}/4 \\
c_7/4 + c_9/4
\end{pmatrix}
$$

Here the $c$'s are the values of $\psi$ on the boundary of the region. This example was a small "toy" problem. In a real problem we must make $\delta$ small enough that the finite differences are a reasonable approximation to the original differential equation. This makes the resulting linear equations very large. For instance in example 5.1 on page 258 above, if we make $\delta = .1$, we will almost get 10000 equations in 10000 unknowns. It is not practical to solve such equations by the usual numerical methods (like Gaussian elimination, for instance). We must generally use *iterative methods* like those discussed in § 1.2.2 (on page 167). We will consider how to set up the Jacobi method to solve the numerical form of Laplace's equation.

Suppose we have a finite-difference form of Laplace's equation defined on a grid, as described above. The function $\psi$ is now a *vector* over all of the grid-points — for instance in example 5.1 on page 258 with $\delta = .1$, $\psi$ has 10000 components. Let $A_{\text{average}}$ be the matrix such that

$$A_{\text{average}} \cdot \psi(\text{at a grid-point})$$

is equal to the average of the values of $\psi$ at the four adjacent grid-points. We will not have to write down what this huge array *is* explicitly (equation (62) on page 261 gives it in a simple case). We only need to know that each row of $A_{\text{average}}$ has at most 4 non-zero entries, each equal to $1/4$. Laplace's equation says that

$$\nabla^2\psi \approx \frac{4}{\delta^2}\left(A_{\text{average}} \cdot \psi - \psi\right) = 0$$

or
$$A_{\text{average}} \cdot \psi - \psi = 0$$
or
$$I \cdot \psi - A_{\text{average}} \cdot \psi = 0$$
(where $I$ is the identity matrix) or
$$A \cdot \psi = 0$$
where $A = I - A_{\text{average}}$. Now we apply the Jacobi method, described in § 1.2.2 on page 167. The array of diagonal elements, $D(A) = I$, so $D(A)^{-1} = I$, $Z(A) = I - D(A)^{-1}A = A_{\text{average}}$ and the Jacobi iteration scheme amounts to
$$\psi^{(n+1)} = A_{\text{average}} \cdot \psi^{(n)}$$

Our numerical algorithm is, therefore,

ALGORITHM 5.2. The $n$-dimensional Laplace equation can be solved by performing the steps: Digitize the region of solution, as in figure V.24 on page 261 (using an $n$-dimensional lattice).

$\psi \leftarrow 0$ at all grid-points
    (of the digitized region of solution),
    except at boundary points for which $\psi$
    must take on other values
    (for instance points like $(1, 2)$ and $(7, 8)$
    in example/ 5.1 on page 258.)

for $i \leftarrow 1$ until $\psi$ doesn't change appreciably
    do in parallel
        if grid-point $p$ is not a boundary point
        $\psi(p) \leftarrow$ average of values of $\psi$ at $2n$ adjacent
          grid-points.

Our criterion for halting the iteration is somewhat vague. We will come back to this later. Essentially, we can just test whether $\psi$ has changed over a given iteration and halt the iteration when the total change in $\psi$ is sufficiently small. The program on page 263 totals up the absolute value of the change in $\psi$ in each iteration and halts when this total is $\leq 2$.

It is possible to prove that when the original partial differential equation was *self-adjoint*, or if the quantity $\delta$ is *sufficiently small*, this iteration-procedure converges[26] to a solution of finite-difference equations — and an *approximate* solution to the original

---

[26]The question of such convergence was discussed in 1.21 on page 168. In § 5.1.3 on page 275 we explore the rate of convergence of this iteration-scheme. The self-adjointness of the original partial differential equation implies that the criteria in that result are satisfied.

partial differential equation (at least *at* the grid-points). This is known as a *relaxation method* for solving the set of finite-difference equations. The difference-equations that arise as a result of the procedure above turn out to have matrices that are dominated by their main diagonal (a necessary condition for the Jacobi method to converge). The *molecule-structure* depicted above turns out to be very significant in finding consistent-ordering schemes for speeding up the Jacobi iteration technique for solving these finite-difference equations.

The term "relaxation method" came from a *vaguely* similar problem of computing the shape of a membrane hanging from its boundary under its own weight. The solution was found by first assuming a upward force that opposes gravity and makes the membrane flat. This restoring force is then "relaxed", one grid-point at a time until the membrane hangs freely.

In many cases the convergence of this procedure is fairly slow — it depends upon the *spectral radius* of the matrix of the system, when it is written in a form like equation (62) — see 1.22 on page 169.

We will present a C* program for implementing the relaxation algorithm in the two-dimensional case. The region under consideration consists of the rectangle $-2 \leq x \leq +2, -4 \leq y \leq +4$. We assign one processor to each grid-point. The size of the mesh, $\delta$, will be $1/16$ — this is dictated by the number of processors available to perform the computation. The smaller $\delta$ is the more accurate the solution. Each processor has float-variables x and y containing the coordinates of its grid point, float-variables psi, oldpsi and diff containing, respectively, the value of $\psi$ in the current iteration, the value in the last iteration, and the difference between them. There is also and int-variable isfixed that determines whether the iteration is to be performed on a given grid-point. In general, we will *not* perform the calculations on the boundary grid-points — at these points the value of $\psi$ is assigned and *fixed*.

When the program is run, it performs the computation that replaces $\psi$ by $\psi_{average}$ and computes the difference. This difference is totaled into the (mono) variable totaldiff and the computation is stopped when this total difference is less than some pre-determined amount (2 in this case):

```
#include <stdio.h>
shape [64][128]plane;
float:plane x, y, psi, oldpsi, diff;
int:plane isfixed;

void main()
{
    float       totaldiff = 1.0;
    int         iteration = 0;
```

```
int          i, j;
/* Initialize coordinates and psi. */
with(plane) {
    x = pcoord(0);  /* Tell each processor its row
                     * number. */
    y = pcoord(1);  /* Tell each processor its
                     * column number. */
    x = x / 16.0 − 2.0;    /* Compute the
                           * x−coordinate of a
                           * processor. */
    y = y / 16.0 − 4.0;    /* Compute the
                           * y−coordinate of a
                           * processor. */
    psi = 0.0;     /* Initialize the psi−function
                   * at 0. */
    diff = 0.0;    /* Initialize the
                   * difference−function at 0. */
    isfixed = 0;   /* Set all of the isfixed flags
                   * to false. */
    where(pcoord(0) == 0 | pcoord(0) == 64) {
        /* Initialize boundary conditions. */
        psi = x + y;
        /* Mark these points as fixed. */
        isfixed = 1;
    }
    where(pcoord(1) == 0 | pcoord(1) == 128) {
        /* Mark these boundary points as fixed. */
        isfixed = 1;
    }
}
with(plane) oldpsi = psi;      /* Save the previous
                               * values of the
                               * psi−function. */
while (totaldiff > 2.0) {
    iteration++;
    with(plane)
        where(isfixed == 0) {   /* Only update the
                                * values of psi at
                                * points that are not
                                * declared to be fixed
```

```
                                  * in value. */
            oldpsi = psi;   /* Save the values of
                              * the psi—function
                              * prior to the update. */
            psi =
                ([.+ 1][.] psi
                +[.− 1][.] psi
                +[.][.+ 1] psi
                +[.][.− 1] psi) / 4.0;
            diff = psi − oldpsi;   /* Calculate how much
                                    * the psi—function has
                                    * changed. */
            totaldiff = 0.0;       /* Set 'totaldiff' to 0
                                    * prior to adding in
                                    * the squares of the
                                    * values of 'diff'. */
            totaldiff += fabs(diff);
            printf("Iteration %d, Difference = %g\n",
                iteration, totaldiff);
        }
    }
    for (i = 0; i < 16; i++) {
        for (j = 0; j < 10; j++)
            printf("%g ",[i][j] psi);
        printf("\n");
    }
    printf("%d Iterations were required\n", iteration);
}
```

We will want to speed up this iteration-scheme. We can do so by using the SOR or consistently-ordered methods described in § 1.2.4 on page 173. Our iteration formula is now:

$$\psi^{(n+1)} = (1 - \mu) \cdot \psi^{(n)} + \mu \cdot A_{\text{average}} \cdot \psi^{(n)}$$

where $\mu$ is a suitable *over-relaxation coefficient*. We will delay considering the question of what value we should use for $\mu$ for the time being. Recall that the SOR method *per se* is a sequential algorithm. We must find a consistent-ordering scheme in order to parallelize it.

At each *grid-point* the two-dimensional form of equation (61) relates each grid-point with its four neighbors. We express this fact by saying the *molecule* of the finite-difference equations is what is depicted in figure V.25:

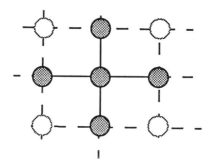

FIGURE V.25. Molecule of the first-order approximation to elliptic differential equations

The interpretation of consistent-ordering schemes in terms of graph-coloring in 1.31 on page 177 is very significant here. Our main result is:

PROPOSITION 5.3. Given finite-difference equations representing elliptic differential equations, with the molecule shown in figure V.25, there exists a consistent-ordering with only *two* sets.

These two sets turn out to be

1. $S_1$ = grid-points for which the sum of indices is *even*; and
2. $S_2$ = grid-points for which the sum of indices is *odd*.

PROOF. Suppose that our digitized region of solution is an $n \times m$ grid, as discussed above, and we have $b$ boundary points (at which the values of $\psi$ are *known*. The fact that the *molecule* of the finite-difference equations is what is depicted in V.25 implies that *however* we map the digitized grid into linear equations, two *indices* (i.e., rows of the matrix of the linear equations) will be *associated* in the sense of statement 1 of 1.27 on page 174, if and only if they are neighbors in the original grid. We want to find a *consistent ordering vector* in the sense of statement 3 of that definition. Recall that such an ordering vector $\gamma$ must satisfy:

1. $\gamma_i - \gamma_j = 1$ if $i$ and $j$ are associated and $i > j$;
2. $\gamma_i - \gamma_j = -1$ if $i$ and $j$ are associated and $i < j$;

Let us map the grid-points of our digitized region into the array in the following way.

- Map all grid-points with the property that the sum of the coordinates is *even* into the lower half of the dimensions of the array.
- Map all grid-points with the property that the sum of the coordinates is *odd* into the upper half of the dimensions of the array.

Figure V.26 shows this in the case where $n = 5$ and $m = 7$.

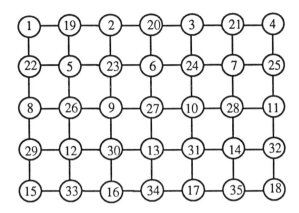

FIGURE V.26. Numbering the grid-points for the odd-even ordering vector

The numbers in circles are the row-number of the that grid-point in the matrix that represent the difference equations. If we go back to our original *grid*, we see that each grid point is *only associated* (in the sense of statement 1 of 1.27 on page 174) with its *four neighbors*. This implies that the following is a valid consistent-ordering vector for our problem:

$$\gamma_i = \begin{cases} 1 & \text{if the sum of grid-coordinates is } even \\ 2 & \text{if the sum of grid-coordinates is } odd \end{cases}$$

□

We can now apply 1.28 on page 175 to solve this as a consistently-ordered system of linear equations. We essentially perform the update-computations on the processors whose coordinates have the property that their sum is *even* (these turn out to be the elements for which the order-vector is equal to 1) and then (in a separate phase) on the processors such that the sum of the coordinates is *odd* — these represent the elements of the array with an ordering vector[27] that is equal to 2.

There is an extensive theory on how one determines the best value of $\mu$ to use. The Ostrowski-Reich Theorem and the theorem of Kahan states that this overrelaxation parameter must satisfy the inequality $0 < \mu < 2$, in order for this SOR-algorithm to converge — see 1.26 on page 174. For the Laplace Equation over a region $\Omega$ with the kind of boundary conditions we have discussed (in which values of $\psi$ are specified on the boundary), Garabedian (see [54]) has given a formula that *estimates* the optimal value to use for $\mu$:

(63)
$$\mu \approx \frac{2}{1 + \pi\delta/\sqrt{A(\Omega)}}$$

---

[27]See statement 3 on page 175 of 1.27 for a definition of this term.

where $A(\Omega)$ is the area of the region $\Omega$. Compare this formula of Garabedian with the formula in 1.29 on page 176. Compare this to equation (75) on page 281 — in that section we derive estimates of the rate of convergence of the Jacobi method. The second statement of 1.29 implies that the spectral radius of the effective matrix used in the SOR method is $\mu - 1$ or

$$\rho(\mathcal{L}_\mu) = \frac{\sqrt{A(\Omega)} - \pi\delta}{\sqrt{A(\Omega)} + \pi\delta}$$

The *rate of convergence* is

(64)
$$-\ln(\rho(\mathcal{L}_\mu)) = \frac{2\pi\delta}{\sqrt{A(\Omega)}} + O(\delta^3)$$

This implies that the execution time of the algorithm (in order to achieve a given degree of accuracy) is $O(1/\delta)$.

In the case of the problem considered in the sample program above, we get $\mu = 1.932908695\ldots$

```
#include <stdio.h>
shape [64][128]plane;
float        lambda = 1.0;
float:plane x, y, psi, oldpsi, diff;
int:plane isfixed;
void update()
{
    psi =
    lambda * ([.+ 1][.] psi
            +[.- 1][.] psi
            +[.][.+ 1] psi
            +[.][.- 1] psi) / 4.0
    + (1.0 - lambda) * oldpsi;
}
void main()
{
    float        totaldiff = 1.0;
    int          iteration = 0;
    int          i, j;
    printf("Enter the over-relaxation coefficient:\n");
    scanf("%g", &lambda);
    /* Initialize coordinates and psi. */
```

```
with(plane) {
    x = pcoord(0);  /* Tell each processor its row
                * number. */
    y = pcoord(1);  /* Tell each processor its
                * column number. */
    x = x / 16.0 − 2.0;   /* Compute the
                    * x−coordinate of a
                    * processor. */
    y = y / 16.0 − 4.0;   /* Compute the
                    * y−coordinate of a
                    * processor. */
    psi = 0.0;    /* Initialize the psi−function
                * at 0. */
    diff = 0.0;   /* Initialize the
                * difference−function at 0. */
    isfixed = 0;   /* Set all of the isfixed flags
                * to false. */
    where((pcoord(0) == 0 | pcoord(0) == 63) |
        ((pcoord(1) == 0) | (pcoord(1) == 127))) {
        /* Initialize boundary conditions. */
        psi = x * y;
        /* Mark these points as fixed. */
        isfixed = 1;
    }
}
with(plane) oldpsi = psi;     /* Save the previous
                    * values of the
                    * psi−function. */
while (totaldiff > 0.1) {
    iteration++;
    with(plane)
        where(isfixed == 0) {   /* Only update the
                        * values of psi at
                        * points that are not
                        * declared to be fixed
                        * in value. */
        oldpsi = psi;   /* Save the values of
                    * the psi−function
                    * prior to the update. */
        /*
```

```
        * Perform the updates on all even
        * grid-points.
        */
        where(((pcoord(0) + pcoord(1)) % 2) == 0)
            update();
        else
        /*
        * Now perform the updates on odd
        * grid-points.
        */
        update();
        diff = psi - oldpsi;/* Calculate how much
                        * the psi-function has
                        * changed. */
        totaldiff = 0.0;      /* Set 'totaldiff' to 0
                        * prior to adding in
                        * the squares of the
                        * values of 'diff'. */
        totaldiff += fabs(diff);
        printf("Iteration %d, Difference = %g\n",
            iteration, totaldiff);
        }
    }
    for (i = 0; i < 16; i++) {
        for (j = 0; j < 10; j++)
            printf("%g ",[i][j] psi);
        printf("\n");
    }
    printf("%d Iterations were required\n", iteration);
    with(plane) {
        diff = psi - x * y;
        totaldiff = 0.0;
        totaldiff += diff;
    }
    printf("Total error=%g\n", totaldiff);
}
```

This program simply prints out some of the data-points produced by the program. One can use various graphics-packages to generate better displays. Figure V.27 shows the result of using matlab to produce a 3 dimensional graph in the case where $\psi$ was

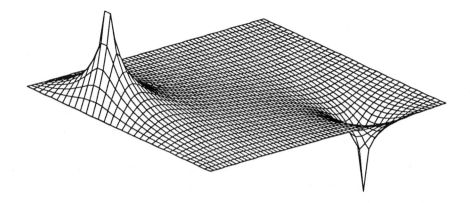

FIGURE V.27. Solution of Laplace's Equation

constrained on the boundary to 0 and set to a positive and negative value at some interior points.

The following routine was used to generate the data in a form for processing by matlab:

```
#include <stdio.h>
extern int    NROWS, NCOLS;
void graphout(FILE * outf, float:current data)
/*
 * Produces an output file in a format that is suitable for
 * input to 'matlab'. We assume global variables NROWS, NCOLS
 * and the definition of a shape shape [NROWS][NCOLS]plane;
 */
{
    int      i, j;
    int      rows = NROWS / 2;
    int      cols = NCOLS / 2;
    fprintf(outf, " [");
    for (i = 0; i < rows; i++) {
        for (j = 0; j < cols − 1; j++)
            fprintf(outf, "%g ",[2 * i][2 * j] data);
        fprintf(outf, "%g",[2 * i][2 * (cols − 1)] data);
        if (i < rows − 1)
            fprintf(outf, "; \n");
        else
            fprintf(outf, "] ");
    }
```

}

Here, the file *outf* must be declared in the file containing the main program that calls *graphout*.

The most efficient implementations of this algorithm involve so-called *multigrid methods*. These techniques exploit the fact that the algorithm converges quickly if the mesh-size is large, but is more accurate (in the sense of giving a solution to the original differential equation) if the mesh-size is small. In multigrid methods, the solution-process is begun with a large mesh-size, and the result is plugged into[28] a procedure with a smaller mesh-size. For instance, we could halve the mesh-size in each phase of a multigrid program. Each change of the mesh-size will generally necessitate a recomputation of the optimal overrelaxation coefficient — see equation (63) on page 267.

As has been mentioned before, even an *exact* solution to the difference equations will not generally be an exact solution to the *differential* equations. To get some idea of how accurate our approximation is, the previous program has boundary conditions for which the *exact solution* is known: $\psi(x, y) = x + y$. Other *exact* solutions of the Laplace equation are: $\psi(x, y) = xy$, $\psi(x, y) = x^2 - y^2$, and $\psi(x, y) = \log((x - a)^2 + (y - b)^2)$, where $a$ and $b$ are arbitrary constants. One important way to get solutions to Laplace's equation is to use the fact (from the theory of complex analysis) that the *real* and the *imaginary* parts of any analytic complex-valued function satisfy the two dimensional Laplace equation. In other words if $f(z)$ is a complex analytic function (like $e^z$ or $z^3$, for instance) and we write $f(x + iy) = u(x, y) + iv(x, y)$, then $u(x, y)$ and $v(x, y)$ each satisfy Laplace's equation.

If the accuracy of these solutions is not sufficient there are several possible steps to take:

1. Use a smaller mesh size. In the sample programs given we would have to increase the number of *processors* — the Connection Machine is currently limited to 64000 processors. It is also possible to use *virtual* processors, in which physical processors each emulate several processors. This latter step slows down the execution time.

2. Use a better finite-difference approximation of the partial derivatives.

5.1.2. *Self-adjoint equations.* Now we will consider how to convert many different elliptic differential equations into self-adjoint equations. Our solution-techniques in this case will be minor variations on the methods described in the previous section.

Throughout this section, we will assume given an elliptic partial differential equations:

---

[28]I.e., used as the initial approximation to a solution.

$$(65) \qquad A_1 \frac{\partial^2 \psi}{\partial x_1^2} + \cdots + A_n \frac{\partial^2 \psi}{\partial x_2^2} + B_1 \frac{\partial \psi}{\partial x_1} + \cdots + B_n \frac{\partial \psi}{\partial x_n} + C\psi = 0$$

where $A_1, \ldots, A_n$, $B_1, \ldots, B_n$, and $C$ are functions of the $x_1, \ldots, x_n$ that all satisfy the conditions $A_i \geq m$, $B_i \geq M$, for some positive constant $M$ and $C \leq 0$. Our integrating factor is a function $\Phi(x_1, \ldots, x_n)$, and we will write $\psi(x_1, \ldots, x_n) = \Phi(x_1, \ldots, x_n) \cdot u(x_1, \ldots, x_n)$.

This equation is *self-adjoint* if

$$\frac{\partial A_i}{\partial x_i} = B_i$$

for all $i = 1, \ldots, n$. If these conditions are *not* satisfied, we can *sometimes* transform the original equation into a self-adjoint one by multiplying the entire equation by a function $\Phi(x_1, \ldots, x_n)$, called an *integrating factor*. Elliptic differential equations that can be transformed in this way are called *essentially self adjoint*. If we multiply (53) by $\Phi$, we get:

$$(66) \qquad A_1 \Phi \frac{\partial^2 \psi}{\partial x_1^2} + \cdots + A_n \Phi \frac{\partial^2 \psi}{\partial x_2^2} + B_1 \Phi \frac{\partial \psi}{\partial x_1} + \cdots + B_n \Phi \frac{\partial \psi}{\partial x_n} + C\Phi\psi = 0$$

and *this* equation is self-adjoint if

$$\frac{\partial \Phi A_i}{\partial x_i} = \Phi B_i$$

It is straightforward to *compute* $\Phi$ — just set:

$$\frac{\partial \Phi A_i}{\partial x_i} = \Phi B_i$$
$$\frac{\partial \Phi}{\partial x_i} A_i + \Phi \frac{\partial A_i}{\partial x_i} = \Phi B_i$$

$$(67) \qquad \frac{\partial \Phi}{\partial x_i} A_i = \Phi \left\{ B_i - \frac{\partial A_i}{\partial x_i} \right\}$$

and temporarily "forget" that it is a *partial* differential equation. We get

$$\frac{\partial \Phi}{\Phi} = \partial x_i \frac{B_i - \dfrac{\partial A_i}{\partial x_i}}{A_i}$$

Now integrate both sides to get

$$\log \Phi = \int \frac{B_i - \dfrac{\partial A_i}{\partial x_i}}{A_i} \, dx_i + C(x_1, \ldots, x_{i-1}, x_{i+1}, \ldots, x_n)$$

Note that our "arbitrary constant" must be a *function* of the other variables, since our original equation was a *partial* differential equation. We get an equation for $\Phi$ that has an *unknown part*, $C(x_2, \ldots, x_n)$, and we plug this into equation (67) for $i = 2$ and solve for $C(x_2, \ldots, x_n)$. We continue this procedure until we have *completely determined* $\Phi$.

A function, $\Phi$, with these properties exists[29] if and only if the following conditions are satisfied:

$$(68) \qquad \frac{\partial}{\partial x_i}\left(\frac{B_j - \partial A_j/\partial x_j}{A_j}\right) = \frac{\partial}{\partial x_j}\left(\frac{B_i - \partial A_i/\partial x_i}{A_i}\right)$$

for all pairs of *distinct* indices $i, j = 1, \ldots, n$. If this condition is satisfied, the original partial differential equation is called *essentially self adjoint*, and can be re-written in the self-adjoint form

$$(69) \qquad \frac{\partial}{\partial x_1}\left(A_1\Phi\frac{\partial\psi}{\partial x_1}\right) + \cdots + \frac{\partial}{\partial x_n}\left(A_n\Phi\frac{\partial\psi}{\partial x_n}\right) + C\Phi\psi = 0$$

Our method for solving this equation is essentially the same as that used for the Laplace equation except that we use the approximations in equation (50) on page 256 for the partial derivatives. The self-adjointness of the original differential equation guarantees that the numerical approximations can be solved by iterative means.

### Exercises:

1. Run the program with boundary conditions $\psi(x, y) = x^2 - y^2$ and determine how accurate the entire algorithm is in this case. (This function is an exact solution to Laplace's Equation).

2. Determine, by experiment, the best value of $\mu$ to use. (Since the formula given above just *estimates* the best value of $\mu$).

3. Is the algorithm described in the program on page 268 EREW (see page 20)? Is it calibrated (see page 62)?

4. Is the partial differential equation

$$\frac{\partial^2\psi}{\partial x^2} + \frac{2}{x}\frac{\partial\psi}{\partial x} + \frac{\partial^2\psi}{\partial y^2} = 0$$

   self-adjoint? If not, how can it be put into a self-adjoint form?

5. Determine whether the following equation is essentially self-adjoint:

$$\frac{\partial^2\psi}{\partial x^2} + \frac{2}{x+y}\frac{\partial\psi}{\partial x} + \frac{\partial^2\psi}{\partial y^2} + \frac{2}{x+y}\frac{\partial\psi}{\partial y} = 0$$

---

[29]This is also a condition for our technique for computing $\Phi$ to be well-defined.

If it *is*, convert it into a self-adjoint form.

6. It is possible to approximate the Laplacian operator (equation (51) on page 257) by using the following approximation for a second partial derivative:

$$\frac{\partial^2 f(x)}{\partial x^2}$$
$$\approx \frac{35f(x-2\delta) - 104f(x-\delta) + 114f(x) - 56f(x+\delta) + 11f(x+2\delta)}{12\delta^2}$$

If $f(x)$ is a smooth function, this is more accurate than the approximation used in equation (50) on page 256 — for instance, this equation is exact if $f(x)$ is a polynomial whose degree is $\leq 4$.

a. What is the *molecule* of this approximation for the Laplacian?

5.1.3. *Rate of convergence.* In this section we will determine how fast the iterative methods in the preceding sections converge in some cases. We will do this by computing the spectral radius (and 2-norm) of the matrices involved in the iterative computations. We will restrict our attention to the simplest case: the two-dimensional form of Laplace's equation for which the domain of the problem is a rectangle. This computation will make extensive use of the material in § 2.4 on page 218, and in particular the computation of the eigenvalues of the matrices $\mathcal{Z}(n)$ defined there.

We will begin by defining matrices $\mathcal{E}(n)$ by

DEFINITION 5.4. Let $n > 0$ be an integer and define the $n \times n$ matrix $\mathcal{E}(n)$ by

$$\mathcal{E}(n) = \begin{cases} 1 & \text{if } |i - j| = 1 \\ 0 & \text{otherwise} \end{cases}$$

For instance $\mathcal{E}(5)$ is the matrix

$$\begin{pmatrix} 0 & 1 & 0 & 0 & 0 \\ 1 & 0 & 1 & 0 & 0 \\ 0 & 1 & 0 & 1 & 0 \\ 0 & 0 & 1 & 0 & 1 \\ 0 & 0 & 0 & 1 & 0 \end{pmatrix}$$

Note that $\mathcal{E}(n)$ looks very similar to $\mathcal{Z}(n)$ defined in 2.10 on page 219 — they only differ in the extra 1's that $\mathcal{Z}(n)$ has in the upper right and lower left corners. The matrices $\mathcal{E}(n)$ turn out to be closely related to the matrices that appear in the Jacobi iteration scheme for solving the finite-difference approximations of elliptic differential equations. We will calculate their eigenvalues and eigenvectors. It is tempting to try use our knowledge of the eigenvalues and eigenvectors of the matrices $\mathcal{Z}(n)$ to perform these computations. This is indeed possible, as is clear by recalling that $\mathcal{Z}(n)$ has a set of eigenvectors:

$$w'(j) = \{0, \sin(2\pi j/n), \sin(4\pi j/n), \dots, \sin(2\pi j(n-1)/n)\}$$

(see equation (30) on page 221). The corresponding eigenvalue is $2\cos(2\pi jk/n)$ and the equation these satisfy is

$$\mathcal{Z}(n) \cdot w'(j) = 2\cos(2\pi jk/n)w'(j)$$

The significant aspect of this equation is that the first component of $w'(j)$ is zero — this implies that the first row and column of that matrix effectively does not exist:

(70) $\quad \mathcal{Z}(n+1) = \begin{pmatrix} 0 & 1 & 0 & \cdots & 0 & 1 \\ \hline 0 & & & & & \\ \vdots & & & \mathcal{E}(n) & & \\ 0 & & & & & \\ 1 & & & & & \end{pmatrix} \cdot \begin{pmatrix} 0 \\ \sin(2\pi k/(n+1)) \\ \vdots \\ \sin(2\pi kn/(n+1)) \end{pmatrix}$

$$= 2\cos 2\pi k/(n+1) \begin{pmatrix} 0 \\ \sin(2\pi k/(n+1)) \\ \vdots \\ \sin(2\pi kn/(n+1)) \end{pmatrix}$$

so

$$\mathcal{E}(n) \cdot \begin{pmatrix} \sin(2\pi k/(n+1)) \\ \vdots \\ \sin(2\pi kn/(n+1)) \end{pmatrix} = 2\cos 2\pi k/(n+1) \begin{pmatrix} \sin(2\pi k/(n+1)) \\ \vdots \\ \sin(2\pi kn/(n+1)) \end{pmatrix}$$

and $\{2\cos 2\pi k/(n+1)\}$ are eigenvalues of $\mathcal{E}(n)$. Unfortunately, we don't know that we have found *all* of the eigenvalues — there are at most $n$ such eigenvalues and we have found $\lceil n/2 \rceil$ of them[30]. In fact, it turns out that we can find additional eigenvalues and eigenvectors. To see this, consider the eigenvalue equation like equation (70) for

---

[30]In the case of computing the eigenvalues of $\mathcal{Z}(n)$ we had other ways of knowing that we found all of the eigenvalues, since the $\mathcal{Z}(n)$ are cyclic matrices.

$\mathcal{Z}(2(n+1))$:

$$\mathcal{Z}(2(n+1)) = \begin{pmatrix} 0 & 1 & 0 & \cdots & 0 & \cdots & 0 & 1 \\ 1 & & & & & & & 0 \\ \vdots & & \mathcal{E}(n) & & 0 & \phantom{0} & & \vdots \\ 0 & & & & 1 & \mathbf{0} & & 0 \\ 0 & & \cdots & 1 & 0 & 1 & \cdots & 0 \\ & & & & 1 & & & \\ \vdots & & \mathbf{0} & & \vdots & & \mathcal{E}(n) & \\ 1 & & & & & & & \end{pmatrix} \cdot \begin{pmatrix} 0 \\ \sin(2\pi k/2(n+1)) \\ \vdots \\ \sin(2\pi kn/2(n+1)) \\ 0 \\ \vdots \\ \sin(2\pi k(2n+1)/2(n+1)) \end{pmatrix}$$

$$= 2\cos 2\pi k/2(n+1) \begin{pmatrix} 0 \\ \sin(2\pi k/2(n+1)) \\ \vdots \\ \sin(2\pi kn/2(n+1)) \\ 0 \\ \vdots \\ \sin(2\pi k(2n+1)/2(n+1)) \end{pmatrix}$$

where the second 0 in the $w'(2(n+1))$ occurs in the row of $\mathcal{Z}(2(n+1))$ directly above the second copy of $\mathcal{E}(n)$. This implies that the large array $\mathcal{Z}(2(n+1))$ can be regarded as *effectively* splitting into two copies of $\mathcal{E}(n)$ and two rows and columns of zeroes — in other words

$$\begin{pmatrix} 0 & 1 & 0 & \cdots & 0 & \cdots & 0 & 1 \\ 1 & & & & & & & 0 \\ \vdots & & \mathcal{E}(n) & & 0 & \mathbf{0} & & \vdots \\ 0 & & & & 1 & & & 0 \\ 0 & & \cdots & 1 & 0 & 1 & \cdots & 0 \\ & & & & 1 & & & \\ \vdots & & \mathbf{0} & & \vdots & & \mathcal{E}(n) & \\ 1 & & & & & & & \end{pmatrix} \cdot \begin{pmatrix} 0 \\ \sin(2\pi k/2(n+1)) \\ \vdots \\ \sin(2\pi kn/2(n+1)) \\ 0 \\ \vdots \\ \sin(2\pi k(2n+1)/2(n+1)) \end{pmatrix}$$

$$= \begin{pmatrix} 0 & 0 & 0 & \cdots & 0 & \cdots & 0 & 0 \\ 0 & & & & & & & 0 \\ \vdots & & \mathcal{E}(n) & & 0 & \mathbf{0} & & \vdots \\ 0 & & & & 0 & & & 0 \\ 0 & & \cdots & 0 & 0 & 0 & \cdots & 0 \\ & & & & 0 & & & \\ \vdots & & \mathbf{0} & & \vdots & & \mathcal{E}(n) & \\ 0 & & & & & & & \end{pmatrix} \cdot \begin{pmatrix} 0 \\ \sin(2\pi k/2(n+1)) \\ \vdots \\ \sin(2\pi kn/2(n+1)) \\ 0 \\ \vdots \\ \sin(2\pi k(2n+1)/2(n+1)) \end{pmatrix}$$

This implies that the values $\{2\cos 2\pi k/2(n+1) = 2\cos \pi k/(n+1)\}$ are also eigenvalues of $\mathcal{E}(n)$. Since there are $n$ of these, and they are all distinct, we have found all of them. We summarize this:

THEOREM 5.5. If the arrays $\mathcal{E}(n)$ are defined by

$$\mathcal{E}(n) = \begin{cases} 1 & \text{if } |i - j| = 1 \\ 0 & \text{otherwise} \end{cases}$$

then the eigenvalues are $\{\lambda_k = 2\cos \pi k/(n+1)\}$ for $k = 1, \ldots, n$, and with corresponding eigenvectors:

$$v_k = (\sin \pi k/(n+1), \sin 2\pi k/(n+1), \ldots, \sin n\pi k/(n+1))$$

Now we will relate this with the numerical solutions of elliptic differential equations. We will assume that the domain of the solution (i.e., the region in which the function $\psi$ is defined) is a *rectangle* defined by $0 \le x < a$ and $0 \le y < b$. The numeric approximation to the two dimensional Laplace equation is equation (61) on page 259:

$$\psi(x,y) = \frac{\psi(x+\delta, y) + \psi(x-\delta, y) + \psi(x, y+\delta) + \psi(x, y-\delta)}{4}$$

As a system of linear equations, this has $\left\lfloor \dfrac{a}{\delta} \right\rfloor \times \left\lfloor \dfrac{b}{\delta} \right\rfloor$ equations, and the same number of unknowns (the values of $\psi(j\delta, k\delta)$, for integral values of $j$ and $k$).

DEFINITION 5.6. We will call the *matrix* of this linear system $E(r, s)$, where $r = \left\lfloor \dfrac{a}{\delta} \right\rfloor$ and $s = \left\lfloor \dfrac{b}{\delta} \right\rfloor$.

In greater generality, if we have a Laplace equation in $n$ dimensions whose domain is an $n$-dimensional rectangular solid with $s_i$ steps in coordinate $i$ (i.e., the total range of $x_i$ is $s_i\delta$), then

$$E(s_1, \ldots, s_n)$$

denotes the matrix for

$$\frac{\psi(x_1+\delta, \ldots, x_n) + \psi(x_1-\delta, \ldots, x_n) + {} \atop {}+\psi(x_1, \ldots, x_n+\delta) + \psi(x_1, \ldots, x_n-\delta)}{2n}$$

Equation (62) on page 261 gives an example of this matrix. We will try to compute the eigenvalues of this matrix — this involves solving the system:

$$\lambda\psi(x,y) = \frac{\psi(x+\delta, y) + \psi(x-\delta, y) + \psi(x, y+\delta) + \psi(x, y-\delta)}{4}$$

Now we make an assumption that will allow us to express this linear system in terms of the arrays $\mathcal{E}(n)$, defined above (in 5.4 on page 275): we assume that $\psi(x, y)$ can be expressed as a product of functions $u(x)$ and $v(y)$[31]. The equation becomes:

$$\lambda u(x) \cdot v(y) = \frac{u(x+\delta) \cdot v(y) + u(x-\delta) \cdot v(y) + u(x) \cdot v(y+\delta) + u(x) \cdot v(y-\delta)}{4}$$

$$= \frac{(u(x+\delta) + u(x-\delta)) \cdot v(y) + u(x) \cdot (v(y+\delta) + v(y-\delta))}{4}$$

If we divide this by $u(x) \cdot v(y)$ we get

(71)
$$\lambda = \frac{\dfrac{u(x+\delta) + u(x-\delta)}{u(x)} + \dfrac{v(y+\delta) + v(y-\delta)}{v(y)}}{4}$$

(72)
$$= \frac{1}{4}\frac{u(x+\delta) + u(x-\delta)}{u(x)} + \frac{1}{4}\frac{v(y+\delta) + v(y-\delta)}{v(y)}$$

Now we notice something kind of interesting:

**We have a function of $x$ (namely $\dfrac{1}{4}\dfrac{u(x+\delta) + u(x-\delta)}{u(x)}$) added to a function of $y$ (namely $\dfrac{1}{4}\dfrac{v(y+\delta) + v(y-\delta)}{v(y)}$)and the result is a constant.**

It implies that a function of $x$ can be expressed as constant minus a function of $y$. How is this possible? A little reflection shows that the *only* way this can happen is for both functions (i.e., the one in $x$ and the one in $y$) to *individually* be constants. Our equation in two variables splits into two equations in one variable:

$$\lambda = \lambda_1 + \lambda_2$$
$$\lambda_1 = \frac{1}{4}\frac{u(x+\delta) + u(x-\delta)}{u(x)}$$

or

$$\lambda_1 u(x) = \frac{1}{4}(u(x+\delta) + u(x-\delta))$$

and

$$\lambda_2 v(y) = \frac{1}{4}(v(y+\delta) + v(y-\delta))$$

[31]This is a very common technique in the theory of partial differential equations, called *separation of variables*.

Incidentally, the thing that makes this separation of variables legal is the fact that the region of definition of $\psi(x, y)$ was a *rectangle*. It follows that eigenvalues of the original equation are sums of eigenvalues of the two equations in one variable. Examination of the arrays that occur in these equations show that they are nothing but $\frac{1}{4}\mathcal{E}(r)$ and $\frac{1}{4}\mathcal{E}(s)$, respectively, where $r = \left\lfloor \frac{a}{\delta} \right\rfloor$ and $s = \left\lfloor \frac{b}{\delta} \right\rfloor$. Consequently, 5.5 on page 277 implies that

1. The possible values of $\lambda_1$ are $\{2\cos\pi j/(r+1)/4 = \cos\pi j/(r+1)/2\}$;
2. The possible values of $\lambda_2$ are $\{\dfrac{2\cos\pi k/(s+1)}{4} = \dfrac{\cos\pi k/(s+1)}{2}\}$;
3. The eigenvalues of $E(r,s)$ include the set of values $\{\dfrac{\cos\pi j/(r+1)}{2} + \dfrac{\cos\pi k/(s+1)}{2}\}$, where $j$ runs from 1 to $r$ and $k$ runs from 1 to $s$.
4. Since $\psi(x,y) = u(x)v(y)$ the eigenfunctions corresponding to these eigenvalues are, respectively,

$$(\sin(\pi j/(r+1))\sin(\pi k/(s+1)), \ldots, \sin(\pi jn/(r+1))\sin(\pi km/(s+1)),$$
$$\ldots, \sin(\pi j/(r+1))\sin(\pi k/(s+1)))$$

where $m$ runs from 1 to $r$ and $n$ runs from 1 to $s$.

It is possible to give a *direct* argument to show that these eigenfunctions are all linearly independent. Since there are $r \times s$ of them, we have found them *all*. To summarize:

THEOREM 5.7. Suppose $\delta > 0$ is some number and $r \geq 1$ and $s \geq 1$ are integers. Consider the system of linear equations

$$\lambda\psi(x,y) = \frac{\psi(x+\delta, y) + \psi(x-\delta, y) + \psi(x, y+\delta) + \psi(x, y-\delta)}{4}$$

where:

1. $0 \leq x < a = r\delta$ and $0 \leq y < b = s\delta$, and
2. $\psi(x, y)$ is a function that is well-defined at points $x = n\delta$ and $y = m\delta$, where $n$ and $m$ are integers;
3. $\psi(0, x) = \psi(x, 0) = \psi(x, s\delta) = \psi(r\delta, y) = 0$.

Then nonzero values of $\psi$ only occur if

$$\lambda = \left\{ \frac{1}{2}\left(\cos\frac{\pi j}{r+1} + \cos\frac{\pi k}{s+1}\right) = 1 - \frac{\pi^2 j^2}{4(r+1)^2} - \frac{\pi^2 k^2}{4(s+1)^2} + O(\delta^4) \right\}$$

where $j$ and $k$ are integers running from 1 to $r$ and 1 to $s$, respectively.

PROOF. The only new piece of information in this theorem is the estimate of the values of the eigenvalues:

$$\lambda = \left\{ 1 - \frac{\pi^2 j^2}{4(r+1)^2} - \frac{\pi^2 k^2}{4(s+1)^2} + O(\delta^4) \right\}$$

This is a result of using the power series expansion of the cosine function:

$$\cos(x) = 1 - \frac{x^2}{2!} + \frac{x^4}{4!} \cdots$$

$\square$

We can compute the rate of convergence of the basic iterative solution to Laplace's equation using 1.22 on page 169:

THEOREM 5.8. Consider the numeric approximation to Laplace's equation in two dimensions, where $\delta > 0$ is the step-size, and $0 < x < a$ and $0 < y < b$. The basic Jacobi iteration method has an error that is reduced by a factor of

$$\rho(E(r,s)) = \frac{1}{2} \left( \cos \frac{\pi}{r+1} + \cos \frac{\pi}{s+1} \right)$$

$$= 1 - \frac{\pi^2}{4(r+1)^2} - \frac{\pi^2}{4(s+1)^2} + O(\delta^4)$$

$$\approx 1 - \frac{\pi^2 \delta^2}{4} \left( \frac{1}{a^2} + \frac{1}{b^2} \right)$$

where $r = \left\lfloor \frac{a}{\delta} \right\rfloor$ and $s = \left\lfloor \frac{b}{\delta} \right\rfloor$.

PROOF. The spectral radius of $E(r, s)$ is equal to the largest value that occurs as the absolute value of an eigenvalue of $E(r, s)$. This happens when $j = k = 1$ in the equation in 5.7 above.  $\square$

We can estimate the optimum relaxation coefficient for the SOR method for solving Laplace's equation, using the equation in 1.29 on page 176. We get:

$$(73) \qquad \omega = \frac{2}{1 + \sqrt{1 - \rho(Z(A))^2}}$$

$$(74) \qquad = \frac{2}{1 + \sqrt{1 - \left(1 - \frac{\pi^2 \delta^2}{4} \left( \frac{1}{a^2} + \frac{1}{b^2} \right)\right)^2}}$$

$$(75) \qquad \approx \frac{2}{1 + \pi \delta \sqrt{\frac{1}{2a^2} + \frac{1}{2b^2}}}$$

If we assume that $a = b$ and $a^2 = b^2 = A$, the area of the region, then equation (75) comes to resemble Garabedian's formula (equation (63) on page 267).

### Exercises:

1. Compute the eigenvalues and eigenvectors of the matrix $E(r, s, t)$ define in 5.6 on page 278, where $r$, $s$, and $t$ are integers.

## 5.2. Parabolic Differential Equations.

5.2.1. *Basic Methods.* These equations frequently occur in the study of *diffusion* phenomena in physics, and in quantum mechanics.

The simplest parabolic differential equation [32] is called the Heat Equation. We have a function $\psi(x_1, \ldots, x_n, t)$, where $x_1, \ldots, x_n$ are spatial coordinates and $t$ is *time*:

$$(76) \qquad \nabla^2 \psi = \frac{1}{a^2} \frac{\partial \psi}{\partial t}$$

Here $a$ is a constant called the *rate of diffusion*. In a real physical problem $\psi$ represents *temperature*, and the heat equation describes how heat flows to equalize temperatures in some physical system.

Another common parabolic equation is the Schrödinger Wave Equation for a particle in a force field:

$$(77) \qquad -\frac{\hbar^2}{2m} \nabla^2 \psi + V(x, y, z)\psi = i\hbar \frac{\partial \psi}{\partial t}$$

Here $\hbar$ is Planck's Constant$/2\pi = 1.054592 \times 10^{-27} g\ cm^2/sec$, $m$ is the mass of the particle, $i = \sqrt{-1}$, and $V(x, y, z)$ is potential energy, which describes the force field acting on the particle. Although this equation looks much more complicated than the

---

[32]For anyone who is interested, the most general parabolic equation looks like:

$$\sum_{i,k}^{n} a_{ik} \frac{\partial^2 \psi}{\partial x_i \partial x_k} = f(x, \psi, \frac{\partial \psi}{\partial x})$$

where $a_{ik} = a_{ki}$ and $\det(a) = 0$.

Heat Equation, its overall behavior (especially from the point of view of numerical solutions) is very similar. Since $\hbar$, $i$, and $m$ are *constants* they can be removed by a suitable change of coordinates. The only real additional complexity is the function $V(x, y, z)$. If this vanishes we get the Heat equation exactly. Unfortunately, physically interesting problems always correspond to the case where $V(x, y, z)$ is nonzero.

Incidentally, the *physical interpretation* of the (complex-valued) function $\psi(x, y, z, t)$ that is the solution of the Schrödinger Wave Equation is something that even physicists don't completely agree on. It is generally thought that the absolute value of $\psi$ (in the complex sense i.e., the sum of the squares of the real and imaginary parts) is the probability of detecting the particle at the given position in space at the given time.

As with the elliptic differential equations, we assume that the differential equation is valid in some region of space (or a plane) and we specify what must happen on the boundary of this region. When the differential equation contains time as one of its variables the boundary conditions usually take the form

> **We specify the value of $\psi(x_i, t)$ completely at some initial time $t_0$ over the domain $\Omega$ of the problem, and specify the behavior of $\psi$ on the boundary of $\Omega$ at all later times.**

Boundary conditions of this sort are often called *initial conditions*. The solution of the partial differential equation then gives the values of $\psi(x_i, t)$ at all times *later* than $t_0$, and over all of $\Omega$.

Here is an example:

EXAMPLE 5.9. Suppose we have an iron sphere of radius 5, centered at the origin. In addition, suppose the sphere is heated to a temperature of 1000 degrees K at time 0 and the boundary of the sphere is kept at a temperature of 0 degrees K. What is the temperature-distribution of the sphere at later times, as a function of $x$, $y$, $z$, and $t$?

Here we make $\Omega$ the sphere of radius 5 centered at the origin and we have the following boundary conditions:

- $\psi(x, y, z, 0) = 1000$ for $(x, y, z) \in \Omega$;
- $\psi(x, y, z, t) = 0$, if $(x, y, z) \in \partial\Omega$ ($\partial\Omega$ denotes the boundary of $\Omega$).

With these boundary conditions, it turns out that the 3-dimensional heat equation solves our problem:

$$\frac{\partial^2 \psi}{\partial x^2} + \frac{\partial^2 \psi}{\partial y^2} + \frac{\partial^2 \psi}{\partial z^2} = \frac{1}{a^2} \frac{\partial \psi}{\partial t}$$

where we must plug a suitable value of $a$ (the thermal conductivity of iron).

Our basic approach to a numerical solution remains the same as in the previous section — we replace $\nabla^2 \psi$ and now $\partial \psi / \partial t$ by *finite differences* and solve the resulting *linear equations*. We will work this out for the Heat equation. We get

$$\frac{\begin{aligned}\psi(x_1 + \delta, \ldots, x_n, t) + \psi(x_1 - \delta, x_2, \ldots, x_n, t) + \cdots + \psi(x_1, \ldots, x_n + \delta t) \\ + \psi(x_1, \ldots, x_n - \delta, t) - 2n\psi(x_1, \ldots, x_n, t)\end{aligned}}{\delta^2}$$

$$= \frac{1}{a^2} \frac{\psi(x_1, \ldots, x_n, t + \delta t) - \psi(x_1, \ldots, x_n, t)}{\delta t}$$

Although this equation is rather formidable-looking we can rewrite it in the form (multiplying it by $a^2$ and $\delta t$, and replacing the numerical $\nabla^2 \psi$ by $\frac{2n}{\delta^2}(\psi_{\text{average}} - \psi)$):

(78) $\qquad \psi(x_1, \ldots, x_n, t + \delta t) = \psi(x_1, \ldots, x_n, t) + \frac{2na^2\delta t}{\delta^2}\left(\psi_{\text{average}} - \psi\right)$

where $\psi_{\text{average}}$ is defined in equation (52) on page 257.

This illustrates one important difference between parabolic and elliptic equations:

1. The value of $\psi$ at time $t + \delta t$ at a given point depends upon its value at time $t$ at that point and at neighboring points (used to compute $\psi_{\text{average}}(t)$). It is not hard to see that these, in turn, depend upon the values of $\psi$ at time $t - \delta t$ over a larger set of neighboring points. In general, as we go further back in time, there is an expanding (in spatial coordinates) set of other values of $\psi$ that determine this value of $\psi$ — and as we go back in time and plot these points they trace out a parabola[33] (very roughly).

2. We use a *different* mesh-sizes for the time and space coordinates, namely $\delta t$ and $\delta$, respectively. It turns out that the iteration will diverge (wildly!) unless $2na^2\delta t/\delta^2 < 1$ or $\delta t < \delta^2/2na^2$. See proposition 5.10 on page 288 for a detailed analysis. This generally means that large numbers of iterations will be needed to compute $\psi$ at late times.

In the parabolic case, we will specify the values of $\psi$ over the *entire region* at time 0, and on the boundaries of the region at all times. We can then solve for the values of $\psi$ at later times using equation 78. This can easily be translated into a C* program. Here is a simple program of this type:

```
#include <stdio.h>
shape[64][128] plane;
float       delta = 1. / 16.;
float       delta_t, thratio;      /* We use a smaller
                                    * delta—value for time
                                    * to minimize round off
                                    * error. */
```

---

[33]In this discussion, it looks like they trace out a *cone*, but in the limit, as $\delta t \rightarrow 0$ it becomes a parabola.

```
float:plane x, y, psi, psi_sum;
void main()
{
    int         iteration = 0;  /* Records the number of
                                 * iterations. */
    int         i, j;
    float       time;       /* Records actual time. */
    /* Initialize coordinates and psi. */
    with(plane) {
        x = pcoord(0);  /* Tell each processor its row
                         * number. */
        y = pcoord(1);  /* Tell each processor its
                         * column number. */
        x = x / 16.0 - 2.0;    /* Compute the
                               * x-coordinate of a
                               * processor. */
        y = y / 16.0 - 4.0;    /* Compute the
                               * y-coordinate of a
                               * processor. */
        where((pcoord(0) != 0) & (pcoord(1) != 0)
            & (pcoord(0) != 63) & (pcoord(1) != 127))
            psi = 5.;       /* We initially set the
                            * temperature of all
                            * points except the
                            * boundary equal to 5
                            * -- the boundary
                            * points are set (and
                            * fixed at) 0. When the
                            * program runs, heat
                            * will drain off the
                            * boundary. */
        else
        psi = 0.;
    }
    delta_t = delta * delta / 10.;
    thratio = delta_t / (delta * delta);
                    /* This is the ratio
                     * used in the formula.
                     * delta_t has been
                     * chosen so that the
```

```
                              * coefficient of each
                              * psi-term will be <1.
                              * In fact, in this case
                              * it will be 4/10. */
    time = 0.;
    for (iteration = 0; time < 1.; iteration++) {
        time = iteration;
        time = time * delta_t;
        with(plane)    /* We only carry out the
                              * iterations on the interior
                              * points.  This has the effect
                              * of FIXING the boundary points
                              * at 0. */
            where((pcoord(0) != 0) & (pcoord(1) != 0)
          & (pcoord(0) != 63) & (pcoord(1) != 127)) {
            psi_sum =
                [.+ 1][.] psi
                +[.- 1][.] psi
                +[.][.+ 1] psi
                +[.][.- 1] psi;
                    /* We use psi_sum
                     * instead of
                     * psi_average to
                     * decrease round off
                     * error. */
            psi = psi + thratio * (psi_sum - 4. * psi);
            }
        }
    for (i = 0; i < 64; i++) {
        int        val;
        for (j = 0; j < 64; j++) {
            val =[i][j] psi;
            printf("%d ", val);
        }
        printf("\n");
    }
}
```

Figure V.28 shows what the output looks like after 1 time unit has passed:
The height of the surface over each point represents the temperature at that point.

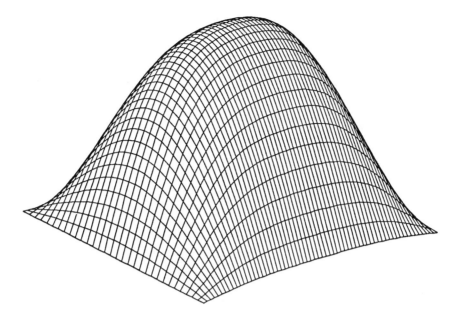

FIGURE V.28. Solution of the heat equation

You can see the heat "draining away" along the boundary.

5.2.2. *Error Analysis*. It is very important to analyze the behavior of numerical solutions of differential equations with respect to errors. These methods are necessarily approximations to the true solutions to the problems they solve, and it turns out that certain variations in the parameters to a given problem have a dramatic effect upon the amount of error that occurs.

Each iteration of one of these algorithms creates some finite amount of error, this error is carried forward into future iterations. It follows that there are two basic sources of error:

1. Stepwise error
2. Propagation of errors

The first source is what we expect since we are approximating a continuous equation by a discrete one. The second source of error is somewhat more problematic — it implies that the quality of the solution we obtain degrades with the number of iterations we carry out. Since each iteration represents a time-step, it means that the quality of our solution degrades with time. It turns out that we will be able to adjust certain parameters of our iterative procedure to guarantee that propagation of error is minimized. In fact that total amount of error from this source can be bounded, under suitable conditions.

Consider the numeric form of the basic heat equation, (78) on page 284. It turns out that the factor $A = \frac{4na^2\delta t}{\delta^2}$ is crucial to the question of how errors propagate in the solution.

PROPOSITION 5.10. In the $n$-dimensional heat equation, suppose that the act of re-placing the partial derivatives by a finite difference results in an error bounded by a number $E$ — i.e.,

$$\left| \frac{\partial \psi}{\partial t} - \frac{\psi(t + \delta t) - \psi(t)}{\delta t} \right| \leq E$$

$$\left| \frac{2n(\psi_{\text{average}} - \psi)}{\delta^2} - \nabla^2 \psi \right| \leq E$$

over the spatial region of the solution and the range of times that we want to study. Then the cumulative error of the solution in (78) on page 284 is $\leq \delta t E \left(1 + \frac{1}{|a|^2}\right)\left(1 + A + A^2 + \cdots + A^k\right)$ in $k$ time-steps of size $\delta t$. Consequently, the total error is $O(\delta t E \cdot A^{k+1})$ if $A > 1$, and $O(\delta t E \cdot (1 - A^{k+1})/(1 - A))$ if $A < 1$.

It is interesting that if $A < 1$ the total error is *bounded*. In this analysis we haven't taken into account the dependence of $E$ on $\delta t$. If we assume that $\psi$ depends upon time in a fashion that is almost linear, then $E$ may be proportional to $\delta t^2$.

We will assume that all errors due to flaws in the numerical computations (i.e., roundoff error) are incorporated into $E$.

PROOF. We prove this by induction on $k$. We begin by showing that the error of a single iteration of the algorithm is $(1 + a^2)\delta t E$.

$$\tilde{E}_1 = \frac{\partial \psi}{\partial t} - \frac{\psi(t + \delta t) - \psi(t)}{\delta t}$$

$$\tilde{E}_2 = \frac{4(\psi_{\text{average}} - \psi)}{\delta^2} - \nabla^2 \psi$$

so the hypotheses imply that

$$\left| \tilde{E}_1 \right| \leq E$$

$$\left| \tilde{E}_2 \right| \leq E$$

Suppose that $\psi$ is an *exact* solution of the Heat equation. We will plug $\psi$ into our approximate version of the Heat Equation and determine the extent to which it satisfies the approximate version. This will measure the amount of error in the approximate equation, since numerical solutions to those equations satisfy them exactly. Suppose

(79)
$$\psi(t + \delta t) = \delta t \frac{\psi(t + \delta t) - \psi(t)}{\delta t} + \psi(t) \text{ exactly}$$

(80)
$$= \delta t \frac{\partial \psi}{\partial t} - \delta t \tilde{E}_1 + \psi(t)$$

(81)
$$= \frac{\delta t}{a^2} \nabla^2 \psi - \delta t \tilde{E}_1 + \psi(t) \text{ because } \psi \text{ satisfies the Heat equation}$$

(82)
$$= \frac{\delta t}{a^2} \frac{4(\psi_{\text{average}} - \psi)}{\delta^2} - \frac{\delta t}{a^2} \tilde{E}_2 - \delta t \tilde{E}_1 + \psi(t)$$

This implies that the error in one iteration of the algorithm is $\leq \delta t E \left(1 + \frac{1}{|a|^2}\right)$. Now we prove the conclusion of this result by induction. Suppose, after $k$ steps the total error is $\leq \delta t E \left(1 + \frac{1}{|a|^2}\right)\left(1 + A + A^2 + \cdots + A^k\right)$. This means that that

$$\left|\hat{\psi} - \psi\right| \leq \delta t E \left(1 + \frac{1}{|a|^2}\right)\left(1 + A + A^2 + \cdots + A^k\right)$$

where $\hat{\psi}$ is the calculated value of $\psi$ (from the algorithm). One further iteration of the algorithm multiplies this error-vector by a matrix that resembles $\frac{2na^2\delta t}{\delta^2}(E(r,s) - I)$, where $I$ is the identity-matrix with $rs$ rows — using the notation of 5.8 on page 281. (We are also assuming that the region of definition of the original Heat equation is rectangular). The eigenvalues of this matrix are $\frac{2na^2\delta t}{\delta^2}$(eigenvalues of $E(r,s) - 1$) and the maximum absolute value of such an eigenvalue is $\leq 2\frac{2na^2\delta t}{\delta^2} = A$. The result-vector has an 2-norm that is $\leq A \cdot \delta t E \left(1 + \frac{1}{|a|^2}\right)\left(1 + A + A^2 + \cdots + A^k\right) = (1 + a^2)\delta t E \left(A + A^2 + \cdots + A^{k+1}\right)$. Since this iteration of the algorithm also adds an error of $\delta t E \left(1 + \frac{1}{|a|^2}\right)$, we get a total error of $\delta t E \left(1 + \frac{1}{|a|^2}\right)\left(1 + A + A^2 + \cdots + A^{k+1}\right)$, which is what we wanted to prove. $\square$

**Exercises:**

1. Analyze the two-dimensional Schrödinger Wave equation (equation (77) on page 282) in the light of the discussion in this section. In this case it is necessary to make some assumptions on the values of the potential, $V(x, y)$.

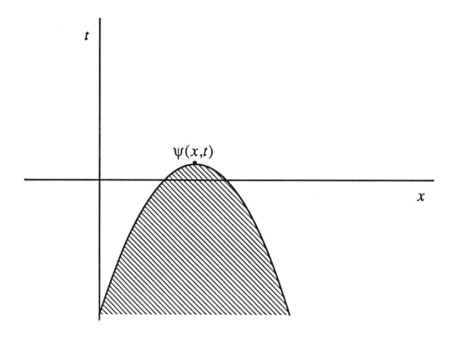

FIGURE V.29. Domain of dependence of a parabolic partial differential equation

5.2.3. *Implicit Methods.* The finite difference approximation methods described in the previous section have a basic flaw. Consider a parabolic partial differential equation in one dimension:

$$\frac{\partial^2 \psi}{\partial x^2} = \frac{\partial \psi}{\partial t}$$

In order to understand the behavior of solutions of this equation, it turns out to be useful to consider the *domain of dependence* of a point $(x, t)$. This is the set of points $(x', t')$ at *earlier times* with the property that the values of $\psi(x', t')$ influences the value of $\psi(x, t)$. It turns out that the domain of dependence of parabolic partial differential equations is a *parabola* in the $x$-$t$ plane, as shown in figure V.29 (this is why these equations are called *parabolic*). In fact this doesn't tell the whole story — it turns out that the value of $\psi(x, t)$ depends upon points *outside* the parabola in figure V.29 so some extent (although the "amount" of dependence falls off rapidly as one moves away from that parabola) — we should really have drawn a "fuzzy" parabola[34].

On the other hand, the finite difference equations like (78) on page 284 have a *conical* domain of dependence — see figure V.30. Furthermore, unlike parabolic differential equations, *this* domain of dependence is *sharp*. The finite-difference approximation of parabolic differential equations *loses* a significant amount of information. It turns out that there is another, slightly more complex, way of approximating parabolic differential equations that capture much of this lost information. We call these approximation-

---

[34]This is because a fundamental solution of the Heat equation is $e^{-x^2/4t}$

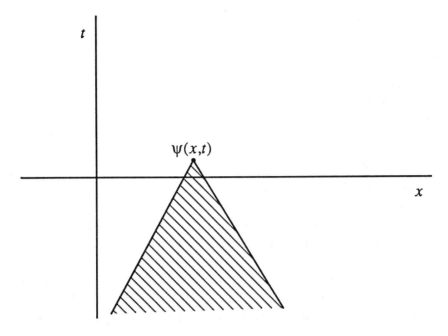

FIGURE V.30. Domain of dependence of a finite-difference equation

techniques, *implicit methods*. They are based upon a simple observation about how to approximate $\partial\psi/\partial t$. In formula (78) we used the expression

$$\frac{\partial\psi}{\partial t} = \lim_{\delta t\to 0}\frac{\psi(x_1,\ldots,x_n,t+\delta t)-\psi(x_1,\ldots,x_n,t)}{\delta t}$$

Implicit methods are based upon the fact that it is *also* true that

(83)
$$\frac{\partial\psi}{\partial t} = \lim_{\delta t\to 0}\frac{\psi(x_1,\ldots,x_n,t)-\psi(x_1,\ldots,x_n,t-\delta t)}{\delta t}$$

Note that the right sides of these equations are *not* the same, but they approach the same limit as $\delta t \to 0$. Using this approximation, we get the following equation for the parabolic partial differential equation:

$$\psi(x_1,\ldots,x_n,t)-\psi(x_1,\ldots,x_n,t-\delta t) = \frac{2na^2\delta t}{\delta^2}\left(\psi_{\text{average}}-\psi\right)$$

Gathering together all terms of the form $\psi(x_1,\ldots,x_n,t)$ gives:

$$\left(1+\frac{2na^2\delta t}{\delta^2}\right)\psi(x_1,\ldots,x_n,t)-\frac{2na^2\delta t}{\delta^2}\psi_{\text{average}}(x_1,\ldots,x_n,t)$$
$$= \psi(x_1,\ldots,x_n,t-\delta t)$$

We usually write the formula in a manner that computes $\psi(x_1, \ldots, x_n, t + \delta t)$ from $\psi(x_1, \ldots, x_n, t)$:

$$(84) \quad \left(1 + \frac{2na^2\delta t}{\delta^2}\right)\psi(x_1, \ldots, x_n, t + \delta t) - \frac{2na^2\delta t}{\delta^2}\psi_{\text{average}}(x_1, \ldots, x_n, t + \delta t)$$

$$= \psi(x_1, \ldots, x_n, t)$$

There are several important differences between this formula and (78):

- Each iteration of this algorithm is *much more complex* than iterations of the algorithm in (78). In fact, each iteration of this algorithm involves a type of computation comparable to numerically solving a the Poisson equation. We must solve a system of linear equations for $\psi(x_1, \ldots, x_n, t + \delta t)$ in terms of $\psi(x_1, \ldots, x_n, t)$.
- Formula (78) explicitly expresses the dependence of $\psi(x_1, \ldots, x_n, t + \delta t)$ upon $\psi(*, t + \delta t)$ at neighboring points. It would appear, intuitively, that the domain of dependence of $\psi$ is more like the parabola in figure V.29. This turns out to be the case.
- It turns out (although this fact might not be entirely obvious) that this algorithm is usually much more numerically stable than that in equation (78). If fact it remains numerically stable even for fairly large values of $\delta t$. See 5.14 on page 295 for a precise statement.

On balance, it is usually advantageous to use the implicit methods described above. Although the computations that must be performed for each iteration of $\delta t$ are more complicated, the number of iterations needed is generally much less since the present algorithm is more accurate and stable. See § 5.2.4 on page 293 for a detailed analysis. As remarked above, we must solve a system of linear equations in order to compute it in terms of $\psi(x_1, \ldots, x_n, t)$. We can use iterative methods like the Jacobi method discussed in § 1.2.2 on page 167. Our basic iteration is:

$$\left(1 + \frac{2na^2\delta t}{\delta^2}\right)\psi^{(k+1)}(x_1, \ldots, x_n, t + \delta t) - \frac{2na^2\delta t}{\delta^2}\psi^{(k)}_{\text{average}}(x_1, \ldots, x_n, t + \delta t)$$

$$= \psi(x_1, \ldots, x_n, t)$$

where $\psi^{(k+1)}(x_1, \ldots, x_n, t + \delta t)$ is the $k + 1^{\text{st}}$ approximation to $\psi(x_1, \ldots, x_n, t + \delta t)$, given $\psi^{(k)}(x_1, \ldots, x_n, t + \delta t)$ and $\psi(x_1, \ldots, x_n, t)$. Note that $\psi(x_1, \ldots, x_n, t)$ plays the part of $b$ in the linear system $Ax = b$. We assume the values of the $\psi(x_1, \ldots, x_n, t)$ are *known*. We may have to perform many iterations in order to get a *reasonable* value of $\psi(x_1, \ldots, x_n, t + \delta t)$. We rewrite this equation to isolate $\psi^{(k+1)}(x_1, \ldots, x_n, t + \delta t)$:

(85) $\psi^{(k+1)}(x_1, \ldots, x_n, t + \delta t)$

$$= \frac{\psi(x_1, \ldots, x_n, t) + \dfrac{2na^2\delta t}{\delta^2}\psi_{\text{average}}^{(k)}(x_1, \ldots, x_n, t + \delta t)}{\left(1 + \dfrac{2na^2\delta t}{\delta^2}\right)}$$

It turns out that the *molecule* of these finite-difference equations is the same as that depicted in figure V.25 on page 266. It follows that we can speed up the convergence of the iteration by using the consistently-ordered methods (see § 1.2.4 on page 173), with the odd-even ordering vector discussed in 5.3 on page 266. Unfortunately we have no *simple* way to determine the optimal overrelaxation coefficient.

Once we have found a reasonable value of $\psi(x_1, \ldots, x_n, t + \delta t)$, we are in a position to compute $\psi(x_1, \ldots, x_n, t + 2\delta t)$ and so on. We may also use multigrid methods like those described on page 272 in each phase of this algorithm.

**Exercises:**

1. Formulate the Schrödinger wave equation (equation (77) on page 282) in an implicit iteration scheme.

5.2.4. *Error Analysis.* In this section we will show that implicit methods are far better-behaved than the explicit ones with regards to propagation of errors. In fact, regardless of step-size, it turns out that the total propagation of error in implicit methods is strictly bounded. We will perform an analysis like that in § 5.2.2 on page 287.

As in § 5.2.2 we will assume that

1. $0 \leq x < a = r\delta$ and $0 \leq y < b = s\delta$, and
2. $\psi(x, y)$ is a function that is well-defined at points $x = n\delta$ and $y = m\delta$, where $n$ and $m$ are integers, and for values of $t$ that are integral multiples of a quantity $\delta t > 0$;

We will examine the behavior of the linear system (84) on page 292 in the case where the original differential equation was two dimensional:

$$\left(1 + \frac{4a^2\delta t}{\delta^2}\right)\psi(x, y, t + \delta t) - \frac{4a^2\delta t}{\delta^2}\psi_{\text{average}}(x, y, t + \delta t) = \psi(x, y, t)$$

We will regard this as an equation of the form

(86) $$\qquad\qquad Z\psi(t + \delta t) = \psi(t)$$

and try to determine the relevant properties of the matrix $Z$ (namely the eigenvalues) and deduce the corresponding properties of $Z^{-1}$ in the equation:

(87) $$\qquad\qquad \psi(t + \delta t) = Z^{-1}\psi(t)$$

CLAIM 5.11. The matrix $Z$ in equation (86) is equal to

$$\left(1 + \frac{4a^2\delta t}{\delta^2}\right)I_{rs} - \frac{4a^2\delta t}{\delta^2}E(r, s)$$

where $I_{rs}$ is an $rs \times rs$ identity matrix.

This follows from the fact that the first term of (84) on page 292 simply multiplied its factor of $\psi$ by a constant, and the second term simply subtracted a multiple of $\psi_{\text{average}}$ — whose matrix-form is equal to $E(r, s)$ (see 5.6 on page 278). This, and exercise 8 on page 167 and its solution on page 481 imply:

CLAIM 5.12. The eigenvalues of the matrix $Z$ defined above, are

$$\left\{1 + \frac{4a^2\delta t}{\delta^2} - \frac{4a^2\delta t}{2\delta^2}\left(\cos\frac{\pi j}{r + 1} + \cos\frac{\pi k}{s + 1}\right)\right\}$$

where $1 \leq j \leq r$ and $1 \leq k \leq s$ are integers.

Now 1.14 on page 162 implies that

PROPOSITION 5.13. The eigenvalues of the matrix $Z^{-1}$ in equation (87) above are

$$\frac{1}{1 + \dfrac{4a^2\delta t}{\delta^2} - \dfrac{4a^2\delta t}{2\delta^2}\left(\cos\frac{\pi j}{r+1} + \cos\frac{\pi k}{s+1}\right)}$$

The eigenvalue with the largest absolute value is the one for which the cosine-terms are as close as possible to 1. This happens when $j = k = 1$. The spectral radius of $Z^{-1}$ is, consequently,

$$\rho(Z^{-1}) = \frac{\delta^2}{\delta^2 + 4\,a^2\delta t - 2\,a^2\delta t\cos(\frac{\pi}{r+1}) - 2\,a^2\delta t\cos(\frac{\pi}{s+1})} < 1$$

since $\cos(\frac{\pi}{r+1}) < 1$ and $\cos(\frac{\pi}{s+1}) < 1$. This means that the implicit methods are always numerically stable. We can estimate the influence of $r$, $s$ and $\delta$ on the degree of numerical stability:

$$\rho(Z^{-1}) = 1 - \frac{a^2\delta t\pi^2}{\delta^2}\left(\frac{1}{r^2} + \frac{1}{s^2}\right) + O(\frac{\delta^4}{r^u s^v})$$

where $u + v \geq 4$.

In analogy with 5.10 on page 288, we get:

PROPOSITION 5.14. In the 2-dimensional heat equation, suppose that the act of replacing the partial derivatives by a finite difference results in a bounded error:

$$\left|\frac{\partial\psi}{\partial t} - \frac{\psi(t+\delta t) - \psi(t)}{\delta t}\right| \leq E_1(\delta t)$$

$$\left|\frac{4(\psi_{\text{average}} - \psi)}{\delta^2} - \nabla^2\psi\right| \leq E_2(\delta)$$

over the spatial region of the solution and the range of times that we want to study. We have written $E_1$ and $E_2$ as functions of $\delta t$ and $\delta$, respectively, to represent their dependence upon these quantities. Then the cumulative error of using the implicit iteration scheme is

$$\leq \frac{\delta^2}{a^2\pi^2}\left(E_1(\delta_t) + \frac{\delta^2}{a^2}E_2(\delta)\right)\frac{r^2 s^2}{r^2 + s^2}$$

As before, we assume that roundoff error is incorporated into $E_1$ and $E_2$. Compare this result with 5.10 on page 288.

PROOF. First we compute the error made in a single iteration. Suppose

$$\tilde{E}_1 = \frac{\partial\psi}{\partial t} - \frac{\psi(t+\delta t) - \psi(t)}{\delta t}$$
$$\tilde{E}_2 = \frac{4(\psi_{\text{average}} - \psi)}{\delta^2} - \nabla^2\psi$$

so the hypotheses imply that

$$\left|\tilde{E}_1\right| \leq E_1(\delta t)$$
$$\left|\tilde{E}_2\right| \leq E_2(\delta)$$

As in the explicit case we will assume that $\psi$ is an exact solution of the Heat Equation. We will plug this exact solution into our finite-difference approximation and see how closely it satisfies the finite-difference version. This will measure the error in using finite differences. We will get equations very similar to equations (79) through (82) on page 289:

$$\psi(t + \delta t) = \delta t \frac{\psi(t + \delta t) - \psi(t)}{\delta t} + \psi(t) \text{ exactly}$$

$$= \delta t \frac{\partial \psi}{\partial t}(t + \delta t) - \delta t \tilde{E}_1 + \psi(t)$$

$$= \frac{\delta t}{a^2} \nabla^2 \psi(t + \delta t) - \delta t \tilde{E}_1 + \psi(t) \text{ because } \psi \text{ satisfies the Heat equation}$$

$$= \frac{\delta t}{a^2} \frac{4(\psi_{average}(t + \delta t) - \psi(t + \delta t))}{\delta^2} - \frac{\delta t}{a^2} \tilde{E}_2 - \delta t \tilde{E}_1 + \psi(t)$$

This implies that the total error made in a *single iteration* of the algorithm is

$$\le E = \left| -\frac{\delta t}{a^2} \tilde{E}_2 - \delta t \tilde{E}_1 \right| \le \delta t \left( E_1 + \frac{1}{a^2} E_2 \right)$$

In the next iteration this error contributes a cumulative error $EA = \rho(Z^{-1})$, and $k$ iterations later its effect is $\le EA^k$. The total error is, consequently

$$E \left( 1 + A + A^2 \ldots \right) \le E = \frac{E}{1 - A}$$

The estimate

$$\rho(Z^{-1}) = 1 - \frac{a^2 \delta t \pi^2}{\delta^2} \left( \frac{1}{r^2} + \frac{1}{s^2} \right) + O(\frac{\delta^4}{r^u s^v})$$

now implies the conclusion. □

**Exercises:**

1. Show that the implicit methods always work for the Schrödinger wave equation (equation (77) on page 282). (In this context "work" means that total error due to propagation of errors is bounded.)

## 5.3. Hyperbolic Differential Equations.

5.3.1. *Background.* The simplest hyperbolic partial differential equation is the Wave Equation[35]. We have a function $\psi(x_1, \ldots, x_n, t)$, where $x_1, \ldots, x_n$ are spatial coordinates and $t$ is *time*:

$$(88) \qquad \nabla^2 \psi = \frac{1}{a^2} \frac{\partial^2 \psi}{\partial t^2}$$

As its name implies, this equation is used to describe general wave-motion. For instance, the 1-dimensional form of the equation can be used the describe a vibrating string:

$$(89) \qquad \frac{\partial^2 \psi}{\partial x^2} = \frac{1}{a^2} \frac{\partial^2 \psi}{\partial t^2}$$

Here, the string lies on the $x$-axis and $\psi$ represents the *displacement* of the string from the $x$-axis as a function of $x$ and $t$ (time). The Wave Equation in two dimensions can be used to describe a vibrating drumhead. Another example is

$$(90) \qquad \frac{\partial^2 \psi}{\partial t^2} - c^2 \frac{\partial^2 \psi}{\partial x^2} - 2 \frac{\partial \psi}{\partial t} = 0$$

This is know as the *telegraph equation* — it describes the propagation of electric current in a wire with leakage.

It is possible to give a closed-form expression for the general solution of the one-dimensional wave equation (89):

$$\psi(x, t) = \frac{\psi_0(x + at) + \psi(x - ai)}{2} + \frac{1}{2} \int_{x-at}^{x+at} \psi_1(u) \, du$$

where $\psi_0 = \psi(x, 0)$, and $\psi_1 = \partial \psi / \partial t | t = 0$. This is known as d'Alembert's solution. Similar (but more complex) solutions exist for higher-dimensional wave equations — these generally involve complex integrals.

Now we discuss the issue of boundary conditions for hyperbolic partial differential equations. They are similar to but slightly more complicated than boundary conditions for parabolic equations because one usually must specify not only the *value* of $\psi$ at some time, but also its *time-derivative*:

> **We specify the values of $\psi(x_i, t)$ and $\dfrac{\partial \psi(x_i, t)}{\partial t}$ completely at some initial time $t_0$ over the domain $\Omega$ of the problem, and specify the behavior of $\psi$ on the boundary of $\Omega$ at all later times.**

---

[35]Not to be confused with the Schrödinger Wave Equation (77).

The additional complexity of these boundary conditions have a simple physical interpretation in the case of the wave equation. We will consider the one-dimensional wave equation, which essentially describes a vibrating string (in this case, $\psi(x,t)$ represents the *displacement* of the string, as a function of time and $x$-position).

- Specifying the value of $\psi$ at time 0 specifies the initial shape of the string. If this is nonzero, but $\dfrac{\partial \psi(x_i,t)}{\partial t}\bigg|_{t=0} = 0$, we have a *plucked string*.

- Specifying $\dfrac{\partial \psi(x_i,t)}{\partial t}\bigg|_{t=0}$ specifies the initial motion of the string. If $\dfrac{\partial \psi(x_i,t)}{\partial t}\bigg|_{t=0} \neq 0$ but $\psi(x,0) = 0$ we have a "struck" string (like in a piano).

Note that we can also have all possible combinations of these conditions.

Generally speaking, the best approach to finding numerical solutions to hyperbolic partial differential equations is via the *method of characteristics* (related to the integral-formulas for solutions mentioned above). See [6] for a discussion of the method of characteristics.

5.3.2. *Finite differences.* We will use the same technique as before — we convert the equation into a finite difference equation. The fact that the equation is hyperbolic gives rise to some unique properties in the finite difference equation.

$$(91)\quad \frac{\psi(x_1,\ldots,x_n,t+2\delta t) - 2\psi(x_1,\ldots,x_n,t+\delta t) + \psi(x_1,\ldots,x_n,t)}{\delta t^2}$$
$$= \frac{2na^2}{\delta^2}\left(\psi_{\text{average}} - \psi\right)$$

which means that

$$(92)\quad \psi(x_1,\ldots,x_n,t+2\delta t)$$
$$= 2\psi(x_1,\ldots,x_n,t+\delta t) - \psi(x_1,\ldots,x_n,t) + \frac{2na^2\delta t^2}{\delta^2}\left(\psi_{\text{average}} - \psi\right)$$

where $\psi_{\text{average}}$ has the meaning it had in 5.2.1 — it is given in equation (52) on page 257.

Note that this equation presents its own unique features: initially we can specify not only the *value* of $\psi$ at an initial time, but we can also specify the value of $\partial\psi/\partial t$.

As mentioned in the discussion on page 298, there are two possibilities for the boundary conditions:

1. The "plucked" case — here $\psi$ is initially nonzero, but $\partial\psi/\partial t = 0$, initially. In this case we set $\psi(x_1,\ldots,x_n,\delta t) = \psi(x_1,\ldots,x_n,0)$ and we continue the numerical solution from $\psi(x_1,\ldots,x_n,2\delta t)$ on.

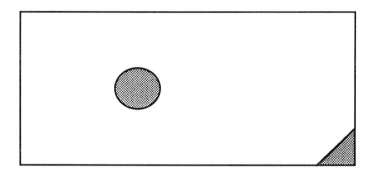

FIGURE V.31. Boundary conditions

2. The "struck" case. In this case $\psi = 0$ initially, but we specify other initial values for $\partial\psi/\partial t$. This is done *numerically* by setting $\psi(x_1, \ldots, x_n, \delta t) = (\partial\psi/\partial t)_{\text{Initial}}\delta t$ and solving for $\psi$ at later times, using 91.

The numeric algorithm tends to be unstable unless the value of $\Delta\psi/\Delta t$ is bounded. In general, we must set up the initial conditions of the problem so that $\psi$ doesn't undergo sudden changes over the spatial coordinates.

Excellent *analytic* solutions can be found in many cases where the boundary has some simple geometric structure. For instance, in the case where the drumhead is *rectangular* the exact analytic solution can be expressed in terms of series involving sines and cosines (Fourier series). In the case where it is *circular* the solution involves *Bessel* functions. Since analytic solutions are preferable to numeric ones whenever they are available, we won't consider such regular cases of the wave equation. We will consider a situation in which the boundary is fairly irregular: it is a rectangular region with a corner chopped off and a fixed disk — i.e., a disk in the rectangle is held fixed. Figure V.31 shows what this looks like.

Here is a C* program that implements this algorithm in the "plucked" case:

```
#include <stdio.h>
int         NROWS = 64;
int         NCOLS = 128;
FILE        *outf;
void        graphout(FILE *, float:current);
shape[64][128] plane;
float       h = 1. / 16.;
float       th = 1. / 64., thratio;
float:plane x, y, psi, oldpsi, psi_del, psi_average,
            psi_rhs;
int:plane canmove;
void
```

```
main()
{
    int         iteration = 0;
    int         i, j;
    float       time;
    /* Initialize coordinates and psi. */
    with(plane) {
        x = pcoord(0);  /* Tell each processor its row
                         * number. */
        y = pcoord(1);  /* Tell each processor its
                         * column number. */
        x = x / 16.0 − 2.0;    /* Compute the
                               * x−coordinate of a
                               * processor. */
        y = y / 16.0 − 4.0;    /* Compute the
                               * y−coordinate of a
                               * processor. */
        /* Set psi to some initial values. */
        psi = 1. / (1. + 2. * x * x + 2. * y * y);
        canmove = (x + y < 5.); /* Cut off corner
                                * 'x+y>=5' */
        /* Cut out small disk centered at (0,−.5) */
        canmove = canmove &&
            ((x * x + (y + .5) * (y + .5)) > .125);
        where(     /* Now constrain the boundary. */
            (pcoord(0) == 0)
            | (pcoord(0) == 63)
            | (pcoord(1) == 0)
            | (pcoord(1) == 127))
              canmove = 0;
        where(canmove == 0) psi = 0.;/* Set psi to 0 where it
                                     * is constrained. */
        else
        /*
         * Now smooth out psi in the neighborhood of the
         * regions cut out.
         */
        psi = psi
            * ((y + .5)
              * (y + .5) − .125)
```

```
                * (5. − x − y)
                * (x + 12)
                * (2. − x)
                * (y + 4.)
                * (4. − y);
        oldpsi = psi;
    }
    thratio = th * th / (h * h);
    time = 0.;
    for (iteration = 0; time < 4.; iteration++) {
        time = iteration;
        time = time * th;
        with(plane)
            where(canmove == 1) {
            psi_average =
                    [.+ 1][.] oldpsi
                    +[.− 1][.] oldpsi
                    +[.][.+ 1] oldpsi
                    +[.][.− 1] oldpsi;
            psi_del = thratio *
                    (psi_average − 4. * oldpsi);
            psi_rhs = psi_del − oldpsi;
            oldpsi = psi;
            psi = 2. * psi + psi_rhs;
            if (iteration % 40 == 0)
                    with(plane) {
                    char        fname[20];
                    printf("Time= %g\n", time);
                    sprintf(fname, "w%d.dat", iteration);
                    outf = fopen(fname, "w");
                    printf("Iteration=%d\n", iteration);
                    graphout(outf, psi);
                    fclose(outf);
                    }
        }
    }
}
```

Here canmove is a parallel variable that determines whether the iteration steps are carried out in a given cycle through the loop. It defines the shape of the region over which the calculations are performed. The initial values of $\psi$ were computed in such a way that it tends to 0 *smoothly* as one approaches the boundary.

The change in $\psi$ as time passes is plotted in figures V.32 through V.34.

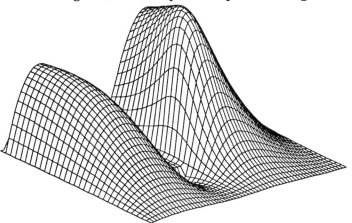

Figure V.32. Initial configuration.

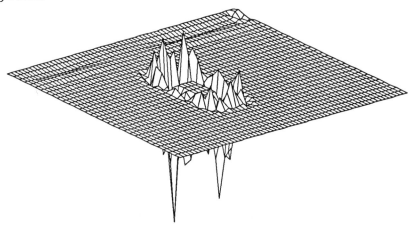

Figure V.33. After 80 iterations.

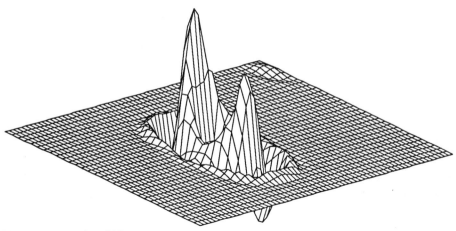

Figure V.34. After 200 iterations.

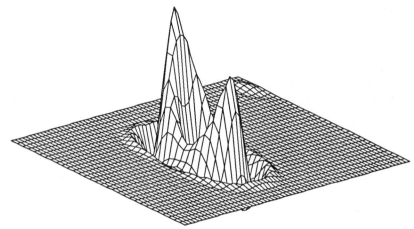

Figure V.35. After 240 iterations.

**5.4. Further reading.** Theoretical physics has provided much of the motivation for the development of the theory of partial differential equations. See [116] and [37] for compendia of physics problems and the differential equations that arise from them.

We have not discussed *mixed* partial differential equations. In general the question of whether a partial differential equation is elliptic, parabolic, or hyperbolic depends on the point at which one tests these properties. Partial differential equations of mixed type have characteristics that actually change from one point to another — in some regions they may be elliptic and in others they may be parabolic, for instance. Here is an example

$$(93) \qquad \left(1 - \frac{u^2}{c^2}\right)\frac{\partial^2 \psi}{\partial x^2} - \frac{2uv}{c^2}\frac{\partial^2 \psi}{\partial x \partial y} + \left(1 - \frac{v^2}{c^2}\right)\frac{\partial^2 \psi}{\partial y^2} = 0$$

This the the equation of 2-dimensional stationary flow without rotation, of a compressible fluid without viscosity. Here $\psi$ is the velocity potential and

$$u = \frac{\partial \psi}{\partial x}$$
$$v = \frac{\partial \psi}{\partial y}$$

are the actual components of the velocity. The number $c$ is the local speed of sound within the fluid — this is some known function of $q = \sqrt{u^2 + v^2}$ (it depends upon the problem). Equation (93) is elliptic if $q < c$ and the flow is said to be *subsonic*. If $q > c$ the flow is *supersonic* and the equation is hyperbolic. As this discussion implies, these equations are important in the study of supersonic fluid flows — see [36] and [14] for applications of mixed differential equations to supersonic shock waves.

We have not touched upon the use of *finite element* methods to numerically solve partial differential equations. See [119] for more information on finite-element methods.

Another method for solving partial differential equations that is gaining wider usage lately is the *boundary element method*. This is somewhat like the methods discussed on page 298 for solving hyperbolic equations. The basic idea of boundary element methods is that (for a linear differential equation) any linear combination of solutions is also a solution. Boundary element methods involve finding *fundamental solutions* to differential equations and expressing arbitrary solutions as sums or integrals of these fundamental solutions, using numerical integration algorithms like those in § 4 on page 245.

# VI
# A Survey of Symbolic Algorithms

In this section we will present a number of symbolic algorithms for the P-RAM computer (since we now know that it can be efficiently simulated by bounded-degree network computers).

## 1. Doubling Algorithms

**1.1. General Principles.** In this section we will present a number of P-RAM algorithms that are closely related. They may be regarded as generalizations of the simple algorithm for adding numbers presented in the introduction, and the sample program in §2 of chapter IV. There are various names for this family of algorithms: doubling algorithms, parallel-prefix algorithms and cyclic reduction algorithms. The different name reflect different applications of this general technique.

The example on page 69 (and the C* program in § 2) of chapter IV shows how to add $n$ numbers in $O(\lg n)$-time. The Brent Scheduling Principle (7.7 on page 52) immediately enables the same computations to be carried out with fewer processors — see 1.5 and 1.6 on page 307. It is not hard to see that this same procedure also works to multiply $n$-numbers in $O(\lg n)$-time. We can combine these to get:

PROPOSITION 1.1. A degree-$n$ polynomial can be evaluated in $O(\lg n)$time on a PRAM computer with $O(n)$ processors.

**Exercises:** Write a C* routine to evaluate a polynomial, given an array of coefficients.

We also get:

PROPOSITION 1.2. Two $n \times n$ matrices $A$ and $B$ can be multiplied in $O(\lg n)$-time using $O(n^3)$ processors.

PROOF. The idea here is that we form the $n^3$ products $A_{ij}B_{jk}$ and take $O(\lg n)$ steps to sum over $j$.   □

Since there exist algorithms for matrix multiplication that require fewer than $n^3$ multiplications (the best current asymptotic estimate, as of as of 1991, is $n^{2.376}$ multiplications — see [33]) we can generalize the above to:

COROLLARY 1.3. If multiplication of $n \times n$ matrices can be accomplished with $M(n)$ multiplications then it can be done in $O(\lg n)$ time using $M(n)$ processors.

This algorithm provides some motivation for using the sample program in §2 of chapter IV: Adding up $n$ numbers efficiently can be done easily by writing

**int** total;

      with(example)

          total += number;

— the C* compiler will automatically use an efficient (i.e $O(\lg n)$-time) algorithm because we are adding instance variables to a mono quantity.

The algorithm for adding up $n$ numbers, given in the introduction, works for *any* *associative* operation.

PROPOSITION 1.4. Let $A$ denote some algebraic system with an associative composition-operation $\star$, i.e. for any $a, b, c \in A$, $(a \star b) \star c = a \star (b \star c)$. Let $a_0, \dots a_{n-1}$ be $n$ elements of $A$. If the computation of $a \star b$ for any $a, b \in A$, can be done in time $T$ then the computation of $\{a_0, a_0 \star a_1, \dots, a_0 \star \cdots \star a_{n-1}\}$ can be done in time $O(T \lg n)$ on a PRAM computer with $O(n)$ processors.

If the operation wasn't associative the (non-parenthesized) composite $a_0 \star \cdots \star a_{n-1}$ wouldn't even be *well-defined*.

PROOF. We follow the sample algorithm in the introduction (or the C* program in §2 in chapter IV) exactly. In fact we present the algorithm as a pseudo-C* program:

```
shape [N]computation;
class pdata:computation {
        datatype an;/* Some data structure
                containing
                an. */
        int PE_number;
        int lower(int);
        void operate(int);
        };
int:computation lower (int iteration)
```

```
{
  int next_iter=iteration+1;
  int:computation PE_num_reduced = (PE_number >>
              next_iter)<<next_iter;
  return PE_num_reduced + (1<<iteration) − 1;
}
void pdata::operate(int iteration)
{
  where (lower(iteration)<PE_number)
  an = star_op([lower(iteration)]an,an);
}
```

Here star_op(a,b) is a procedure that computes $a \star b$. It would have to declared via something like:

```
datatype:current starop(datatype:current,
              datatype:current)
```

We initialize [i]an with $a_i$ and carry out the computations via:

```
for(i=0,i<=log(N)+1;i++)
    with (computation) operate(i);
```

At this point, PE[i].an will equal $a_0 \star \cdots \star a_i$.  □

The Brent Scheduling Principle implies that the number of processors can be reduced to $O(n/\lg n)$:

The following result first appeared in [26]:

ALGORITHM 1.5. Let $\{a_0, \ldots, a_{n-1}\}$ be $n$ elements, let $\star$ be an associative operations. Given $K$ processors, the quantity $A(n) = a_0 \star \cdots \star a_{n-1}$ can be computed in $T$ parallel time units, where

$$T = \begin{cases} \lceil n/K \rceil - 1 + \lg K & \text{if } \lfloor n/2 \rfloor > K \\ \lg n & \text{if } \lfloor n/2 \rfloor \leq K \end{cases}$$

This is a direct application of 7.7 on page 52.

In the first case, we perform *sequential* computations with the $K$ processors until the number of data-items is reduced to $2K$. At that point, the parallel algorithm 2 it used. Note that the last step is possible because we have reduced the number of terms to be added to the point where the original algorithm works.

This immediately gives rise to the algorithm:

COROLLARY 1.6. Let $\{a_0, \ldots, a_{n-1}\}$ be $n$ elements, let $\star$ be an associative operations. Given $n/\lg n$ processors, the quantity $A(n) = a_0 \star \cdots \star a_{n-1}$ can be computed in $O(\lg n)$ parallel time units.

**1.2. Recurrence Relations.** Next, we consider the problem of solving recurrence-relations. These are equations of the form:

$$x_i = x_{i-1} - 2x_{i-2} + 1$$

and the problem is to solve these equations for all the $\{x_i\}$ in terms of some finite set of parameters — in this particular case $x_0$ and $x_1$. Failing this we can try to compute $x_0, \ldots, x_n$ for some large value of $n$.

Recurrence-relations occur in a number of areas including:

- Markov-processes.
- numerical solutions of ordinary differential equations, where a differential equation is replaced by a finite-difference equation. See [163] — this is an entire book on recurrence relations.
- Series solutions of differential equations, in which a general power-series is plugged into the differential equation, usually give rise to recurrence relations that can be solved for the coefficients of that power-series. See [142] for more information on these topics.
- The Bernoulli Method for finding roots of algebraic equations. We will briefly discuss this last application.

Suppose we want to find the roots of the polynomial:

$$f(x) = x^n + a_1 x^{n-1} + \cdots + a_n = 0$$

Let $S_0 = S_1 = \cdots = S_{n-2} = 0$ and let $S_{n-1} = 1$. Now we define the higher terms of the sequence $\{S_i\}$ via the recurrence relation

$$S_k = -a_1 S_{k-1} - a_2 S_{k-2} - \cdots - a_n S_{k-n}$$

Suppose the roots of $f(x) = 0$ are $\alpha_1, \alpha_2, \ldots, \alpha_n$ with $|\alpha_1| \geq |\alpha_2| \geq \cdots \geq |\alpha_n|$. It turns out that, if $\alpha_1$ is a real, simple root, then

$$\lim_{k \to \infty} \frac{S_k}{S_{k-1}} = \alpha_1$$

If $\alpha_1, \alpha_2$ are a pair of complex conjugate roots set $\alpha_1 = Re^{i\theta}, \alpha_2 = Re^{-i\theta}$. If $|\alpha_3| < R$, then

$$(94) \qquad L_1 = \lim_{k \to \infty} \frac{S_k^2 - S_{k+1}S_{k-1}}{S_{k-1}^2 - S_k S_{k-2}} = R^2$$

$$(95) \qquad L_2 = \lim_{k \to \infty} \frac{S_k S_{k-1} - S_{k+1} S_{k-2}}{S_{k-1}^2 - S_k S_{k-2}} = 2R \cos \theta$$

This allows us to compute the largest root of the equation, and we can solve for the other roots by dividing the original polynomial by $(x - \alpha_1)$, in the first case, and by $(x - \alpha_1)(x - \alpha_2)$, in the second. In the second case,

$$(96) \qquad R = \sqrt{L_1}$$

$$(97) \qquad \cos \theta = \frac{L_2}{2\sqrt{L_1}}$$

$$(98) \qquad \sin \theta = \sqrt{1 - (\cos \theta)^2}$$

and

$$(99) \qquad \alpha_1 = R(\cos \theta + i \sin \theta)$$

$$(100) \qquad \alpha_2 = R(\cos \theta - i \sin \theta)$$

Now we turn to the problem of solving recurrence-equations. We first choose a data-structure to represent such a recurrence. Let

$$x_i = \sum_{j=1}^{k} B_{i,j} x_{i-j} + Z_i$$

denote a general $k$-level recurrence relation — here $Z_i$ is a *constant*. We will represent it via a triple $(1, L_i, Z)$, where $L_i$ is the list of numbers

$$[1, -B_{i,1}, \dots, -B_{i,k}]$$

and $Z$ is the list $[Z_1, \dots]$. We will also want to define a few simple operations upon lists of numbers:

1. If $L_1$ and $L_2$ are two lists of numbers we can define the elementwise sum and difference of these lists $L_1 + L_2$, $L_1 - L_2$ — when one list is shorter than the other it is extended by 0's on the right.
2. If $z$ is some real number, we can form the elementwise product of a list, $L$, by $z$: $zL$;
3. We can define a right shift-operation on lists. If $L = [a_1, \dots, a_m]$ is a list of numbers, then $\Sigma L = [0, a_1, \dots, a_m]$.

Given these definitions, we can define an operation $\star$ on objects $(t, L, Z)$, where $t$ is an integer $\geq 1$, $L$ is a list of numbers $[1, 0, \ldots, 0, at, \ldots, a_k]$, where there are at least $t-1$ 0's following the 1 on the left, and $Z$ is a number. $(t_1, L_1, Z_1) \star (t_2, L_2, Z_2) = (t_1 + t_2, L_1 - \sum_{j=j_1}^{t_1+t_2-1} a_j \Sigma^j L_2, Z')$, where $L_1 = [1, 0, \ldots, 0, a_k, \ldots]$, and $Z' = Z_1 - \sum_{j=j_1}^{t_1+t_2-1} a_j$.

Our algorithm for adding up $n$ numbers in $O(\lg n)$-time implies that this composite can be computed in $O(\lg n)$-time on a PRAM, where $n$ is the size of the lists $Z_1$, $Z_2$, and $Z'$. This construct represents the operation of substituting the recurrence-relation represented by $L_2$ into that represented by $L_1$. This follows from how we associate integer-sequences with recurrence-relations: if the recurrence-relation is true, the integer-sequence is a linear form that is identically equal to the number on the right in the triple. It follows that $(t_1, L_1, Z_1) \star (t_2, L_2, Z_2)$ also contains a linear-form that is identically equal to $Z'$ if all of the constituent recurrence-relations are satisfied. It follows that the result of translating $(t_1, L_1, Z_1) \star (t_2, L_2, Z_2)$ back into a recurrence-relation is a logical consequence of the recurrence-relations that gave rise to $(t_1, L_1, Z_1)$ and $(t_2, L_2, Z_2)$. One interesting property of $(t_1, L_1, Z_1) \star (t_2, L_2, Z_2)$ is that it has a run of at least $k_1 + k_2$ zeroes. It is not hard to see that, if $(1, L_i, Z)$, where $L_i = [1, -B_i, 1, \ldots, -B_{i,k}]$ represents the recurrence-relation $x_i = B_{i,j}x_{i-j} + Z_i$ then $(1, L_1, Z) \star \cdots \star (1, L_{n-k}, Z)$ represents a formula expressing $x_n$ in terms of $x_1, \ldots, x_k$ and the $\{Z_i\}$ — this is a solution of the recurrence-relation. Proposition 1.4 implies that:

THEOREM 1.7. Given the recurrence-relation $x_i = \sum_{j=1}^{k} B_{i,j}x_{i-j} + Z_i$ the value of $x_n$ can be computed in time $O(\lg^2(n-k))$ on a PRAM-computer, with $O(kn^2)$ processors.

1. It is only necessary to verify that the composition $(t_1, L_1, Z_1) \star (t_2, L_2, Z_2)$ is *associative*. This is left as an exercise for the reader.

2. The *most general* recurrence-relation has each $x_i$ and $Z_i$ a *linear array* and each $B_j$ a *matrix*. The result above, and 1.4 imply that *these* recurrence-relations can be solved in time that is $O(\lg^3 n)$ — the definitions of the triples $(t, L, Z)$ and the composition-operation $(t_1, L_1, Z_1) \star (t_2, L_2, Z_2)$ must be changed slightly.

Exercises:

1. Consider the Bernoulli Method on page 308. Note that $S_i$ is asymptotic to $\alpha_1^i$ and so, may either go off to $\infty$ or to 0 as $i \to \infty$. How would you deal with situation?

**1.3. Deterministic Finite Automata.** Next we consider an algorithm for the simulation of a deterministic finite automaton. See chapter 2 of [71] for a detailed treatment. We will only recall a few basic definitions.

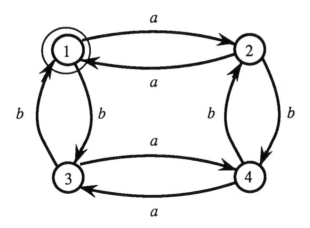

FIGURE VI.1. A DFA

DEFINITION 1.8. A deterministic finite-state automaton (or DFA), $D$, consists of a triple $(A, S, T)$ where:

A is a set of symbols called the alphabet of $D$;

**S** is a set of states of $D$ with a distinguished **Start state**, $S_0$, and **stop states** $\{P_0, \ldots, P_k\}$;

$T$ is a function $T: S \times A \to S$, called the **transition function** of $D$.

A string of symbols $\{a_0, \ldots, a_n\}$ taken from the alphabet, $A$, is called a **string** of the **language** determined by $D$ if the following procedure results in one of the stop states of $D$: Let $s = S_0$, the start state of $D$. **for** i=0 **to** n **do** s=T(s,$a_i$);

DFA's are used in many applications that require simple pattern-recognition capability. For instance the lexical analyzer of a compiler is usually an implementation of a suitable DFA. The following are examples of languages determined by DFA's:

1. All strings of the alphabet $\{a, b, c\}$ whose length is not a multiple of 3.
2. All C comments: i.e. strings of the form '/*' followed by a string that doesn't contain '*/' followed by '*/'.

DFA's are frequently represented by *transition diagrams* — see figure VI.1.

This is an example of a DFA whose language is the set of string in $a$ and $b$ with an *even* number of $a$'s and an *even* number of $b$'s. The *start* state of this DFA is State 1 and this is also its *one stop state*.

We now present a parallel algorithm for testing a given string to see whether it is in the language determined by a DFA. We assume that the DFA is specified by:

1. a $k \times t$-table, $T$, where $k$ is the number of states in the DFA (which we assume to be numbered from 0 to $k - 1$) and $t$ is the number of letters in the alphabet. Without loss of generality we assume that the start state is number 0;
2. A list of numbers between 0 and $k - 1$ denoting the stop states of the DFA.

The *idea* of this algorithm is that:

1. each letter, $a$, of the alphabet can be regarded as a function,
   $f_a: \{0, \ldots, k-1\} \to \{0, \ldots, k-1\}$ — simply define $f_a = T(*, a)$.
2. the operation of *composing functions* is *associative*.

In order to give the formal description of our algorithm we define the following operation:

DEFINITION 1.9. Let $A$ and $B$ be two arrays of size $k$, and whose entries are integers between 0 and $k-1$. Then $A \star B$ is defined via: $(A \star B)_i = B[A[i]]$, for all $0 \le i \le k-1$.

Our algorithm is the following:

ALGORITHM 1.10. Given a character-string of length $n$, $s = \{c_0, \ldots, c_{n-1}\}$, and given a DFA, $D$, with $k$ states and transition function $T$, the following procedure determines whether $s$ is in the language recognized by $D$ in time $O(\lg n)$ with $O(kn)$ processors:

1. Associate a linear array $A(c_j)$ with each character of the string via: $A(c_j)_i = T(i, c_j)$. Each entry of this array is a number from 0 to $k - 1$.
2. Now compute $A = A(c_0) \star \cdots \star A(c_j)$. The string $s$ is in the language recognized by $D$ if and only if $A_0$ is a valid stop state of $D$.

□

1. The composition-operation can clearly be performed in constant time with $k$ processors. Consequently, the algorithm requires $O(\lg n)$ with $O(kn)$ processors.

2. DFA's are frequently described in such a way that some states have *no* transitions on certain characters. If this is the case, simply add an extra "garbage state" to the DFA such that a character that had no transition defined for it in the original DFA is given a transition to this garbage state. All transitions from the garbage state are back to itself, and it is never a valid stop state.

3. This algorithm could be used to implement the front end of a compiler. DFA's in such circumstances usually have actions associated with the various stop states and we are usually more interested in these actions than simply recognizing strings as being in the language defined by the DFA.

Exercises:

1. Write a C* program to recognize the language defined by the DFA in figure VI.1.
2. Consider algorithm 1.4 on page 306. Suppose you are only given only $n/\lg n$ processors (rather than $n$). Modify this algorithm so that it performs its computations in $O(T \lg^2 n)$ time, rather than $O(T \lg n)$-time. (So the "price" of using only $n/\lg n$ processors is a slowdown of the algorithm by a factor of $O(\lg n)$).

3. Same question as exercise 1 above, but give a version of algorithm 1.4 whose execution time is still $O(T \lg n)$ — here the constant of proportionality may be larger than before, but *growth* in time-complexity is no greater than before. Note that this version of algorithm 1.4 gives optimal parallel speedup — the parallel execution time is proportional to the sequential execution time divided by the number of processors. (Hint: look at §7.2 in chapter II)

## 2. Graph Algorithms

One area where a great deal of work has been done in the development of parallel algorithms is that of graph algorithms. Throughout this section $G = (V, E)$ will be assumed to be an undirected graph with vertex set $V$ and edge set $E$ — recall the definitions on page 98. We will also need the following definitions:

DEFINITION 2.1. Let $G = (V, E)$ be a connected graph. Then:

1. $G$ is a *tree* if removal of any vertex with more than one incident edge disconnects $G$. See figure VI.2, part a.
2. $G$ is an *in-tree* if it is a tree, and a directed graph, and there exists one vertex, called the *root*, with the property that there exists a directed path from every vertex in $G$ to the root. See figure VI.2, part b.
3. $G$ is an *out-tree* if it is a tree, and a directed graph, and there exists one vertex, called the *root*, with the property that there exists a directed path from the root to every vertex in $G$. See figure VI.2, part c.

**2.1. The Euler Tour Algorithm.** We will begin with a very simple and ingenious algorithm that is often used as a kind of subroutine to other algorithms. We have already seen an application of this algorithm — see § 3.

The input to this algorithm is an *undirected tree*. The algorithm requires these edges to be ordered in *some* way, but makes no additional requirements. The various versions of this algorithm all compute functions of the tree including:

1. the $\begin{Bmatrix} \text{preorder} \\ \text{postorder} \\ \text{inorder} \end{Bmatrix}$ rank of the vertices.
2. various other functions associated with the graph like distance from the root; number of direct descendants, etc.

In addition we will need to assume that:

1. the edges incident upon each vertex are *ordered* in some way.
2. each node has at least two children.

Applications of the algorithm may require that the ordering of the edges have some other special properties. For instance, the algorithm in § 3 requires that the edges be ordered in a way that is compatible with the terms in the original expression. Figure

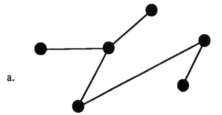

a.

Removal of any interior vertex disconnects this graph

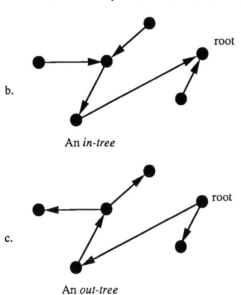

b.

An *in-tree*

c.

An *out-tree*

FIGURE VI.2. Examples of trees

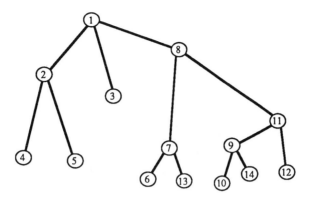

FIGURE VI.3

FIGURE VI.3

VI.3 is a sample tree that we will use in our discussion. We will assume that the root of this tree is the top, and that the edges connecting a vertex with its children are ordered from left to right.

ALGORITHM 2.2. **Euler Tour.**

1. Convert the undirected tree into a directed tree, rooted at some vertex. Mark this root-vertex. Now perform $\lg n$ iterations of the following operation:

   All undirected edges incident upon a marked vertex are directed *away* from it. Now mark all of the vertices at the *other ends* of these new directed edges. This is illustrated in figure VI.4.

   Since the process of directing edges can be done in constant time on a PRAM computer, this entire procedure requires $O(\lg n)$ time. We basically use this step so we can determine the *parent* of a vertex.

2. Replace each directed edge of this tree by two *new* directed edges going in opposite directions between the same vertices as the original edge. We get the result in figure VI.5

3. At each vertex link each directed edge with the next higher directed edge whose direction is compatible with it. For instance, if a directed edge is *entering* the vertex, link it with one that is *leaving* the vertex. The result is a linked list of directed edges and vertices. Each vertex of the original tree gives rise to *several* elements of this linked list. At this point, applications of the Euler Tour technique usually carry out additional operations. In most cases they associate a *number* with each vertex in the *linked list*. Figure VI.6 shows the Euler Tour that results — the darkly-shaded disks represent the vertices of the Euler Tour and the larger circles containing them represent the corresponding vertices of the original tree.

These steps can clearly be carried out in *unit* time with a SIMD computer. The result will be a linked list, called the *Euler Tour* associated with the original tree. What is done

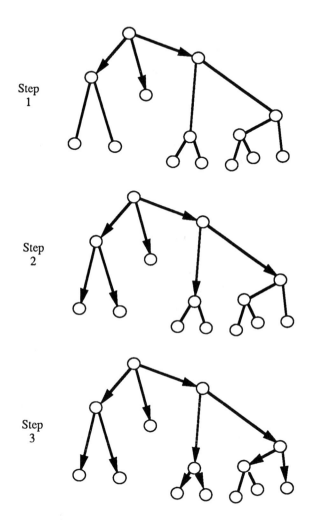

Step
1

Step
2

Step
3

FIGURE VI.4. Converting the input tree into a directed tree

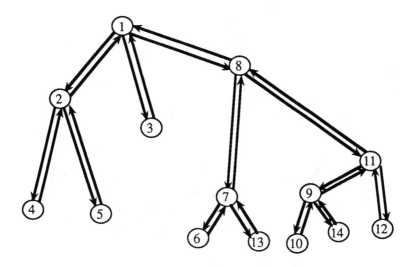

FIGURE VI.5. Graph with doubled edges

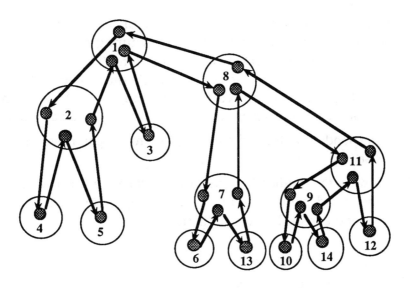

FIGURE VI.6. The Euler Tour

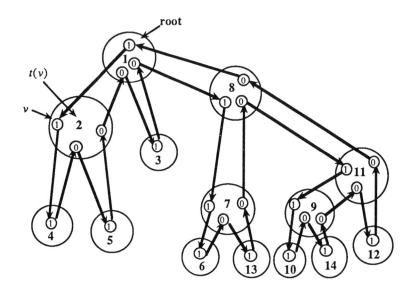

FIGURE VI.7. The preorder-numbered Euler Tour

next depends upon how we choose to apply this Euler Tour.

We will give some sample applications of the Euler Tour technique:

1. Computation of the ordering of the vertices of the original graph in an preorder-traversal.

As remarked above, each vertex of the original graph appears several times in the Euler Tour: once when the vertex is first encountered during a preorder traversal, and again each time the Euler Tour backtracks to that vertex. In order to compute the preorder numbering of the vertices, we simply modify the procedure for forming the Euler Tour slightly. Suppose $v$ is a vertex of the Euler Tour, and $t(v)$ is the corresponding vertex of the original tree that gives rise to $t$. Then $v$ is assigned a value of $a(v)$, where

- 1 if $v$ corresponds to a directed edge coming from a parent-vertex;
- 0 otherwise.

Figure VI.7 illustrates this numbering scheme. We take the list of vertices resulting from breaking the closed path in figure VI.5 at the top vertex (we will assume that vertex 1 is the *root* of the tree). This results in the sequence:

$$\{(1,1), (2,1), (3,1), (4,1), (2,0), (5,1), (2,0),$$
$$(1,0),(3,0), (1,0), (8,1), (7,1), (6,1), (7,0),$$
$$(13,1), (8,0), (11,1), (9,1), (10,1), (9,0),$$
$$(14,1), (11,0), (12,1), (11,0), (8,0), (1,0)\}$$

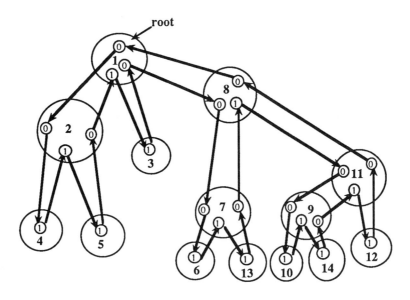

FIGURE VI.8. inorder-numbered Euler Tour

Now compute the cumulative sum of the second index in this list, and we get the rank of each vertex in a preorder traversal of the original tree. This cumulative sum is easily computed in $O(\lg n)$ time via the algorithm on page 69.

$\{(1,1), (2,2), (3,3), (4,4), (5,5), (8,6), (7,7),$
$(6,8), (13,9), (11,10), (9,11), (10,12), (14,13), (12,14)\}$

2. We can also compute an *inorder traversal* of the graph using another variant of this algorithm. We use the following numbering scheme: In this case $a(v)$ is defined as follows:

- if $t(v)$ has no children, $a(v) = 1$;
- if $t(v)$ has children, but $v$ arose as a result of a directed edge entering $t(v)$ from a parent vertex, $a(v) = 0$;
- if $v$ arose as a result of a directed edge entering $t(v)$ from a child, and exiting to another child, *and* it is the lowest-ordered such edge (in the ordering-scheme supplied with the input) then $a(v) = 1$. If it is not the lowest-ordered such edge then $a(v) = 0$
- if $v$ arose as a result of a directed edge entering $t(v)$ from a child, and exiting to a parent, $a(v) = 0$;

Figure VI.8 illustrates this numbering scheme.

Although this version of the algorithm is a little more complicated than the previous one, it still executed in $O(\lg n)$ time.

See [150] for more information on the Euler Tour Algorithm.

**Exercises:**

1. Write a C* program to implement the Euler Tour Algorithm.
2. Given a version of the Euler Tour algorithm that produces the *postfix ordering*.
3. Given a version of the Euler Tour algorithm that counts the *distance* of each vertex from the root.

**2.2. Parallel Tree Contractions.** This is a very general technique for efficiently performing certain types of computations on trees in which each interior vertex has at least two children. It was developed by Miller and Reif in [112], [113], and [114]. Usually, we start with a rooted tree[1], and want to compute some quantity associated with the vertices of the tree. These quantities must have the property that the value of this quantity at a vertex, $v$, depends upon the entire subtree whose root is $v$. In general, the object of the computation using parallel tree contractions, is to compute these quantities for the root of the tree. These quantities might, for instance, be:

- the number of children of $v$.
- the distance from $v$ to the $\left\{\begin{array}{l}\text{closest}\\\text{furthest}\end{array}\right\}$ leaf-vertex, in the subtree rooted at $v$.

The Parallel Tree-Contraction method consists of the following *pruning operation*:

> **Select all vertices of the tree whose immediate children are leaf-vertices. Perform the computation for the selected vertices, and delete their children.**

The requirement that each vertex of the tree has at least two immediate children implies that at least half of all of the vertices of the tree are leaf-vertices, at any given time. This means that at least half of the vertices of the tree are deleted in each step, and at most $O(\lg n)$ pruning operations need be carried out before the entire tree is reduced to its root.

Here is a simple example of parallel tree contraction technique, applied to compute the distance from the root to the nearest leaf:

---

[1]I.e., a tree with some vertex marked as the root.

ALGORITHM 2.3. Let $T$ be a rooted tree with $n$ vertices in which each vertex has at least two and not more than $c$ children, and suppose a number $f$ is associated with each vertex. The following algorithm computes distance from the root to the nearest leaf of $T$ in $O(\lg n)$-time on a CREW-PRAM computer with $O(n)$ processors. We assume that each vertex of $T$ has a data-item $d$ associated with it. When the algorithm completes, the value of $d$ at the root of $T$ will contain the quantity in question.

**while** the root has children **do in parallel**
    Mark all leaf-vertices
    Mark all vertices whose children are leaf-vertices
    **for** all vertices **do in parallel**
        $d \leftarrow 0$
    **endfor**
    **for** each vertex $v$ whose children $z_1, \dots, z_c$ are leaves
        **do in parallel**
            $d(v) \leftarrow 1 + \min(d(z_1), \dots, d(z_c))$
    **endfor**
    Delete the leaf-vertices from $T$
**endwhile**

In order to modify this algorithm to compute the distance from the root to the *furthest* leaf, we would only have to replace the min-function by the max-function.

In order to see that this algorithm works, we use induction on the distance being computed. Clearly, if the root has a child that is a leaf-vertex, the algorithm will compute the correct value — this child will not get pruned until the last step of the algorithm, at which time the algorithm will perform the assignment

$$d(\text{root}) \leftarrow 1$$

If we assume the algorithm works for all trees with the distance from the root to the nearest leaf $\leq k$, it is not hard to see that it is true for trees in which this distance is $k + 1$:

> If $T'$ is a tree for which this distance is $k + 1$, let the leaf nearest the root be $\ell$, and let $r'$ be the child of the root that has $\ell$ as its descendent. Then the distance from $r'$ to $\ell$ is $k$, and the tree-contraction algorithm will correctly compute the distance from $r'$ to $\ell$ in the first few iterations (by assumption).
> The next iteration will set $d(\text{root})$ to $k + 1$.

Parallel tree contractions have many important applications. First significant application was to the parallel evaluation of arithmetic expressions. This application is described in some detail in § 3 on page 370 of this book — also see [113] and [56].

In [114] Miller and Reif develop algorithms for determining whether trees are isomorphic, using parallel tree-contractions.

**2.3. Shortest Paths.** In this algorithm we will assume that the edges of a graph are *weighted*. In other words, they have numbers attached to them. You can think of the weight of each edge as its "length" — in many applications of this work, that is *exactly* what the weights mean. Since each edge of the graph has a length, it makes sense to speak of the length of a path through the graph. One natural question is whether it is possible to travel through the graph from one vertex to another and, if so, what the *shortest* path between the vertices is. There is a simple algorithm for finding the *lengths* of the shortest path and (with a little modification), the paths themselves. We begin with some definitions:

DEFINITION 2.4. Let $G = (V, E)$ be an undirected graph with $n$ vertices.

1. The *adjacency matrix* of $G$ is defined to be an $n \times n$ matrix $A$, such that

$$A_{i,j} = \begin{cases} 1 & \text{if there is an edge connecting vertex } i \text{ and vertex } j \\ 0 & \text{otherwise} \end{cases}$$

2. If $G$ is a directed graph, we define the adjacency matrix by

$$A_{i,j} = \begin{cases} 1 & \text{if there is an edge from vertex } i \text{ to vertex } j \\ 0 & \text{otherwise} \end{cases}$$

3. If $G$ is a weighted graph, the *weight matrix* of $G$ is defined by

$$A_{i,j} = \begin{cases} 0 & \text{if } i = j \\ w(i,j) & \text{if there is an edge from vertex } i \text{ to vertex } j \\ \infty & \text{otherwise} \end{cases}$$

where $w(i,j)$ is the weight of the edge connecting vertex $i$ and $j$ (if this edge exists).

Note that the diagonal elements of all three of these matrices are 0.

PROPOSITION 2.5. Let $G$ be a positively weighted graph with $|V| = n$. There exists an algorithm for the distance between all pairs of vertices of $G$ that executes in $O(\lg^2 n)$ time using $O(n^{2.376})$ processors.

Incidentally, we call a graph *positively weighted* if all of the weights are $\geq 0$.

Our algorithm involves defining a *variant* of *matrix multiplication* that is used for a *dynamic-programming* solution to the distance problem, i.e.

$$(A \times B)_{ij} = \min_k(A_{ik} + B_{kj})$$

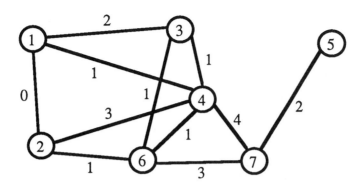

FIGURE VI.9. Weighted graph

Here we assume the distance matrix $D$ is set up in such a way that entries corresponding to missing edges are set to $\infty$ — where $\infty$ is some number that always compares higher than any other number under consideration. Figure VI.9 shows a weighted graph that we might use. The processor-count of $O(n^{2.376})$ is based upon the results of Don Coppersmith Schmuel Winograd in [33], which shows that matrix-multiplication can be done with $n^{2.376}$ processors. Our exotic matrix multiplication algorithm can be implemented in a similar fashion.

The corresponding distance matrix is:

$$D = \begin{pmatrix} 0 & 0 & 2 & 1 & \infty & \infty & \infty \\ 0 & 0 & \infty & 3 & \infty & 1 & \infty \\ 2 & \infty & 0 & 1 & \infty & 1 & \infty \\ 1 & 3 & 1 & 0 & \infty & 1 & 4 \\ \infty & \infty & \infty & \infty & 0 & \infty & 2 \\ \infty & 1 & 1 & 1 & \infty & 0 & 3 \\ \infty & \infty & \infty & 4 & 2 & 3 & 0 \end{pmatrix}$$

As remarked above, matrix multiplication in this sense can be done using $O(n^{2.376})$ processors in $O(\lg n)$ steps. Now "square" the distance matrix $\lceil \lg n \rceil$ times.

Here is a sample C* program for carrying out this "dynamic programming" variant of matrix multiplication

```
#include <stdio.h>
#include <values.h>
#include <stdlib.h>
shape [64][128]mats;
void dynamic_programming(int, int, int, int:current*,
      int:current*, int:current*);
void dynamic_programming(int m, int n, int k,
                int:current *mat1,
                int:current *mat2,
                int:current *outmat)
```

```
{
    shape [32][32][32]tempsh;
    int:tempsh tempv;
    *outmat = MAXINT;
    with(tempsh)
    {
    bool:tempsh region = (pcoord(0) < m) &
            (pcoord(1) < n)
            & (pcoord(2) < k);
        where(region) {
            tempv =[pcoord(0)][pcoord(1)](*mat1) +
                [pcoord(1)][pcoord(2)](*mat2);
            [pcoord(0)][pcoord(2)](*outmat) <?= tempv;
        }
    }
}
```

The number **MAXINT** is defined in the include-file `<values.h>`. It is equal to the largest possible integer — we use it as $\infty^2$. Also note that we have to initialize the *outmat array at **MAXINT** since the reduction-operation <?= has the effect of taking the minimum of the initial value of *outmat and the values of tempv. In other words, it has the effect of repeatedly taking min(*outmat, some value in tempv). The remarks on page 141, regarding a practical program for matrix multiplication apply to this dynamic programming algorithm as well.

The *proof* that this algorithm works is fairly straightforward. Suppose $A$ is the matrix giving the lengths of the edges of a graph and let $A^k$ be the result of carrying out the dynamic programming form of matrix multiplication $k$ times. We claim that $A^k$ gives the length of the shortest paths between pairs of vertices, if those paths have $\leq k + 1$ vertices in them. This is clearly true for $k = 1$. If it is true for some value of $k$ it is not hard to prove it for $A^{k+1}$ — just note that:

1. the shortest path with $k + 1$ vertices in it is the concatenation of its *first edge,* with the shortest path from the end vertex of that edge with $k$ vertices in it, to the destination;

2. we can *find* that shortest path by trying every edge coming out of the starting vertex as a possible candidate for its first edge. Then we pick the combination **(first edge, shortest path with $k$ vertices)** with the *shortest total length.* This variant on the algorithm above is called Floyd's Algorithm.

The second statement basically describes our algorithm for dynamic-programming matrix multiplication. The proof is completed by noting that we don't have to carry out the matrix-multiplication more than $n$ times (if $n$ is the number of vertices in the graph) since no path will have more than $n$ vertices in it.

---

[2]You should use **MAXINT**/2, or **MAXINT**>>1 for $\infty$ in the entries of the A matrix. This is because the addition-step in the dynamic-programming matrix multiplication algorithm could add **MAXINT** to another number, resulting in overflow.

**Exercises:**

1. The algorithm for all-pairs shortest paths requires that the edge-weights be *nonnegative*. What goes wrong if some edge-weights are negative?

2. In the context of the preceding problem, how could the results of this section be generalized to graphs for which some edge-weights are negative.

**2.4. Connected Components.** Another problem for which there is significant interest in parallel algorithms is that of connected components and the closely related problem of minimal spanning trees of graphs.

The first algorithm for connected components was published in 1979. It is due to Hirschberg, Chandra and Sarawate — see [70]. It was designed to run on a CREW-PRAM computer and it executes in $O(\lg n)$ time using $O(n^2/\lg n)$ processors. This algorithm was improved by Willie in his doctoral dissertation so that it used $O(|V| + |E|)$ processors, where $|V|$ is the number of vertices in a graph, and $|E|$ is the number of edges of the graph in question. It was further improved by Chin, Lam, and Chen in 1982 in [27] to use $O(n^2/\lg^2 n)$ processors.

In 1982 Shiloach and Vishkin published (see [141]) an algorithm for connected components that executes in $O(\lg n)$ time using $O(|V| + 2|E|)$ processors. This algorithm is interesting both for its simplicity, and the fact that it requires the parallel computer to be a CRCW-PRAM machine.

2.4.1. *Algorithm for a CREW Computer.* We will discuss a variation on the algorithm of Chin, Lam, and Chen because of its simplicity and the fact that it can be easily modified to solve other graph-theoretic problems, such as that of minimal spanning trees.

Component numbers are equal to the minimum of the vertex numbers over the component.

We regard the vertices of the graph as being partitioned into collections called "super-vertices". In the beginning of the algorithm, each vertex of the graph is regarded as a super-vertex itself. In each phase of the algorithm, each super-vertex is *merged* with at least one other super-vertex to which it is connected by an edge. Since this procedure *halves* (at least) the number of super-vertices in each phase of the algorithm, the total number of phases that are required is $\leq \lg n$. We keep track of these super-vertices by means of an array called $D$. If $v$ is some vertex of the graph, the value of $D(v)$ records the number of the super-vertex containing $v$. Essentially, the number of a super-vertex is equal to the smallest vertex number that occurs within that super vertex — i.e., if a super-vertex contains vertices $\{4, 7, 12, 25\}$, then the *number* of this super-vertex is 4.

At the end of the algorithm $D(i)$ is the number of the lowest numbered vertex that can be reached via a path from vertex $i$.

ALGORITHM 2.6. **Connected Components.**

> **Input:** A graph $G$ with $|V| = n$ described by an $n \times n$ adjacency matrix $A(i, j)$.
> **Output:** A linear array $D(i)$, where $D(i)$ is equal to the component number of the component that contains vertex $i$.
> **Auxiliary memory:** One-dimensional arrays $C$, Flag, and $S$, each with $n$ components.

1. Initialization-step:

   **for all** $i, 0 \leq i < n$ **do in parallel**
   $\quad D(i) \leftarrow i$
   $\quad \text{Flag}(i) \leftarrow 1$
   **endfor**

   The remainder of the algorithm consists in performing

   do steps 2 through 8 lg $n$ times:

2.   a. Construct the set $S$: $S \leftarrow \{i | \text{Flag}(i) = 1\}$.
   b. **Selection.** All vertices record the number of the lowest-numbered neighbor in a parallel variable named $C$:
      **for all pairs** $(i, j), 0 \leq i, j < n$ and $j \in S$ **do in parallel**
      $\quad\quad C(i) \leftarrow \min\{j | A_{i,j} = 1\}$
      $\quad\quad\quad$ **if** $A_{i,j} = 0$ for all $j$, **then** set $C(i) \leftarrow i$
      **endfor**

3. Isolated super-vertices are eliminated. These are super-vertices for which no neighbors were found in the previous step. All computations are now complete for these super-vertices, so we set their flag to zero to prevent them from being considered in future steps.

   **for all** $i \in S$ **do in parallel**
   $\quad$ **if** $C(i) = i$, **then** $\text{Flag}(i) \leftarrow 0$
   **endfor**

4. At the end of the previous step the value $C(i)$ was equal to the smallest super-vertex to which super-vertex $i$ is adjacent. We set the super-vertex number of $i$ equal to this.

   **for all** $i \in S$, **do in parallel**
   $\quad\quad D(i) \leftarrow C(i)$
   **endfor**

5. **Consolidation.** One potential problem arises at this point. The super-vertices might no longer be *well-defined*. One basic requirement is that all vertices within the same super-vertex (set) have the same super-vertex number. We have now updated these super-vertex numbers so they are equal to the number of some *neighboring* super-vertex. This may destroy the consistency requirement, because *that* super-vertex may have been assigned to some third super-vertex. We essentially want $D(D(i))$ to be the same as $D(i)$ for all $i$.

   We restore consistency to the definition of super-vertices in the present step — we perform basically a kind of *doubling* or "folding" operation. This operation never needs to be performed more than $\lg n$ times, since the length of any chain of super-vertex pointers is halved in each step. We will do this by operating on the $C$ array, so that a future assignment of $C$ to $D$ will create well-defined super-vertices.

   **for** $i \leftarrow 0$ **until** $i > \lg n$ **do**
      **for all** $j \in S$ **do in parallel**
         $C(j) \leftarrow C(C(j))$
      **endfor**
   **endfor**

6. Update super-vertex numbers.
   a. Now we update $D$ array again. We only want to update it if the *new* super-vertex values are *smaller* than the *current* ones:
      **for all** $i \in S$ **do in parallel**
         $D(i) \leftarrow \min(C(i), D(C(i)))$
      **endfor**
      This corrects a problem that might have been introduced in step 4 above — if $i$ happens to be the vertex whose number is the same as its super-vertex number (i.e., its number is a minimum among all of the vertices in its super-vertex), then we don't want its $D$-value to change in this iteration of the algorithm. Unfortunately, step 4 *will* have changed its $D$-value to be equal to the minimum of the values that occur among the *neighbors* of vertex $i$. The present step will correct that.
   b. Update $D$ array for *all* vertices (i.e. include those *not* in the currently-active set). This step is necessary because the activity of the algorithm is restricted to a *subset* of all of the vertices of the graph. The present step updates vertices *pointing* to super-vertices that were newly-merged in the present step:
      **for all** $i$ **do in parallel**
         $D(i) \leftarrow D(D(i))$
      **endfor**

7. Here, we clean up the original adjacency matrix $A$ to make it reflect the merging

of super-vertices:

   a. This puts an arc from vertex $i$ to new super-vertex $j$ — in other words $A(i,j) \leftarrow 1$ if there is an edge between $i$ and a vertex merged into $j$.

     **for all** $i \in S$ **do in parallel**
       **for all** $j \in S$ **and** $j = D(i)$ **do in parallel**

$$A(i,j) \leftarrow \bigvee_{\text{for all } k \in S} \{A(i,k) | D(k) = j\}$$

     **endfor**
   **endfor**

Although this step *appears* to require $O(n^3)$ processors, as written here, it actually involves combining certain *rows* of the $A$-matrix. See the C* program at the end of this section (page 332).

   b. This puts an arc from super-vertex vertex $i$ to super-vertex $j$ if there is an arc to $j$ from a vertex merged into $i$.

     **for all** $j \in S$ such that $j = D(j)$ **do in parallel**
       **for all** $i \in S$ **and** $i = D(i)$ **do in parallel**

$$A(i,j) \leftarrow \bigvee_{\text{for all } k \in S} \{A(k,j) | D(k) = i\}$$

     **endfor**
   **endfor**

   c. Remove diagonal entries:

     **for all** $i \in S$ **do in parallel**
       $A(i,i) \leftarrow 0$
     **endfor**

8. One of the ways the algorithm of Chin, Lam and Chen achieves the processor-requirement of only $O(n^2/\lg^2 n)$ is that processors not working on super-vertices are removed from the set of active processors. The original algorithm of Hirschberg, Chandra and Sarawate omits this step.

   **for all** $i \in S$ **do in parallel**
     **if** $D(i) \neq i$ **then**
       Flag$(i) \leftarrow 0$
   **endfor**

As written, and implemented in the most straightforward way, the algorithm above has an execution-time of $O(\lg^2 n)$, using $O(n^2)$ processors. Hirschberg, Chandra and Sarawate achieve a processor-bound of $O(n\lceil n/\lg n\rceil)$ by using a version of 1.5 on page 307 in the steps that compute *minima* of neighboring vertex-numbers[3]. The algorithm of Chin, Lam and Chen in [27] achieves the processor-requirement of $O(n\lceil n/\lg^2 n\rceil)$ by

---

[3]Finding a minimum of quantities is a *census operation* as mentioned on page 117.

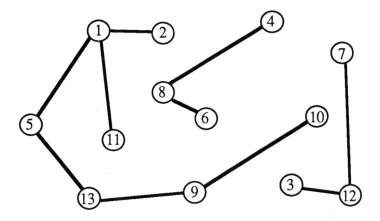

FIGURE VI.10. Initial Graph

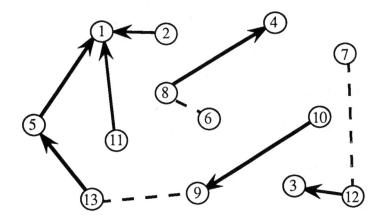

FIGURE VI.11. After the first step

a much more clever (and complex) application of the same technique. Their algorithm makes explicit use of the fact that the only processors that are involved in each step of the computation are those associated with super-vertices (so the number of such processors decreases in each phase of the algorithm).

Here is an example of this algorithm. We will start with the graph in figure VI.10.

After the initial step of the algorithm, the $C$-array will contain the number of the smallest vertex neighboring each vertex. We can think of the $C$-array as defining "pointers", making each vertex point to its lowest-numbered neighbor. We get the graph in figure VI.11.

Vertices that are lower-numbered than any of their neighbors have no pointers coming out of them — vertex 6, for instance.

The $C$-array does *not* define a partitioning of the graph into components at this stage. In order for the $C$-array to define a partitioning of the graph into components,

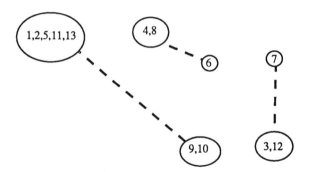

FIGURE VI.12. After the first consolidation

it would be necessary for the C-pointer of the vertex representing all vertices in the same partition to be the *same*. For instance vertex 5 is the target of 13, but couldn't possibly be the vertex that represents a partition of the graph because vertex 5 points to vertex 1, rather than *itself*. We must perform the "folding"-operation to make these partition-pointers consistent. The result is the graph in figure VI.12.

The partitioning of the graph is now well-defined and the number of vertices of the graph has been strictly reduced — each vertex of the present graph is a *super-vertex*, derived from a set of vertices of the original graph. In the next step of the algorithm, the super-vertices will correspond to components of the original graph.

In our C* program we will assume the $A$-matrix is stored in a parallel variable A in a shape named 'graph', and C, D, and Flag are parallel variables stored in a parallel variable in a shape named 'components'.

Here is a C* program that implements this algorithm:

```
#include <values.h>
#include <math.h>
#include <stdio.h>
shape [64][128]graph;
shape [8192]components;
#define N 10
int:graph A, temp;
int:components C, D, Flag,in_S;
int i, j;
FILE *graph_file;
void
    main()
{
    int L = (int) (log((float) N) /
            log(2.0) + 1);

    /* Adjacency matrix stored in a text file. */
    graph_file = fopen("gfile", "r");
    for (i = 0; i < N; i++)
        for (j = 0; j < N; j++)
            {
                int temp;
```

```
            fscanf(graph_file, "%d", &temp);
            [i][j]A = temp;
        }

/*
 * Initialize super-vertex array so that each
 * vertex starts out being a super vertex.
 */
with (components)
    {
    D = pcoord(0);
    Flag=1;
    }


/* Main loop for the algorithm. */
for (i = 0; i <= L; i++)
    with (graph)
    where ((pcoord(0) < N) &
        (pcoord(1) < N))
    {
        int i;

        /* This is step 2. */

        with(components) in_S=Flag;

        /*
         * Locate smallest-numbered
         * neighboring super vertex:
         */

        with(components)
          where ([pcoord(0)]in_S == 1)
            C=pcoord(0);

        where ([pcoord(0)]in_S == 1)
          where (A == 1)
            {
            [pcoord(0)]C <?= pcoord(1);
            }

        /* This is step 3 */
        where ([pcoord(0)]in_S == 1)
          where ([pcoord(0)]C == pcoord(0))
            [pcoord(0)]Flag = 0;

        /* This is step 4 */

        with(components)
          where (in_S == 1)
            D=C;

        /* This is step 5 */
```

```
        for (i = 0; i <= L; i++)
            with (components)
                C = [C]C;

        /* This is step 6a */

        with(components)
            where (in_S == 1)
                D=(C <? [D]C);

        /* This is step 6b */

        with(components)
            D=[D]D;


        /* Step 7a */
        where ([pcoord(0)]in_S == 1)
            where ([pcoord(1)]in_S == 1)
                [[pcoord(1)]D][pcoord(1)]A |= A;

        /* Step 7b */
        where ([pcoord(0)]in_S == 1)
            where ([pcoord(1)]in_S == 1)
                [pcoord(0)][[pcoord(1)]D]A |= A;

        /* Step 7c */
        with(components)
            where ([pcoord(0)]in_S == 1)
                [pcoord(0)][pcoord(0)]A = 0;

        /* Step 8 */
        with(components)
            where ([pcoord(0)]in_S == 1)
                where ([pcoord(0)]D != pcoord(0))
                    [pcoord(0)]Flag = 0;
    }               /* End of big for-loop. */

    for  (i = 0; i < N; i++)
        printf("%d \n",[i]D);
}
```

A few comments are in order here:

In step 7a, the original algorithm, 2.6 on page 326 does

**for all** $i \in S$ **do in parallel**
    **for all** $j \in S$ and $j = D(i)$ **do in parallel**
        $A(i,j) \leftarrow \bigvee_{\text{for all } k \in S} \{A(i,k)|D(k) = j\}$
    **endfor**
**endfor**

This pseudocode in step 7a of 2.6 states that we are to compute the **OR** of all of the vertices in each super-vertex and send it to the vertex that represents it.

The C\* language doesn't lend itself to an explicit implementation of this operation. Instead, we implement an operation that is logically equivalent to it:

> **where** ([pcoord(0)]in_S == 1)
>   **where** ([pcoord(1)]in_S == 1)
>     [[pcoord(1)]D][pcoord(1)]A |= A;

Here, we are using a census-operation in C\* (see page 117) to route information in selected portions of the $A$-array to vertices numbered by D-array values (which represent the super-vertices), and to combine these values by a logical **OR** operation.

*2.4.2. Algorithm for a CRCW computer.* Now we will discuss a faster algorithm for connected components, that takes advantage of concurrent-write operations available on a CRCW-PRAM computer. It is due to Shiloach and Vishkin and it runs in $O(\lg n)$ time using $|V| + 2|E|$ processors, where $|V|$ is the number of vertices in the graph and $|E|$ is the number of edges. It also takes advantage of a changed format of the input data. The input graph of this algorithm is represented as a list of edges rather than an adjacency-matrix.

ALGORITHM 2.7. The following algorithm computes the set of connected components of a graph $G = (V, E)$, where $|V| = n$. The algorithm executes in $O(\lg n)$ steps using $O(|V| + 2|E|)$ processors.

> **Input:** A graph, $G = (V, E)$, represented as a list of edges $\{(v_1, v_2), (v_2, v_1), \dots\}$. Here each edge of $G$ occurs twice — once for each ordering of the end-vertices. Assign one processor to each vertex, and to each entry in the edge-list.
>
> **Output:** A 1-dimensional array, $D_{2 \lg n}$, with $n$ elements. For each vertex $i$, $D(i)$ is the number of the component containing $i$.
>
> **Auxiliary memory:** One dimensional arrays $s$, $D_i$, and $s'$, each of which has $n$ elements. Here $i$ runs from 0 to $2 \lg n$.

**Initialization.** Set $D_0(i) \leftarrow i$ and $Q(i) \leftarrow 0$, for all vertices $i \in V$. Store the $D$-array in the first $n$ processors, where $n = |V|$. Store the list of edges in the next $2m$ processors, where $m = |E|$.

0. **while** $s = s'$ **do in parallel:**
1. Shortcutting:

> **for** $i$, $1 \le i \le n$ **do in parallel**
>   $D_s(i) \leftarrow D_{s-1}(D_{s-1}(i))$
>     **if** $D_s(i) \ne D_{s-1}(i)$
>       $Q(D(i)) \leftarrow s$

**endfor**

This is like *one iteration* of step 5 of algorithm 2.6 — see page 327. We will use the $Q$ array to keep track of whether an array entry was changed in a given step. When none of the array entries are changed, the algorithm is complete.

2. Tree-Hooking

> **for** all processors holding an edge $(u, v)$) **do in parallel**
> > **then if** $D_s(u) = D_{s-1}(u)$
> > > **then if** $D_s(v) < D_s(u)$
> > > > **then** $D_s(D_s(u)) \leftarrow D_s(v)$
> > > > $Q(D_s(v)) \leftarrow u$

**endfor**

This is essentially the *merging* phase of algorithm 2.6. Note that:

1. the present algorithm "pipelines" consolidation and merging, since we only perform *one* step of the consolidation-phase (rather than $\lg n$ steps) each time we perform this step.

2. Only processors that point to super-vertex representatives participate in this step. This limitation is imposed precisely because the consolidation and merging phases are pipelined. Since we have only performed a *single* consolidation step, not all vertices are properly consolidated at this time. We do not want to merge vertices that are not consolidated into super-vertices.

If $D_s(u)$ hasn't been changed (so it pointed to a representative of the current supervertex), then the processor checks to see whether vertex $v$ is contained in a smaller-numbered supervertex. If so, it puts $D_s(u)$ into that supervertex (starting the merge-operation of this supervertex). *Many* processors carry out this step at the *same time* — the CRCW property of the hardware enables only one to succeed.

3. Stagnant Supervertices

> **if** $i > n$ and processor $i$ contains edge $(u, v)$
> > **then if** $D_s(u) = D_s(D_s(u))$ and $Q(D_s(u)) < s$
> > > **then if** $D_s(u) \neq D_s(v)$
> > > > **then** $D_s(D_s(u)) \leftarrow D_s(v)$

These are supervertices that haven't been changed by the first two steps of this iteration of the algorithm — i.e. they haven't been hooked onto any other super-vertex, and no other super-vertex has been hooked onto them. (This fact is determined by the test $Q(D_s(u)) < s$ — $Q(D_s(u))$ records the iteration in which the super-vertex containing $u$ was last updated) The fact that a super-vertex is stagnant implies that:

1. The super-vertex is fully consolidated (so no short-cutting steps are taken).

2. None of the vertices *in* this super-vertex is adjacent to any lower-numbered super-vertex. It follows that every vertex of this super-vertex is adjacent to either:

    a. *Another* vertex of the same super-vertex.

    b. *Higher* numbered super-vertices. This case can *never* occur in algorithm 2.6 (so that stagnant super-vertices also never occur), because that algorithm *always* connects super-vertices to their *lowest numbered* neighbors. In the merging step of the present algorithm, one (random[4]) processor succeeds in updating the $D$-array — it might not be the "right" one.

Stagnant super-vertices have numbers that are *local minima* among their neighbors. The present step arbitrarily merges them with any neighbor.

4. Second Shortcutting

**for** $i, 1 \leq i \leq n$ **do in parallel**
    **then** $D_s(i) \leftarrow D_s(D_s(i))$
**endfor**

This has two effects:

– It performs a *further consolidation* of non-stagnant super-vertices.

– It *completely consolidates* stagnant super-vertices that were merged in the previous step. This is due to the definition of a stagnant super-vertex, which implies that it is already completely consolidated within itself (i.e., all vertices in it point to the root). Only *one* consolidation-step is required to incorporate the vertices of this super-vertex into its neighbor.

5. Completion criterion

**if** $i \leq n$ and $Q(i) = s$
    **then** $s' \leftarrow s' + 1$
        $s \leftarrow s + 1$

We conclude this section with an example.

EXAMPLE 2.8. We use the graph depicted in figure VI.10 on page 329. Initially $D(i) \leftarrow i$.

1. The first short-cutting step has no effect.
2. The first tree-hooking step makes the following assignments:
    a. $D(D(2)) \leftarrow D(1)$;
    b. $D(D(11)) \leftarrow D(1)$
    c. $D(D(5)) \leftarrow D(1)$;

---

[4]At least, it is unspecified. *However* the hardware decides which store-operation succeeds, it has nothing to do with the present algorithm.

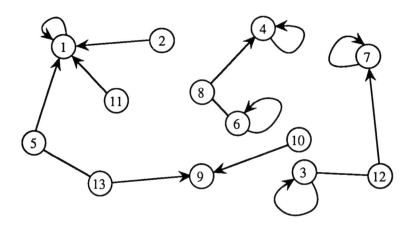

FIGURE VI.13. Result of the Tree-Hooking step of the Shiloach-Vishkin Algorithm

d. $\begin{cases} D(D(13)) \leftarrow D(5) \\ D(D(13)) \leftarrow D(9) \end{cases}$ — this is a CRCW assignment. We assume that

$D(D(13)) \leftarrow D(9)$ actually takes effect;

e. $D(D(10)) \leftarrow D(9)$;

f. $\begin{cases} D(D(8)) \leftarrow D(4) \\ D(D(8)) \leftarrow D(6) \end{cases}$ — this is a CRCW assignment. We assume that

$D(D(8) \leftarrow D(4)$ actually takes effect;

g. $\begin{cases} D(D(12)) \leftarrow D(7) \\ D(D(12)) \leftarrow D(3) \end{cases}$ — this is a CRCW assignment. We assume that

$D(D(12) \leftarrow D(7)$ actually takes effect;

The result of this step is depicted in figure VI.13.

3. Vertices 3 and 6 represents supervertices that weren't changed in any way in the first two steps of the algorithm. The are, consequently, *stagnant*. The second tree-hooking step hooks them onto neighbors. The result is depicted in figure VI.14.

4. The second consolidation-step combines all of the stagnant vertices into their respective super-vertices to produce the graph in figure VI.15.

5. The next iteration of the algorithm completely processes all of the components of the graph.

**Exercises:**

1. The Connection Machine is a CRCW computer. Program the Shiloach-Vishkin algorithm in C*.

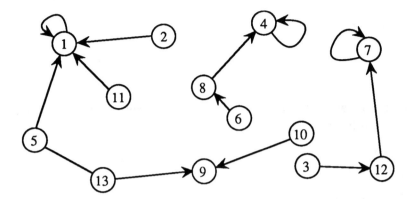

FIGURE VI.14. Tree-Hooking of Stagnant Vertices

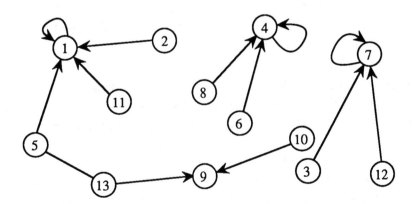

FIGURE VI.15. Second Short-cutting

2. Modify the Chin-Lam-Chen algorithm to accept its input as a list of edges. This reduces the execution-time of the algorithm somewhat. Does this modification allow the execution-time to become $O(\lg n)$?

**2.5. Spanning Trees and Forests.** Another interesting problem is that of finding *spanning trees* of undirected graphs.

DEFINITION 2.9. Let $G = (V, E)$ be a connected graph. A *spanning tree* of $G$ is a subgraph $T$ such that:

1. All of the vertices of $G$ are in $T$;
2. $T$ is a *tree* — i.e. $T$ has *no cycles*.

If $G$ is not connected, and consists of components $\{G_1, \ldots, G_k\}$, then a *spanning forest* of $G$ is a set of trees $\{T_1, \ldots, T_k\}$, where $T_i$ is a spanning tree of $G_i$.

In [133], Carla Savage showed that the connected components algorithm, 2.6 on page 326, actually computes a spanning forest of $G$.

Whenever a vertex (or super-vertex) of the algorithm is merged with a lower-numbered neighboring vertex (or super-vertex) we select the edge connecting them to be in the spanning tree. It is clear that the set of selected edged will form a subgraph of the original graph that *spans* it — every vertex (and, in the later stages, every super-vertex) participated in the procedure. It is not quite so clear that the result will be a *tree (or forest)* — we must show that it has no closed cycles.

We show this by contradiction:

> Suppose the set of selected edges (in some phase of the algorithm) has a cycle. Then this cycle has a vertex (or super-vertex) whose numbers is maximal. That vertex can only have a single edge incident upon it since:
>
> - No other vertex will select it;
> - It will only select a single other vertex (its *minimal* neighbor).
>
> This contradicts the assumption that this vertex was in a cycle.

We can modify algorithm 2.6 by replacing certain steps, so that whenever it merges two super-vertices along an edge, it records that edge. The result will be a spanning tree of the graph. We augment the data-structures in that algorithm with four arrays:

- $\mathrm{Edge}(1, *)$, $\mathrm{Edge}(2, *)$, where $\{(\mathrm{Edge}(1, i), \mathrm{Edge}(2, i)) | i$ such that $D(i) \neq i\}$, is the set of edges of the spanning tree.

- $B(1, i, j)$, $B(2, i, j)$ record the endpoints of the edge connecting super-vertices $i$ and $j$, when those super-vertices get merged. This reflects a subtlety of the spanning-tree problem: Since we have collapsed *sets* of vertices into *super-vertices* in each iteration, we need some mechanism for recovering information about the original vertices and edges in the graph. When we merge two super-vertices in our algorithm, we use the $B$-arrays to determine which edge in the original graph was used to accomplish this merge.

The execution-time of the present algorithm is, like the connected-components algorithm, equal to $O(\lg^2 n)$. Like that algorithm, it can be made to use $O(n^2/\lg^n)$ processors, though a clever (and somewhat complicated) application of the Brent Scheduling Principle. The modified algorithm is

ALGORITHM 2.10. **Spanning Forest.**

**Input:** A graph $G$ with $|V| = n$ described by an $n \times n$ adjacency matrix $A(i, j)$.
**Output:** A $2 \times n$ array Edge, such that $\text{Edge}(1, i)$ and $\text{Edge}(2, i)$ are the end-vertices of the edges in a spanning tree.
**Auxiliary memory:** A one-dimensional arrays $C$ and Flag, each with $n$ elements. A $2 \times n \times n$ array $B$.

1. Initialization.
   a.
   **for all** $i, 0 \le i < n - 1$ **do in parallel**
      $D(i) \leftarrow i$
      $\text{Flag}(i) \leftarrow 1$
      $\text{Edge}(1, i) \leftarrow 0$
      $\text{Edge}(2, i) \leftarrow 0$
   **endfor**
   b. We initialize the $B$-arrays. $B(1, i, j)$ and $B(2, i, j)$ will represent the end-vertices that will connect super-vertices $i$ and $j$.
   **for all** $i, j, 0 \le i, j \le n - 1$ **do in parallel**
      $B(1, i, j) \leftarrow i$
      $B(2, i, j) \leftarrow j$
   **endfor**

The remainder of the algorithm consists in

do steps 2 through 9 $\lg n$ times:

Construct the set $S$: $S \leftarrow \{i | \text{Flag}(i) = 1\}$.

2. Selection. As in algorithm 2.6, we choose the lowest-numbered super-vertex, $j_0$, adjacent to super-vertex $i$. We record the edge involved in $\text{Edge}(1, i)$ and $\text{Edge}(2, i)$. It is necessary to determine which *actual edge* is used to connect these super-vertices, since the numbers $i$ and $j$ are only *super-vertex* numbers. We use the $B$-arrays for this.

   **for all** $i \in S$ **do in parallel**
      Choose $j_0$ such that $j_0 = \min\{j | A(i,j) = 1; j \in S\}$
      **if none then** $j_0 \leftarrow j$
      $C(i) \leftarrow j_0$
      $\text{Edge}(1,i) \leftarrow B(1,i,j_0)$
      $\text{Edge}(2,i) \leftarrow B(2,i,j_0)$
   **endfor**

3. Removal of isolated super-vertices.

   **for all** $i$ such that $i \in S$ **do in parallel**
      **if** $C(i) = i$, **then** $\text{Flag}(i) \leftarrow 0$

4. Update $D$.

   **for all** $i \in S$, **do in parallel**
      $D(i) \leftarrow C(i)$
   **endfor**

5. Consolidation.

   **for** $i \leftarrow 0$ **until** $i > \lg n$
      $j \in S$ **do in parallel**
         $C(j) \leftarrow C(C(j))$
   **endfor**

6. Final update.

      a.
      **for all** $i \in S$ **do in parallel**
         $D(i) \leftarrow \min(C(i), D(C(i)))$
      **endfor**
      b. Propagation of final update to previous phase.
      **for all** $i$ **do in parallel**
         $D(i) \leftarrow D(D(i))$
      **endfor**

7. Update the incidence-matrix and $B$-arrays.

      a. We update the $A$-array, as in algorithm 2.6. We must also update the $B$ arrays to keep track of which actual vertices in the original graph will be adjacent to vertices in other super-vertices. We make use of the *existing* $B$-arrays in this procedure. This step locates, for each vertex $i$, the super-vertex $j$ that contains a vertex $j_0$ adjacent to $i$. We also record, in the $B$-arrays, the edge that is used.
      **for all** $i \in S$ **do in parallel**
         **for all** $j \in S$ such that $j = D(j)$ **do in parallel**
            Choose $j_0 \in S$ such that $D(j_0) = j$ AND $A(i,j_0) = 1$
            **if none then** $j_0 \leftarrow j$

   **endfor**
   **endfor**
   $A(i,j) \leftarrow A(i, j_0)$
   $B(1, i, j) \leftarrow B(1, i, j_0)$
   $B(2, i, j) \leftarrow B(2, i, j_0)$
   b. This step locates, for each super-vertex $j$, the super-vertex $i$ that contains a vertex $i_0$ adjacent to $j$. We make explicit use of the results of the
   previous step. This step completes the merge of the two super-vertices.
   The $A$-array now reflects adjacency of the two *super*-vertices (in the last
   step we had vertices of one super-vertex being adjacent to the other super-
   vertex). The $B$-arrays for the pair of super-vertices now contain the edges
   found in the previous step.
   **for all** $j \in S$ such that $j = D(j)$ **do in parallel**
       **for all** $i \in S$ such that $i = D(i)$ **do in parallel**
           Choose $i_0 \in S$ such that $D(i_0) = i$ AND $A(i_0, j) = 1$
           **if** none **then** $i_0 \leftarrow i$
       **endfor**
   **endfor** $A(i,j) \leftarrow A(i_0, j)$
   $B(1, i, j) \leftarrow B(1, i_0, j)$
   $B(2, i, j) \leftarrow B(2, i_0, j)$
   c.
   **for all** $i \in S$ **do in parallel**
       $A(i, i) \leftarrow 0$
   **endfor**
8. Select only current super-vertices for remaining phases of the
   algorithm.
   **for all** $i \in S$ **do in parallel**
       **if** $D(i) \neq i$ **then**
           Flag$(i) \leftarrow 0$
   **endfor**
9. Output selected edges.
   **for all** $i, 0 \leq i \leq n - 1$ **do in parallel**
       **if** $i \neq D(i)$ **then** output
           $(\text{Edge}(1, i), \text{Edge}(2, i))$
   **endfor**

The handling of the $B$-arrays represents one of the added subtleties of the present
algorithm over the algorithm for connected components. Because of this, we will give
a detailed example:

EXAMPLE 2.11. We start with the graph depicted in figure VI.16. Its incidence matrix

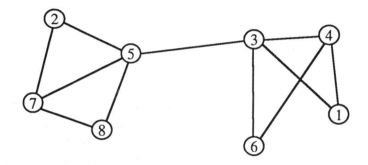

FIGURE VI.16

is:

(101)
$$A = \begin{pmatrix} 1 & 0 & 1 & 1 & 0 & 0 & 0 & 0 \\ 0 & 1 & 0 & 0 & 1 & 0 & 1 & 0 \\ 1 & 0 & 1 & 1 & 0 & 1 & 0 & 0 \\ 1 & 0 & 1 & 1 & 0 & 1 & 0 & 0 \\ 0 & 1 & 0 & 0 & 1 & 0 & 1 & 1 \\ 0 & 0 & 1 & 1 & 0 & 1 & 0 & 0 \\ 0 & 1 & 0 & 0 & 1 & 0 & 1 & 1 \\ 0 & 0 & 0 & 0 & 1 & 0 & 1 & 1 \end{pmatrix}$$

Now we run the algorithm. At the start of phase 1 we have:

- $B(1, i, j) = i$, $B(2, i, j) = j$, for $1 \leq i, j \leq 8$;
- $\text{Flag}(i) = 1$, for $1 \leq i \leq 8$;
- $\text{Edge}(1, i) = \text{Edge}(2, i) = 0$ for $1 \leq i \leq 8$;
- $S = \{1, 2, 3, 4, 5, 6, 7, 8\}$;

Step 2 sets the $C$-array and updates the $\text{Edge}(1, *)$ and $\text{Edge}(2, *)$ arrays:

$$C = \begin{pmatrix} 3 \\ 5 \\ 1 \\ 1 \\ 2 \\ 3 \\ 2 \\ 5 \end{pmatrix}, \ \text{Edge}(1, *) = \begin{pmatrix} 1 \\ 2 \\ 3 \\ 4 \\ 5 \\ 6 \\ 7 \\ 8 \end{pmatrix} \text{ and, } \text{Edge}(1, *) = \begin{pmatrix} 3 \\ 5 \\ 1 \\ 1 \\ 2 \\ 3 \\ 2 \\ 5 \end{pmatrix}$$

In step 4 we perform the assignment $D \leftarrow C$, and in step 5 we consolidate the

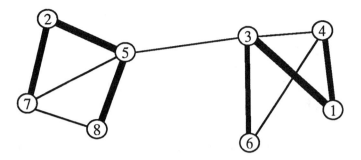

FIGURE VI.17

$C$-array. The outcome of this step is:

$$D = \begin{pmatrix} 3 \\ 5 \\ 1 \\ 1 \\ 2 \\ 3 \\ 2 \\ 5 \end{pmatrix} \text{ and } C = \begin{pmatrix} 1 \\ 2 \\ 3 \\ 3 \\ 5 \\ 1 \\ 5 \\ 2 \end{pmatrix}$$

In step 6 we perform the final update of the $D$-array to get:

$$D = \begin{pmatrix} 1 \\ 2 \\ 1 \\ 1 \\ 2 \\ 1 \\ 2 \\ 2 \end{pmatrix}$$

Note that the construction of the spanning tree is not yet complete — we have two supervertices (numbered 1 and 2), as depicted in figure VI.17, and spanning-trees of each of these super-vertices.

Now we update the incidence matrix and the $B$ arrays, in step 7. Part a:

$i = 1$ No change.

$i = 2$ No change.

$i = 3$ When $j = 1$ there is no change, but when $j = 2$, we get
- $j_0 = 5$;
- $B(1,3,2) \leftarrow 3$;
- $B(2,3,2) \leftarrow B(2,3,5) = 5$.

$i = 5$ When $j = 2$ there is no change, but when $j = 1$, we get
- $j_0 = 3$;
- $B(1,5,1) \leftarrow 5$;
- $B(2,5,1) \leftarrow B(2,5,3) = 3$.

Note that the updating of the arrays is not yet complete. We know that vertex 3 is adjacent to super-vertex 1, but we don't know that super-vertices 1 and 2 are adjacent.

The next phase of the algorithm completes the spanning tree by adding edge $(3,5) = (B(1,1,2), B(2,1,2))$ to it.

Part b:

$j = 1$ If $i = 2$ we get $i_0 = 5$ and
- $A(2,1) = 1$;
- $B(1,2,1) = 5$;
- $B(2,2,1) = 3$.

$j = 2$ If $1 = 2$ we get $i_0 = 3$ and
- $A(1,2) = 1$;
- $B(1,1,2) = 3$;
- $B(2,1,2) = 5$.

The next iteration of this algorithm merges the two super-vertices 1 and 2 and completes the spanning tree by adding edge $(B(1,1,2), B(2,1,2)) = (3,5)$.

**Exercises:**

1. In step 7 the updating of the $A$-matrix and the $B$-matrices is done in two steps because the algorithm must be able to run on a CREW computer. Could this operation be simplified if we had a CRCW computer?
2. Is it possible to convert the Shiloach-Vishkin algorithm (2.7 on page 333) for connected components into an algorithm for a spanning-tree that runs in $O(\lg n)$ time on a CRCW computer? If so, what has to be changed?

2.5.1. *An algorithm for an inverted spanning forest.* The algorithm above can be modified to give an algorithm for an *inverted spanning forest* of the graph in question. This is a spanning tree of each component of a graph that is a *directed tree*, with the edges of the tree pointing toward the root. There are a number of applications for an inverted

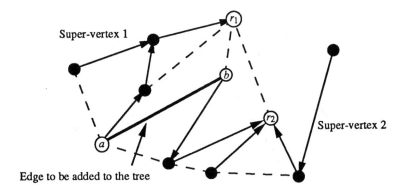

FIGURE VI.18. Forming a directed spanning tree

spanning forest of an undirected graph. We will be interested in the application to computing a *cycle-basis* for a graph in section 2.7 on page 363. A cycle-basis can be used to determine and enumerate the closed cycles of a graph.

Algorithm 2.10 on page 339 *almost* accomplishes this: it finds directed edges that point to the vertex representing a super-vertex. The problem with this algorithm is that, when two super-vertices are merged, the vertices that get joined by the merge-operation may not be the parents of their respective sub-spanning trees. Consequently, the directions of the edges are not compatible, and we don't get a directed spanning tree of the new super-vertex that is formed — see figure VI.18. The two super-vertices in this figure cannot be merged along the indicated edge in such a way that that the directionality of the subtrees are properly respected. The solution to this problem is to *reverse* the directions of the edges connecting one root to a vertex where the merge-operation is taking effect.

Several steps are involved. Suppose we want to merge super-vertex 1 with super-vertex 2 in such a way that super-vertex 1 becomes a subset of super-vertex 2. Suppose, in addition, that this merging operation takes place along an edge $(a, b)$. We must find the *path* in super-vertex 1 that connects vertex $a$ with the *representative* of this super-vertex, $r_1$. Then we reverse the directions of the edges along this path — and we obtain the result shown in VI.19.

In order to carry out this operation we must have an algorithm for computing the path from $a$ to $r_1$. We have an *in-tree* (see 2.1 on page 313 for the definition) on the super-vertex represented by $r_1$ — we only need to *follow* the directed edges to their target.

We will want to regard these directed edges as defining a *function* on the vertices in this super-vertex. The value of this function on any vertex, $v$, is simply the vertex that is at the *other end* of unique directed edge containing $v$. This is simpler than it sounds — the directed edges are given by the arrays $(\text{Edge}(1, i), \text{Edge}(2, i))$ in algorithm 2.18 on page 357. We define $f(\text{Edge}(1, i)) = \text{Edge}(2, i)$.

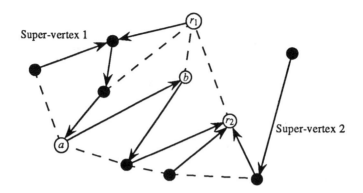

FIGURE VI.19. Directed spanning tree on merged super-vertices

The path from $a$ to $r_1$ is *nothing* but the sequence of vertices that result from *repeated iteration* of $f$: $\{a, f(a), f(f(a)), \ldots, f^{(k)}(a)\}$. Having computed $f$ itself, we can compute the result of repeatedly iterating $f$ in $O(\lg n)$ time by a cyclic-reduction type of algorithm like 1.4 on page 306. We get the following algorithm for computing paths from vertices in an in-tree (defined in 2.1 on page 313) to its root:

ALGORITHM 2.12. Suppose $T$ is an in-tree with root $r$ and let $f$, be the function that maps a vertex to its successor in $T$. Then, given any vertex $a$ in the super-vertex[5] corresponding to $f$, we can compute the sequence $\{a, f(a), f(f(a)), \ldots, f^{(k)}(a)\}$ by the following procedure:

**for** all $i$ such that $1 \leq i \leq n$ **do in parallel**
   $f^0(i) \leftarrow i, f^1(i) \leftarrow f(i)$
      **for** $t \leftarrow 0$ **to** $\lg(n-1) - 1$ **do**
         **for** $s$ such that $1 \leq s \leq 2^t$
         and $i$ such that $1 \leq i \leq n$ **do in parallel**
           $f^{2^t+s}(i) \leftarrow f^{2^t}(f^s(i))$
      **endfor**
**endfor**

We will use the notation $\hat{f}$ to denote the table that results from this procedure. This can be regarded as a function of *two* variables: $\hat{f}(v, i) = f^i(v)$.

We are now in a position to give an algorithm for finding an inverted spanning forest of a graph. We basically perform the steps of the spanning-tree algorithm, 2.10 on page 339, and in each iteration of the main loop:

Each super-vertex is equipped with an inverted spanning tree of *itself* (i.e., as a subgraph of the main graph). Whenever we *merge* two super-vertices, one of the super-vertices (called the *subordinate one* gets merged into the other (namely, the lower

---

[5]Recall that $f$ is a directed spanning-tree defined on a super-vertex of the original graph.

numbered one). We perform the following additional steps (beyond what the original spanning-tree algorithm does):

1. We determine which *vertex*, $v$, of the subordinate super-vertex is attached to the other. This is a matter of keeping track of the *edge* being used to join the super-vertices.

2. We compute a path, $p$, from $v$ in the inverted spanning tree of its super-vertex, to its root. We use algorithm 2.12 above.

3. We *reverse* the direction of the edges along this path.

This amounts to modifying several steps of algorithm 2.10 and results in an algorithm that requires $O(\lg^2 n)$ time and uses $O(n^2/\lg n)$ processors. The step in which we apply algorithm 2.12 can be done in $O(\lg n)$-time using $O(n\lceil n/\lg n\rceil)$ processors — it, consequently, dominates the execution-time or the processor-count of the algorithm. In [151], Tsin and Chin present a fairly complicated algorithm for an inverted spanning forest that is a variation upon the present one, but only requires $O(n^2/\lg^2 n)$ processors. Our algorithm for an inverted spanning forest is thus:

ALGORITHM 2.13. **Inverted Spanning Forest.**

>   **Input:** A graph $G$ with $|V| = n$ described by an $n \times n$ adjacency matrix $A(i, j)$.

>   **Output:** A function, $f$, such that for all vertices, $v \in V$, $f(v)$ is the successor of $v$ in an inverted spanning tree of $G$.

>   **Auxiliary memory:** A one-dimensional arrays $C$, Flag, and $\hat{f}$, each with $n$ elements. A $2 \times n \times n$ array $B$.

1. Initialization.

>   a.
>
>   **for** all $i, 0 \leq i < n - 1$ **do in parallel**
>   $\quad D(i) \leftarrow i$
>   $\quad \text{Flag}(i) \leftarrow 1$
>   $\quad f(i) \leftarrow i$
>   **endfor**
>
>   b. We initialize the $B$-arrays. $B(1, i, j)$ and $B(2, i, j)$ will represent the end-vertices that will connect super-vertices $i$ and $j$.
>
>   **for** all $i, j, 0 \leq i, j \leq n - 1$ **do in parallel**
>   $\quad B(1, i, j) \leftarrow i$
>   $\quad B(2, i, j) \leftarrow j$
>   **endfor**

The remainder of the algorithm consists in

do steps 2 through 8 $\lg n$ times:

Construct the set $S$: $S \leftarrow \{i | \text{Flag}(i) = 1\}$.

2. Selection. As in algorithm 2.6, we choose the lowest-numbered super-vertex, $j_0$, adjacent to super-vertex $i$. We record the edge involved in the $f$-function.

It is necessary to determine which *actual edge* is used to connect these super-vertices, since the numbers $i$ and $j$ are only *super-vertex* numbers. We use the $B$-arrays for this.

a.

**for all** $i \in S$ **do in parallel**
    Choose $j_0$ such that $j_0 = \min\{j | A(i,j) = 1; j \in S\}$
    **if** none **then** $j_0 \leftarrow j$
    $C(i) \leftarrow j_0$
**if** $(D(B(1,i,j_0)) = D(i))$ **then**
    $f(B(1,i,j_0)) \leftarrow B(2,i,j_0)$
**endfor**

b. In this step we compute $(\hat{f})$, using algorithm 2.12.
$\hat{f}(1,*) \leftarrow f$
**for** $i \leftarrow 1$ **to** $\lg(n-1) - 1$ **do**
    **for** $j, 1 \le j \le 2^i$ **do in parallel**
        $\hat{f}(2^i + j, *) \leftarrow \hat{f}(2^i, *) \circ \hat{f}(j, *)$
    **endfor**
**endfor**

c. Now we compute the *lengths* of the paths connecting vertices to the roots of their respective super-vertices. We do this by performing a binary search on the sequence $\hat{f}(j,i)$ for each vertex $i$
**for** $j, 1 \le j \le n$ **do in parallel**
    $\text{Depth}(i) \leftarrow \min\{j | \hat{f}(j,i) = D(i)\}$
**endfor**

d. Step 2a above adjoined a new edge, $e$, to the spanning tree. This edge connected two super-vertices, $i$ and $j_0$, where super-vertex $i$ is being incorporated into super-vertex $j_0$. Now we *reverse* the edges of the path that from the end of $e$ that lies in super-vertex $i$ to vertex $i$. The end of $e$ that lies in super-vertex $i$ is numbered $B(1,i,j_0)$.
**for** $k, 1 \le k \le \text{Depth}(B(1,i,j_0))$ **do in parallel**
    $\text{temp1}(i) \leftarrow B(1,i,j_0)$
    $\text{temp2}(i) \leftarrow f(\text{temp1}(i))$
**endfor**

**for** $k, 1 \le k \le \text{Depth}(B(1,i,j_0))$ **do in parallel**
    $f(\text{temp2}(i)) \leftarrow \text{temp1}(i)$
**endfor**

3. Removal of isolated super-vertices.

**for all** $i$ such that $i \in S$ **do in parallel**

if $C(i) = i$, then Flag$(i) \leftarrow 0$
   **endfor**

4. Update $D$.

   **for** $i \in S$, **do in parallel**
      $D(i) \leftarrow C(i)$
   **endfor**

5. Consolidation.

   **for** $i \leftarrow 0$ **until** $i > \lg n$
      $j \in S$ **do in parallel**
         $C(j) \leftarrow C(C(j))$
      **endfor**
   **endfor**

6. Final update.

      a.

         **for** $i \in S$ **do in parallel**
            $D(i) \leftarrow \min(C(i), D(C(i)))$
      **endfor**

      b. Propagation of final update to previous phase.
         **for all** $i$ **do in parallel**
            $D(i) \leftarrow D(D(i))$
         **endfor**

7. Update the incidence-matrix and $B$-arrays.

      a.

      **for all** $i \in S$ **do in parallel**
         **for all** $j \in S$ such that $j = D(j)$ **do in parallel**
            Choose $j_0 \in S$ such that $D(j_0) = j$ AND $A(i, j_0) = 1$
            **if none then** $j_0 \leftarrow j$
            $A(i, j) \leftarrow A(i, j_0)$
            $B(1, i, j) \leftarrow B(1, i, j_0)$
            $B(2, i, j) \leftarrow B(2, i, j_0)$
         **endfor**
      **endfor**

      b.

      **for all** $j \in S$ such that $j = D(j)$ **do in parallel**
         **for all** $i \in S$ such that $i = D(i)$ **do in parallel**
            Choose $i_0 \in S$ such that $D(i_0) = i$ AND $A(i_0, j) = 1$
            **if none then** $i_0 \leftarrow i$
            $A(i, j) \leftarrow A(i_0, j)$
            $B(1, i, j) \leftarrow B(1, i_0, j)$

$$B(2,i,j) \leftarrow B(2,i_0,j)$$
   **endfor**
**endfor**
c.
**for all** $i \in S$ **do in parallel**
   $A(i,i) \leftarrow 0$
**endfor**

8. Select only current super-vertices for remaining phases of the algorithm.

   **for all** $i \in S$ **do in parallel**
      **if** $D(i) \neq i$ **then**
         $\text{Flag}(i) \leftarrow 0$
   **endfor**

**2.6. Minimal Spanning Trees and Forests.** If $G$ is a *weighted graph* — i.e. there exists a *weight function* $w: E \to \mathbb{R}$, then a *minimal spanning tree* is a spanning tree such that the sum of the weights of the edges is a *minimum* (over *all possible* spanning trees).

1. We will assume, for the time being, that the weights are all *positive*.

2. Minimal spanning trees have many applications. Besides the obvious ones in network theory there are applications to problems like the traveling salesman problem, the problem of determining cycles in a graph, etc.

We will briefly discuss some of these applications:

DEFINITION 2.14. Let $\{c_1, \ldots, c_n\}$ be $n$ points on a plane. A *minimal tour* of these points is a closed path that passes through all of them, and which has the shortest length of all possible such paths.

1. Given $n$ points on a plane, the problem of computing a minimal tour is well-known to be *NP-complete*. This means that there is *probably* no polynomial-time algorithm for solving it. See [55] as a general reference for NP-completeness.

2. In the *original* traveling salesman problem, the points $\{c_1, \ldots, c_n\}$ represented *cities* and the idea was that a salesman must *visit* each of these cities at least once. A solution to this problem would represent a travel plan that would have the least cost. Solutions to this problem have obvious applications to general problems of routing utilities, etc.

There do exist algorithm for finding an *approximate solution* to this problem. One such algorithm makes use of minimal spanning trees.

Suppose we are given $n$ points $\{c_1, \ldots, c_n\}$ on a plane (we might be given the coordinates of these points). Form the *complete graph* on these points — recall that a complete graph on a set of vertices is a graph with edges connecting *every pair* of vertices — see page 99. This means the the complete graph on $n$ vertices has exactly

$\binom{n}{2} = n(n-1)/2$ edges. Now assign a *weight* to each edge of this complete graph equal to the *distance* between the "cities" at its ends. We will call this weighted complete graph the *TSP graph* associated with the given traveling salesman problem.

PROPOSITION 2.15. The total weight of the minimal spanning tree of the TSP graph of some traveling salesman problem is $\leq$ the total distance of the minimal tour of that traveling salesman problem.

PROOF. If we delete one edge from the minimal tour, we get a *spanning tree* of the complete graph on the $n$ cities. The *weight* of this spanning tree must be $\geq$ the weight of the corresponding *minimal spanning tree*. $\square$

Note that we can get a *kind* of tour of the $n$ cities by simply tracing over the minimal spanning tree of the TSP graph — where we traverse each edge *twice*. Although this tour isn't *minimal* the proposition above immediately implies that:

PROPOSITION 2.16. The weight of the tour of the $n$ cities obtained from a minimal spanning tree by the procedure described above is $\leq 2W$, where $W$ is the weight of a minimal tour.

This implies that the tour obtained from a minimal spanning tree is, at *least*, not *worse* than twice as bad as an optimal tour. If you don't like the idea of traversing some edges twice, you can jump directly from one city to the next unvisited city. The triangle inequality implies that this doesn't increase the total length of the tour.

There is a well-known algorithm for computing minimal spanning trees called *Borůvka's Algorithm*. It is commonly known as Soullin's Algorithm, but was actually developed by Borůvka in 1926 — see [19], and [149]. It was developed before parallel computers existed, but it lends itself easily to parallelization. The resulting parallel algorithm bears a striking resemblance to the algorithm for connected components and spanning trees discussed above.

We begin by describing the basic algorithm. We must initially assume that the weights of all edges are *all distinct*. This is not a very restrictive assumption since we can define weights lexicographically.

The idea of this algorithm is as follows:

As with connected components, we regard the vertices of the graph as being partitioned into collections called "super-vertices". In the beginning of the algorithm, each vertex of the graph is regarded as a super-vertex itself. In each phase of the algorithm, each super-vertex is *merged* with at least one other super-vertex to which it is connected by an edge. Since this procedure *halves* (at least) the number of super-vertices in each phase of the algorithm, the total number of phases that are required is $\approx \lg n$. Each phase consists of the following steps:

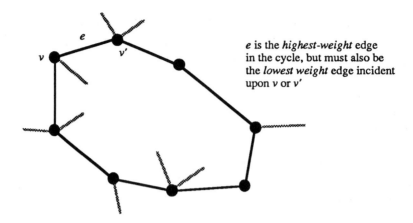

e is the *highest-weight* edge
in the cycle, but must also be
the *lowest weight* edge incident
upon $v$ or $v'$

FIGURE VI.20. A possible cycle in the MST Algorithm

1. Each super-vertex selects the *lowest-weight* edge that is incident upon it. These lowest-weight edges become part of the minimal spanning tree.

   If we consider the super-vertices, equipped *only* with these minimal-weight edges, we get several disjoint graphs. They are subgraphs of the original graph.

2. These subgraphs are collapsed into new super-vertices. The result is a graph with edges from a vertex to itself, and multiple edges between different super-vertices. We eliminate all self-edges and, when there are several edges between the same two vertices, we eliminate all but the one with the lowest weight.

THEOREM 2.17. (Borůvka's Theorem) Iteration of the two steps described above, $\lg n$ times, results in a minimal spanning tree of the weighted graph.

PROOF. 1. We first prove that the algorithm produces a *tree*. This is very similar to the proof that the spanning-tree algorithm of the previous section, produced a tree. This argument makes use of the assumption that the edge-weights are all distinct. If we drop this assumption, it is possible (even if we resolve conflicts between equal-weight edges arbitrarily) to get cycles in the set of selected edges. Consider the graph that is a single closed cycle with all edges of weight 1.

It is clearly possible to carry out step 1 of Borůvka's algorithm (with arbitrary selection among equal-weight edges) in such a way that the selected edges form the original cycle. If the weights of the edges are all *different* the edge with the highest weight is *omitted* in the selection process — so we get a path that is not closed. If we assume that the result (of Borůvka's algorithm) contains a cycle we get a contradiction by the following argument (see figure VI.20 on page 352):

> Consider the edge, $e$, with the *highest weight* of all the edges in the cycle
> — this must be unique since all weights are distinct. This edge must be

the lowest-weight edge incident on one of its end vertices, say $v$. But this leads to a contradiction since there is another edge, $e'$, incident upon $v$ that is also included in the cycle. The weight of $e'$ must be strictly less than that of $e$ since all weights are *distinct*. But this contradicts the way $e$ was selected for inclusion in the cycle — since it can't be the lowest weight edge incident upon $v$.

2. Every vertex of the original graph is in the tree — this implies that it is a *spanning tree*. It also implies that no new edges can be added to the result of Borůvka's algorithm — the result would no longer be a tree. This follows Edge-Selection step of the algorithm.

3. All of the edges *in* a minimal spanning tree will be selected by the algorithm. Suppose we have a minimal spanning tree $T$, and suppose that the edges of the graph *not* contained in $T$ are $\{e_1, \dots, e_k\}$. We will actually show that none of the $e_i$ are ever selected by the algorithm. If we *add* $e_k$ to $T$, we get a graph $T \cup e_k$ with a cycle.

We claim that $e_k$ is the *maximum-weight* edge in this cycle.

If *not*, we could exchange $e_k$ with the maximum-weight edge in this cycle to get a new tree $T'$ whose weight is strictly less than that of $T$. This contradicts the assumption that $T$ was a minimal spanning tree.

Now mark these edges $\{e_1, \dots, e_k\} \in G$ and run Borůvka's algorithm on the original graph. We claim that the $e_i$ are never selected by Borůvka's algorithm. Certainly, this is true in the first phase. In the following phases

1. If one of the $e_i$ connects a super-vertex to itself, it is eliminated.
2. If there is more than one edge between the same super-vertices and one of them is $e_i$, it will be *eliminated*, since it will never be the minimum-weight edge between these super-vertices.
3. Each super-vertex will consist of a connected subtree of $T$. When we collapse them to single vertices, $T$ remains a tree, and a cycle containing an $e_i$ remains a (smaller) cycle — see figure VI.21. The edge $e_i$ remains the maximum-weight edge of this new smaller cycle, and is not selected in the next iteration of the algorithm.

4. The algorithm executes in $O(\lg^2 n)$ time. This follows from exactly the same argument that was used in algorithm 2.6 on page 326. □

We can convert this procedure into a concrete algorithm by making a few small changes to 2.10 on page 339. We eliminate the incidence matrix and replace it by a weight matrix $W$. The fact that *no edge* connects two vertices is represented in $W$ by a weight of $\infty$. We also eliminate *ties* in weights of edges[6] by numbering the edges of the graph, and selecting the lowest-numbered edge with a given weight.

---

[6]Since Borůvka's algorithm required that all edge-weights be *unique*.

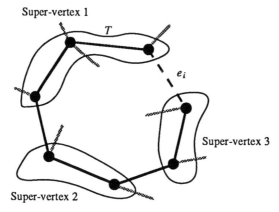

Before super-vertices are collapsed

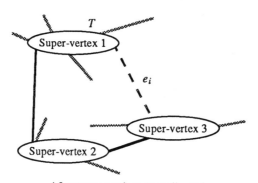

After super-vertices are collapsed

FIGURE VI.21

ALGORITHM 2.18. (Borůvka's Algorithm.)

**Input:** A graph $G$ with $|V| = n$ described by an $n \times n$ weight matrix $W(i, j)$.

**Output:** A $2 \times n$ array Edge, such that $\mathrm{Edge}(1, i)$ and $\mathrm{Edge}(2, i)$ are the end-vertices of the edges in a minimal spanning tree.

**Auxiliary memory:** A one-dimensional arrays $C$ and Flag, each with $n$ elements. A $2 \times n \times n$ array $B$.

1. Initialization.

   a.

   **for all** $i, 0 \le i < n - 1$ **do in parallel**
   $\quad D(i) \leftarrow i$
   $\quad \mathrm{Flag}(i) \leftarrow 1$
   $\quad \mathrm{Edge}(1, i) \leftarrow 0$
   $\quad \mathrm{Edge}(2, i) \leftarrow 0$
   **endfor**

   b.

   **for all** $i, j, 0 \le i, j \le n - 1$ **do in parallel**
   $\quad B(1, i, j) \leftarrow i$
   $\quad B(2, i, j) \leftarrow j$
   **endfor**

   The remainder of the algorithm consists in

   do steps 2 through 8 $\lg n$ times:

   Construct the set $S$: $S \leftarrow \{i | \mathrm{Flag}(i) = 1\}$.

2. Selection.

   **for all** $i \in S$ **do in parallel**
   $\quad$ Choose $j_0$ such that $W(i, j_0) = \min\{W(i, j) | j \in S\}$
   $\quad$ **if** none **then** $j_0 \leftarrow i$
   $\quad C(i) \leftarrow j_0$
   $\quad \mathrm{Edge}(1, i) \leftarrow B(1, i, j_0)$
   $\quad \mathrm{Edge}(2, i) \leftarrow B(2, i, j_0)$
   **endfor**

3. Removal of isolated super-vertices.

   **for all** $i$ such that $i \in S$ **do in parallel**
   $\quad$ **if** $C(i) = i$, **then** $\mathrm{Flag}(i) \leftarrow 0$
   **endfor**

4. Update $D$.

   **for all** $i \in S$, **do in parallel**
   $\quad D(i) \leftarrow C(i)$
   **endfor**

5. Consolidation.

**for** $i = 0$ **until** $i > \lg n$
    $j \in S$ **do in parallel**
      $C(j) \leftarrow C(C(j))$
**endfor**

6. Final update.

    a.
    **for** all $i \in S$ **do in parallel**
      $D(i) \leftarrow \min(C(i), D(C(i)))$
    **endfor**
    b.
    **for** all $i$ **do in parallel**
      $D(i) \leftarrow D(D(i))$
    **endfor**

7. Update the weight-matrix and $B$-arrays.

    a.
    **for** all $i \in S$ **do in parallel**
      **for** all $j \in S$ such that $j = D(j)$ **do in parallel**
        Choose $j_0 \in S$ such that
          $W(i, j_0) = \min\{W(i, k) | D(k) = i, k \in S\}$
        **if** none **then** $j_0 \leftarrow j$
        $W(i, j) = W(i, j_0)$
        $B(1, i, j) \leftarrow B(1, i, j_0)$
        $B(2, i, j) \leftarrow B(2, i, j_0)$
      **endfor**
    **endfor**
    b.
    **for** all $j \in S$ such that $j = D(j)$ **do in parallel**
      **for** all $i \in S$ such that $i = D(i)$ **do in parallel**
        Choose $i_0 \in S$ such that
          $W(i_0, j) = \min\{W(k, j) | D(k) = i, k \in S\}$
        **if** none **then** $i_0 \leftarrow i$
        $W(i, j) = W(i_0, j)$
        $B(1, i, j) \leftarrow B(1, i_0, j)$
        $B(2, i, j) \leftarrow B(2, i_0, j)$
      **endfor**
    **endfor**
    c.
    **for** all $i \in S$ **do in parallel**
      $W(i, i) \leftarrow \infty$

**endfor**

8. Select only current super-vertices for remaining phases of the algorithm.

    **for all** $i \in S$ **do in parallel**
      **if** $D(i) \neq i$ **then**
        Flag$(i) \leftarrow 0$
    **endfor**

9. Output selected edges.

    **for all** $i, 0 \leq i \leq n-1$ **do**
      **if** $i \neq D(i)$ **then** output
        $(\text{Edge}(1,i), \text{Edge}(2,i))$
    **endfor**

Our program for computing minimal spanning trees is very much like that for connected components.

**Input:** A graph $G$ with $|V| = n$ described by an $n-1 \times n-1$ weight matrix $W(i,j)$[7]. (Nonexistence of an edge between two vertices $i$ and $j$ is denoted by setting $W(i,j)$ and $W(j,i)$ to "infinity". In our C* program below, this is the pre-defined constant **MAXFLOAT**, defined in <values.h>).

**Output:** An $n \times n$ array **B**, giving the adjacency matrix of the minimal spanning tree.

1. We will assume the A-matrix is stored in a parallel float variable A in a shape named 'graph', and B is stored in a parallel **int** variable.

2. This algorithm executes in $O(\lg^2 n)$ time using $O(n^2)$ processors.

This operation never combines super-vertices that are not connected by an edge, so it does no harm. This operation never needs to be performed more than $\lg n$ times, since the length of any chain of super-vertex pointers is halved in each step.

Here is a C* program for minimal spanning trees.

```
#include <values.h>
#include <math.h>
#include <stdio.h>
shape [64][128]graph;
shape [8192]components;
#define N 10
int:graph temp, T, B1, B2;   /* T holds the minimal
         * spanning forest. */
float:graph W;
float:components temp2;
int:components C, D, Flag, in_S, Edge1, Edge2,
   i_0, j_0;
int i, j;
FILE *graph_file;
```

---

[7]This is an $n \times n$ weight matrix with the main diagonal *omitted*

```
void main()
{
  int L = (int) (log((float) N) /
          log(2.0) + 1);

  /* Weight matrix stored in a text file. */
  graph_file = fopen("gfile", "r");
  with (graph)
  {
    char temp[100];
    for  (i = 0; i < N; i++)
    {
      char *p;
        fscanf(graph_file, "%[^\n]\n", temp);
        p = temp;
      for  (j = 0; j < N; j++)
        if  (i != j)
        {
          char str[20];
          float value;
          int items_read = 0;
            sscanf(p, "%s", str);
            p += strlen(str);
            p++;
            items_read = sscanf(str, "%g", &value);
          if  (items_read < 1)
          {
            if (strchr(str, 'i') != NULL)
              value = MAXFLOAT;
            else
              printf("Invalid field = %s\n", str);
          }
            [i][j]W = value;
        } else
          [i][j]W = MAXFLOAT;
    }
  }

  /*
   * Initialize super-vertex array so that each
   * vertex starts out being a super vertex.
   */
  with (components)
  {
    D = pcoord(0);
    Flag = 1;
    Edge1 = 0;
    Edge2 = 0;
  }
  with (graph)
  {
    B1 = pcoord(0);
    B2 = pcoord(1);
```

```
}

/* Main loop for the algorithm. */
for  (i = 0; i <= L; i++)
{
  int i;

  /* This is step 2. */

  with (components)
  where (pcoord(0) < N)
      in_S = Flag;
  else
  in_S = 0;

  with (graph)

  where ((([pcoord(0)]in_S == 1) &&
      ([pcoord(1)]in_S == 1))
  {
    [pcoord(0)]temp2 = MAXFLOAT;
    [pcoord(0)]temp2 <?= W;
    [pcoord(0)]j_0 = N + 1;
    where ([pcoord(0)]temp2 == W)
    [pcoord(0)]j_0 <?= pcoord(1);
    where ([pcoord(0)]j_0 == N + 1)
    [pcoord(0)]j_0 = pcoord(0);
    [pcoord(0)]C = [pcoord(0)]j_0;
    [pcoord(0)]Edge1 = [pcoord(0)][[pcoord(0)]j_0]B1;
    [pcoord(0)]Edge2 = [pcoord(0)][[pcoord(0)]j_0]B2;
  }

  /* This is step 3 */
  with (components)
  where ([pcoord(0)]in_S == 1)
  where ([pcoord(0)]C == pcoord(0))
  [pcoord(0)]Flag = 0;

  /* This is step 4 */

  with (components)
  where (in_S == 1)
  D = C;

  /* This is step 5 */

  for (i = 0; i <= L; i++)
    with (components)
    where (in_S == 1)
    C = [C]C;

  /* This is step 6a */

  with (components)
```

```
where (in_S == 1)
D = (C <?[C]D);

/* This is step 6b */

with (components)
where (in_S == 1)
D = [D]D;


/* Step 7a */
with (graph)
where ([pcoord(0)]in_S == 1)
where ([pcoord(1)]in_S == 1)
where (pcoord(1) == [pcoord(1)]D)
{
    [pcoord(0)]temp2 = MAXFLOAT;
    [pcoord(0)]temp2 <?= W;

    [pcoord(0)]j_0 = N + 1;
    where ([pcoord(0)]temp2 == W)
    [pcoord(0)]j_0 <?= pcoord(1);
    where ([pcoord(0)]temp2 == MAXFLOAT)
    [pcoord(0)]j_0 = pcoord(1);

    W = [pcoord(0)][[pcoord(0)]j_0]W;
    B1 = [pcoord(0)][[pcoord(0)]j_0]B1;
    B2 = [pcoord(0)][[pcoord(0)]j_0]B2;
}


/* Step 7b */
with (graph)
where ([pcoord(0)]in_S == 1)
where ([pcoord(1)]in_S == 1)
where (pcoord(0) == [pcoord(0)]D)
where (pcoord(1) == [pcoord(1)]D)
{
    [pcoord(1)]temp2 = MAXFLOAT;
    [pcoord(1)]temp2 <?= W;

    [pcoord(1)]i_0 = N + 1;
    where ([pcoord(1)]temp2 == W)
    [pcoord(1)]i_0 <?= pcoord(0);
  where ([pcoord(1)]temp2 == MAXFLOAT)
    [pcoord(1)]i_0 = pcoord(0);

        W = [[pcoord(1)]i_0][pcoord(1)]W;
        B1 = [[pcoord(1)]i_0][pcoord(1)]B1;
        B2 = [[pcoord(1)]i_0][pcoord(1)]B2;
}


/* Step 7c */
with (components)
```

```
        where ([pcoord(0)]in_S == 1)
        [pcoord(0)][pcoord(0)]W = MAXFLOAT;

        /* Step 8 */
        with (components)
        where ([pcoord(0)]in_S == 1)
        where ([pcoord(0)]D != pcoord(0))
        [pcoord(0)]Flag = 0;

  }          /* End of big for-loop. */
      /* Step 9 */
  for (i = 0; i < N; i++)
    if ([i]D != i)
      printf("Edge1=%d, Edge2=%d\n",[i]Edge1,[i]Edge2);
}
```

Here is a sample run of this program:

The input file is:

```
          1.1    i    1.0    i     i     i     i     i     i
   1.1           3.1    i     i     i     i     i     i     i
    i     3.1           0.0   1.3    i     i     i     i     i
   1.0     i     0.0          1.2    i     i     i     i     i
    i      i     1.3   1.2          3.5    i     i     i     i
    i      i      i     i    3.5          2.1    i    2.4    i
    i      i      i     i     i    2.1          2.2    i     i
    i      i      i     i     i     i    2.2          .9   1.7
    i      i      i     i     i    2.4    i    .9           i
    i      i      i     i     i     i     i    1.7    i
```

here we don't include any *diagonal* entries, so the array is really N-1 ×N-1. The letter *i* denotes the fact that there is *no* edge between the two vertices in question – the input routine enters **MAXFLOAT** into the corresponding array positions of the Connection Machine (representing *infinite* weight).

See figure VI.22 for the graph that this represents.

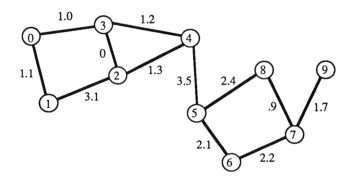

FIGURE VI.22. Input graph for the MST program

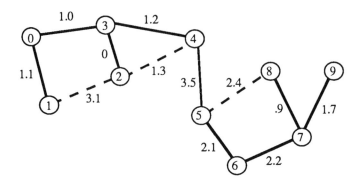

FIGURE VI.23. Minimal spanning tree

The output is

```
0  1  0  1  0  0  0  0  0  0
1  0  0  0  0  0  0  0  0  0
0  0  0  1  0  0  0  0  0  0
1  0  1  0  1  0  0  0  0  0
0  0  0  1  0  1  0  0  0  0
0  0  0  0  1  0  1  0  0  0
0  0  0  0  0  1  0  1  0  0
0  0  0  0  0  0  1  0  1  1
0  0  0  0  0  0  0  1  0  0
0  0  0  0  0  0  0  1  0  0
```

and this represents the minimal spanning tree shown in figure VI.23.


**Exercises:**

1. Find a minimal spanning tree of the graph with weight-matrix

$$
\begin{pmatrix}
 & 2 & 3 & 1 & 0 & 4 & 5 & \infty & \infty \\
2 & & \infty & \infty & 8 & -1 & \infty & -3 & 6 \\
3 & \infty & & 9 & \infty & 7 & \infty & -2 & \infty \\
1 & \infty & 9 & & \infty & -4 & \infty & \infty & 10 \\
0 & 8 & \infty & \infty & & 11 & 12 & \infty & \infty \\
4 & -1 & 7 & -4 & 11 & & \infty & 13 & \infty \\
5 & \infty & \infty & \infty & 12 & \infty & & \infty & 14 \\
\infty & -3 & -2 & \infty & \infty & 13 & \infty & & \infty \\
\infty & 6 & \infty & 10 & \infty & \infty & 14 & \infty &
\end{pmatrix}
$$

How many components does this graph have?

2. Find an algorithm for computing a *maximum weight* spanning tree of an undirected graph.

## 2.7. Cycles in a Graph.

2.7.1. *Definitions.* There are many other graph-theoretic calculations one may make based upon the ability to compute spanning trees. We will give some examples.

DEFINITION 2.19. If $G = (V, E)$ is a graph, a:

1. *path* in $G$ is a sequence of vertices $\{v_1, \ldots, v_k\}$ such that $(v_i, v_{i+1}) \in E$ for all $1 \le i \le k$;
2. *cycle* is a path in which the start point and the end point are the same vertex.
3. *simple cycle* is a cycle in which no vertex occurs more than once (i.e. it does not intersect itself).

We define a notion of *sum* on the edges of an undirected graph. The sum of two distinct edges is their *union*, but the sum of two copies of the *same* edge is the empty set. We can easily extend this definition to *sets* of edges, and to *cycles* in the graph. Given this definition, the set of all cycles of a graph, equipped with this definition of sum form a mathematical system called a *group*. Recall that a group is defined by:

DEFINITION 2.20. A *group*, $G$, is a set of objects equipped with an operation $\star: G \times G \to G$ satisfying the conditions:

1. It has an *identity element*, $e$. This has the property that, for all $g \in G$, $e \star g = g \star e = g$.

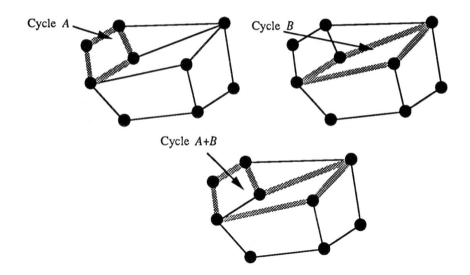

FIGURE VI.24. The sum of two cycles in a graph

2. The operation $\star$ is *associative*: for every set of three elements $g_1, g_2, g_3 \in G$, $g_1 \star (g_2 \star g_3) = (g_1 \star g_2) \star g_3$.

3. Every element $g \in G$ has an inverse, $g^{-1}$ with respect to $\star$: $g \star g^{-1} = g^{-1} \star g = e$.

A group will be called *abelian* if its $\star$-operation is also *commutative* — i.e., for any two elements $g_1, g_2 \in G$, $g_1 \star g_2 = g_2 \star g_1$.

The identity-element in our case, is the *empty set*. The *inverse* of any element is *itself*. Figure VI.24 illustrates this notion of the sum of two cycles in a graph. Notice that the common edge in cycle $A$ and $B$ *disappears* in the sum, $A + B$.

DEFINITION 2.21. A *basis* for the cycles of a graph are a set of cycles with the property that all cycles in the graph can be expressed uniquely as sums of elements of the basis.

Given a graph $G$ with some spanning tree $T$, and an edge $E \in G - T$ the cycle corresponding to this edge is the cycle that results from forming the *union* of $E$ with the path $P$ connecting the *endpoints* of $E$ in $T$.

2.7.2. *A simple algorithm for a cycle basis.* In general, there are *many distinct bases* for the cycles of a graph. It turns out that there is a 1-1 correspondence between the cycles in such a basis, and the edges that do *not* occur in a spanning tree of the graph.
   Given:
   1. algorithm 2.10 on page 339.
   2. Algorithm 2.5 on page 322 for finding shortest paths in the spanning tree.

we can easily compute a basis for the cycles of a graph by:

   1. Finding a spanning tree for the graph, using Borůvka's algorithm (after assigning unique weights to the edges);

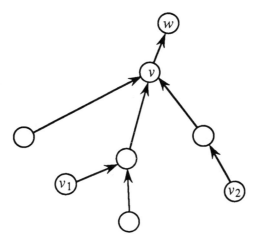

FIGURE VI.25

2. Computing the set of *omitted edges* in the graph — these are edges that appeared in the original graph, and do *not* appear in the spanning tree.

3. Compute the *shortest paths* in the spanning tree, between the end-vertices of these omitted edges — using algorithm 2.5 on page 322. The union of an omitted edge with the shortest path connecting its end-vertices is a basis-cycle of the graph.

This algorithm for a cycle-basis of a graph is not particularly efficient: it requires $O(n^2)$ processors for the computation of the spanning tree and $O(n^{2.376})$ processors for finding the minimal paths in this spanning tree. The path-finding step is the most expensive, and makes the processor requirement for the whole algorithm $O(n^{2.376})$.

2.7.3. *Lowest Common Ancestors.* We will present an algorithm for finding these paths that uses fewer processors. It was developed in [151] by Tsin and Chin and makes use of the algorithm for an inverted spanning forest in section 2.5.1 on page 344. A key step in this algorithm involves finding *lowest common ancestors* of pairs of vertices in a inverted spanning tree.

DEFINITION 2.22. Let $T = (V, E)$ be an in-tree (see 2.1 on page 313).

1. A vertex $v \in V$ is an *ancestor* of a vertex $v' \in V$, written $v \succeq v'$, if there is a *directed path* from $v'$ to $v$ in $T$.

2. Let $v_1, v_2 \in V$ be two vertices. A vertex $v \in V$ is the *lowest common ancestor* of $v_1$ and $v_2$ if $v \succeq v_1$ and $v \succeq v_2$, and for any vertex $w$, $w \succeq v_1$ and $w \succeq v_2 \implies w \succeq v$. See figure VI.25 for an example.

It is fairly simple to compute the lowest common ancestor of two vertices in an inverted spanning tree. We use an algorithm based on algorithm 2.12 on page 346:

ALGORITHM 2.23. Let $T = (V, E)$ be an in-tree described by a function $f: V \to V$ that maps a vertex to its immediate ancestor. Given two vertices $v_1, v_2 \in V$, we can compute their lowest common ancestor $v$ by:

1. Compute the path-function, $\hat{f}$, associated with $f$, using 2.12 on page 346.
2. Perform a binary search of the rows $\hat{f}(v_1, *)$ and $\hat{f}(v_2, *)$ to determine the smallest value of $j$ such that $\hat{f}(v_1, j) = \hat{f}(v_2, j)$. Then the common value, $\hat{f}(v_1, j)$ is the lowest common ancestor of $v_1$ and $v_2$.

This algorithm can clearly be carried out in parallel for many different pairs of vertices in $T$. It is the basis of our algorithm for a cycle-basis of a graph. We use lowest common ancestors to compute a path connecting the endpoints of an *omitted edge* in a spanning tree — recall the algorithm on page 364.

ALGORITHM 2.24. Let $G = (V, E)$ be an undirected graph, with $|V| = n$. We can compute a basis for the cycles of $G$ by:

1. Compute an inverted spanning forest, $T = (V, E')$, of $G$, via algorithm 2.13 on page 347.
2. Compute the set, $\{e_1, \ldots, e_k\}$, of *omitted edges* of $G$ — these are edges of $G$ that *do not occur in* $T$. They are $E - E'$ (where we must temporarily regard the edges in $E'$ as undirected edges).
3. Compute the set of paths from the leaves of $T$ to the root, using algorithm 2.12 on page 346.
4. For each of the $e_i$, **do in parallel:**
   a. Compute the end-vertices $\{v_{i,1}, v_{i,2}\}$;
   b. Compute the lowest-common-ancestor $v_i$ of $\{v_{i,1}, v_{i,2}\}$.
   c. Compute the paths $p_{1,i}$ and $p_{2,i}$ from $v_{i,1}$ and $v_{i,2}$ to $v_i$, respectively.
   d. Output $e_i \cup p_{1,i} \cup p_{2,i}$ as the $i^{\text{th}}$ fundamental cycle of $G$.

Unlike the simple algorithm on page 364, this algorithm requires $O(n^2)$ in its straightforward implementation. The more sophisticated implementations described in [151] require $O(n^2 / \lg^2 n)$ processors. The execution time of both of these implementations is $O(\lg^2 n)$.

**Exercises:**

1. Write a C* program to implement the algorithm described on page 488 for testing whether weighted graphs have negative cycles
2. Suppose $G$ is a complete graph on $n$ vertices (recall that a complete graph has an edge connecting every pair of vertices). Describe a cycle-basis for $G$.

3. A *planar graph*, $G$, is a graph that can be embedded in a plane — in other words, we can draw such a graph on a sheet of paper in such a way that no edge crosses another. If we *delete* the planar graph from the plane into which it is embedded[8], the plane breaks up into a collection of polygon, called the *faces* of the embedding. The faces of such an embedding correspond to simple cycles in the graph — say $\{c_1, \ldots, c_k\}$. Show that one of these cycles can be expressed as a sum of the other, and if we delete it, the remaining cycles form a basis for the cycles of $G$.

**2.8. Further Reading.** Graph theory has many applications to design and analysis of networks. A good general reference on graph theory is [63] by Harary. Research on the design of efficient parallel algorithms in graph theory goes back to the mid 1970's. Many problems considered in this section were studied in Carla Savage's thesis [133]. Chandra investigated the use of graphs in relation to the problem of multiplying matrices — see [23]. *Planar graphs* are graphs that can be embedded in a plane. In 1982 Ja' Ja' and Simon found NC-parallel algorithms for testing whether a given graph is planar, and for embedding a graph in a plane if it *is* — see [75]. This was enhanced by Klein and Reif in 1986 — see [87], and later by Ramachandran and Reif in 1989 — see [126].

One problem we haven't considered here is that of computing a *depth-first search* of a graph. Depth-first search is the process of searching a graph in such a way that the search moves forward[9] until it reaches a vertex whose neighbors have all been searched — at this point it backtracks a minimum distance and continues in a new direction.It is easy to see that the time required for a sequential algorithm to perform depth-first search on a graph with $n$ vertices is $O(n)$. Depth-first search forms the basis of many sequential algorithms — see [147]. In 1977, Alton and Eckstein found a parallel algorithm for depth first search that required $O(\sqrt{n})$-time with $O(n)$ processors — see [5]. Corneil and Reghbati conjectured that depth-first search was inherently sequential in [35] and Reif proved that *lexicographically first* depth-first search[10] is P-complete in [127]. The author found an NC-algorithm for depth-first search of planar graphs —

[8]I.e., if draw the graph on a piece of paper in such a way that no edge crosses over another, and then cut the paper along the edges.

[9]I.e., moves to new vertices

[10]This is the form of depth-first search that always selects the lowest-numbered vertex when it has multiple choices.

see [143]. This was subsequently improved by Hagerup in 1990 — see [61]. The best parallel algorithm (in terms of time and number of processors required) for depth-first search of planar graphs is currently (1992) that of Shannon — see [140]. This was extended to arbitrary graphs that do not contain subgraphs isomorphic to $K_{3,3}$ by Khuller in [85][11].

In 1987 Aggarwal and Anderson found a *random* NC algorithm for depth-first search of general graphs. This is an algorithm that makes random choices during its execution, but has *expected* execution-time that is poly-logarithmic in the number of processors. See chapter VII for a discussion of randomized algorithms in general.

A graph is called *chordal* if every cycle of length $\geq 4$ can be "short-circuited" in the sense that there exists at least one pair of non-consecutive vertices in every such cycle that are connected by an edge. These graphs have the property that there exist NC algorithms for many important graph-theoretic problems, including: optimal coloring, maximum independent set, minimum clique cover. See the paper, [118], of Naor, Naor and Schäffer.

If the maximum degree of a vertex in a graph is $\Delta$, there is an algorithm, due to Luby, for coloring the graph in $\Delta + 1$ colors. This algorithm runs on a CREW PRAM and executes in $O(\lg^3 n \lg \lg n)$ time, using $O(n + m)$ processors — see [106]. In Karchmer and Naor found an algorithm that colors the graph with $\Delta$ colors. this second algorithm runs on a CRCW-PRAM — see [80].

There are several algorithms for finding a maximal independent set in a graph — see page 460 of the present book for a discussion of these results.

In general we haven't considered algorithms for *directed graphs* here. These graphs are much harder to work with than undirected graphs and, consequently, much less is known about them. For instances, the problem of finding even a *sequential* algorithm for spanning in-trees and out-trees of a weighted directed graph is much more difficult than the corresponding problem for undirected graphs. Several *incorrect* papers were published before a valid algorithm was found. See [148], by Robert Tarjan. The first NC-parallel algorithms for this problem were developed by Lovász in 1985 ([103]) and in the thesis of Zhang Yixin ([166]).

Exercises:

1. Write a program that not only finds the *length* of the shortest path between every pair of vertices, but also finds the *actual paths* as well. This requires a variant on the algorithm for lengths of shortest paths presented above. In this variant every entry of the matrix is a *pair*: (distance, list of edges). The algorithm not only adds the distances together (and selects the smallest distance) but also

---

[11] $K_{3,3}$ is the complete bipartite graph on two sets of 3 vertices — see 8.8 on page 104.

*concatenates* lists of edges. The lists of edges in the original A-matrix *initially* are either *empty* (for *diagonal* entries) or have only a single edge.

2. Given an example of a weighted graph for which Borůvka's algorithm finds a minimal spanning tree in a *single step*.

3. Suppose $G = (V, E)$ is a connected undirected graph. An edge $e$ with the property that *deleting* it *disconnects* the graph is called a *bridge* of $G$. It turns out that an edge is a bridge if and only if it is not contained in *any* fundamental cycle of the graph. Give an algorithm for computing all of the bridges of a connected graph. It should execute in $O(\lg^2 n)$ time and require no more than $O(n^2)$ processors.

4. Give an example of a weighted graph with $n$ vertices, for which Borůvka's algorithm requires the full $\lceil \lg n \rceil$ phases.

5. The program for minimal spanning trees can be improved in several ways:
   a. The requirement that all weights be *distinct* can be eliminated by either redefining the method of *comparing* weights so that the weights of all pairs of edges behave as if they are distinct;
   b. The method of storing edge-weights can be made considerably more *memory-efficient* by storing them in a *list of triples* (start-vertex, end-vertex, weight).

   Modify the program for minimal spanning trees to implement these improvements.

6. Write a C* program for finding an approximate solution to the Traveling Salesmen problem

7. What is wrong with the following procedure for finding an inverted spanning forest of a graph?
   a. Find a spanning tree of the graph using Borůvka's algorithm (with some arbitrary distinct values given for the weights. It is easy to modify Borůvka's algorithm to give a minimal undirected spanning forest of a graph with more than one component.
   b. Select an arbitrary vertex to be the root.
   c. Make this *undirected* spanning tree into a *directed* tree via a technique like the first step of 2.2 on page 315. Make the directed edges all point toward the root.

### 3. Parsing and the Evaluation of arithmetic expressions

**3.1. Introduction.** This section develops an algorithm for parsing precedence grammars and the evaluation of an arithmetic expression in $O(\lg n)$ time. Although parts of this algorithm are somewhat similar to the doubling algorithms of the previous section, we put these algorithms in a separate section because of their complexity. They are interesting for a number of reasons.

Parsing is a procedure used in the front-end of a compiler — this is the section that reads and analyzes the source program (as opposed to the sections that generate object code).

The *front end* of a compiler generally consists of two modules: the *scanner* and the *parser*. The scanner reads the input and very roughly speaking checks spelling — it recognizes grammatic elements like identifiers, constants, etc. The parser then takes the sequence of these identified grammatic elements (called tokens) and analyzes their overall syntax. For instance, in a Pascal compiler the scanner might recognize the occurrence of an identifier and the parser might note that this identifier was used as the target of an assignment statement.

In the preceding section we saw how a DFA can be efficiently implemented on a SIMD machine. This essentially amounts to an implementation of a scanner. In the present section we will show how a parser (at least for certain simple grammars) can also be implemented on such a machine. It follows that for certain simple programming languages (ones whose grammar is a operator-precedence grammar), the whole front end of a compiler can be implemented on a SIMD machine We then show how certain operations similar to those in the code-generation part of a compiler can also be efficiently implemented on a SIMD machine. Essentially the algorithm presented here will take a syntax tree for an arithmetic expression (this is the output of the parser) and evaluate the expression, but it turns out not to be much more work to generate code to compute the expression.

Recall that a syntax-tree describes the order in which the operations are performed — it is like a parse tree that has been stripped of some unnecessary verbiage (i.e. unit productions, names of nonterminals, etc.). If the expression in question is $(a + b)/(c - d) + e/f$ the corresponding syntax tree is shown in figure VI.26.

See [2], chapter 2, for more information on syntax trees.

**3.2. An algorithm for building syntax-trees.** The algorithm for building syntax trees of expressions is due to Bar-On and Vishkin in [12]. We begin with a simple algorithm for fully-parenthesizing an expression. A fully parenthesized expression is one in which all precedence rules can be deduced by the patterns of parentheses: for instance $a + b * c$ becomes $a + (b * c)$, since multiplication generally (i.e. in most programming languages) has precedence over addition. Operators within the same level of nesting of parentheses are assumed to be of equal precedence. The algorithm

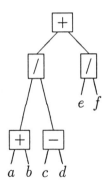

FIGURE VI.26

is:

ALGORITHM 3.1.   1. Assume the original expression was enclosed in parentheses.

2.  for each '+', '-' operator insert two left parentheses to its right and two right parentheses to its left — i.e. '$a + b$' becomes '$a)) + ((b$'.

3.  For each '*', '/' operator insert one left parenthesis to its right and one right parenthesis to its left — i.e. '$a * b$' becomes '$a) * (b$'.

4.  For each left (resp. right) parenthesis, insert two additional left (resp. right parentheses) to its right (resp. left).   □

These four steps, applied to '$a + b * (c - d)$' lead to '$(((a)) + ((b) * ((((c)) - ((d))))))$'. It is intuitively clear that this procedure will respect the precedence rules — *'s will tend to get buried more deeply in the nested parentheses since they only get one layer of opposing parentheses put around them, while +'s get two. Note that this procedure always results in too many parentheses — this does not turn out to create a problem. At least the parentheses are at the proper level of nesting. In the expression with extraneous parentheses computations will be carried out in the correct order if the expressions inside the parentheses are evaluated first. For instance, if we strip away the extraneous parentheses in the sample expression above we get: '$(a + (b * (c - d)))$'. Bar-On and Vishkin remark that this procedure for handling precedence-rules was used in the first FORTRAN compilers.

This little trick turns out to work on general operator-precedence grammars.

This construction can clearly be carried out in unit time with $n$ processors — simply assign one processor to each character of the input. Each processor replaces its character by a pointer to a linked list with the extra symbols (for instance). Although the rest of the algorithm can handle the data in its present form we copy this array of characters and linked lists into a new character-array. This can be done in $O(\lg n)$ time using $n$ processors — we add up the number of characters that will precede each character in the new array (this is just adding up the cumulative lengths of all of the little linked lists formed in the step above) — this gives us the position of each character in the new array. We can use an algorithm like that presented in the introduction for adding up $n$ numbers — i.e. algorithm 3.1.4 with $\star$ replaced by +. We now move the characters in constant time (one step if we have enough processors).

We will, consequently, assume that our expression is the highly parenthesized output of 3.1. We must now match up pairs of parentheses. Basically this means we want to have a pointer associated with each parenthesis pointing to its matching parenthesis — this might involve an auxiliary array giving subscript-values. Now we give an algorithm for matching up parentheses. The algorithm given here is a simplification of the algorithm due to Bar-On and Vishkin — our version assumes that we have $n$ processors to work with, rather that $n/\lg n$. See the appendix for this section for a discussion of how the algorithm must be changed to allow for the smaller number of processors.

ALGORITHM 3.2. We present an algorithm that is slightly different from that of Bar-On and Vishkin — it is simpler in the present context since it makes use of other algorithms in this book. We start by defining an array $A[*]$:

$$A[j] = \begin{cases} -1, & \text{if there is a left parenthesis at position } j; \\ 1, & \text{if there is a right parenthesis at position } j. \end{cases}$$

Now we compute the cumulative sums of this array, using the algorithm in the introduction on page 69 (or 1.4). Let this array of sums be $L[*]$, so $L[i] = \sum_{j=1}^{i} A[j]$. Basically, $L[i]$ indicates the *level of nesting* of parentheses at position $i$ of the character array, if that position has a left parenthesis, and the level of nesting $-1$, if that position has a right parenthesis.

Here is the idea of the binary search: Given any right parenthesis in, say, position $i$ of the character array, the position, $j$, of the matching left parenthesis is the maximum value such that $L[j] = L[i] + 1$. We set up an tree to facilitate a binary search for this value of $j$. Each level of this tree has half as many siblings as the next lower level and each node has two numbers stored in it: **Min-Right**, which is equal to the minimum value of $L[i]$ for all *right* parentheses among that node's descendants and **Min-Left**, which is the corresponding value for *left* parentheses. See figure VI.27.

We can clearly calculate these values for all of the nodes of the tree in $O(\lg n)$ time using $n$ processors (actually we can get away with $n/\lg n$ processors, if we use an argument like that in the solution to exercise 3 at the end of §1). Now given this tree and a right parenthesis at location $i$, we locate the matching left parenthesis at location $j$ by:

1. Moving *up* the tree from position $i$ until we arrive at the right child of a node whose left child, $k$, satisfies **Min-Left** $\leq L[i] + 1$;

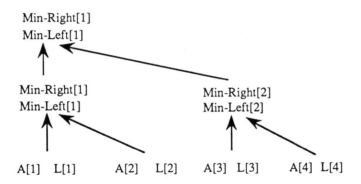

FIGURE VI.27. Parenthesis-matching algorithm

2. Travel *down* the tree from node $k$, always choosing *right children* if possible (and left children otherwise). Choosing a right child is possible if its **Min-Left** value is $\leq L[i] + 1$.

Given this matching of parentheses, we can build the syntax tree, via:

1. Remove extra parentheses: In this, and in the next step processors assigned to *parentheses* are active and all *others* are *inactive*. Every right parenthesis is deleted if there is another right parenthesis to its left, and the corresponding left parentheses are also adjacent Similarly, a left parenthesis is deleted if there is another one to its right and the corresponding right parentheses are adjacent.

2. Each pair of parentheses encloses a subexpression representing a small subtree of the entire syntax tree. The root of this little subtree corresponds to the arithmetic operation being performed in the subexpression. We assign a processor to the left parenthesis and it searches for the operator being executed in the subexpression. There are several possibilities:

    1. the quantity to the right of the left parenthesis is a constant or variable. In this case the operator is the next token, in sequence.

    2. the quantity to the right of the left parenthesis is another left parenthesis, $p$. In this case the left-operand (i.e. child, in the subtree) is another subexpression. The operator (or root of the subtree) is the token following the right parenthesis corresponding to $p$.

3. Pointers are set between the root (operator) and the pair of matching parentheses that enclose the expression.

4. Processors assigned to operators are active, all others are inactive. Suppose the operator is one of '$+, -, *, or /$'. The previous operator is set to be its left child and the operand to its right its right child. The processor finds them as follows: if the entry to the left of the operator contains a constant or a variable, then the previous entry contains the operator. Otherwise the entry contains a right parenthesis. The operator

is in the entry to the left of the left match. The entry on the right contains the operand (if this is a left parenthesis, then the node is the corresponding root).

The syntax tree has been effectively constructed at this point.

**3.3. An algorithm for evaluating a syntax tree.** Given a syntax tree for an expression, we can evaluate the expression using an algorithm due to Gibbons and Rytter in [56].

The statement of their result is:

THEOREM 3.3. There exists an algorithm for evaluating an expression of length $n$ consisting of numbers, and operations $+, -, *$, and $/$ in time $O(\lg n)$ using $n/\lg n$ processors.

3.4.    1. Note that this algorithm is asymptotically optimal since its execution time is proportional to the sequential execution time divided by the number of processors.
2. The methods employed in this algorithm turn out to work for any set of algebraic operations $\{\star_1, \ldots, \star_k\}$ with the property that:
   a. The most general expression involving these operations in one variable and constant parameters is of *bounded size*. Say this expression is: $f(x; c_1, \ldots, c_t)$, where $x$ is the one variable and the $c_i$ are parameters.
   b. the *composite* of two such general expressions can be computed in *bounded time*. In other words, if $f(x; c_1', \ldots, c_t') = f(f(x; c_1, \ldots, c_t); d_1, \ldots, d_t)$, then the $\{c_i'\}$ can be computed from the $\{c_i\}$ and the $\{d_i\}$ in constant time.

In the case of the arithmetic expressions in the statement of the theorem, the most general expression in one variable is $f(x; a, b, c, d) = (ax + b)/(cx + d)$. The composite of two such expressions is clearly also of this type and the new coefficients can be computed in constant time:

If $f(x; a, b, c, d) = (ax + b)/(cx + d)$, then $f(f(x; \hat{a}, \hat{b}, \hat{c}, \hat{d}); a, b, c, d) = f(x; a', b', c', d')$, where $a' = \hat{a}a + \hat{c}b, b' = a\hat{b} + \hat{d}b, c' = \hat{a}c + \hat{c}d, d' = c\hat{b} + d\hat{d}$. In fact it is easy to see that, if we write these sets of four numbers as *matrices*
$$\begin{pmatrix} a' & b' \\ c' & d' \end{pmatrix} = \begin{pmatrix} a & b \\ c & d \end{pmatrix} \begin{pmatrix} \hat{a} & \hat{b} \\ \hat{c} & \hat{d} \end{pmatrix}$$ (matrix product).

This reasoning implies that the algorithms presented here can also evaluate certain kinds of *logical expressions* in time $O(\lg n)$. It turns out (as pointed out by Gibbons and Rytter in [56]) that the problem of parsing bracket languages and input-driven languages can be reduced to evaluating a suitable algebraic expression, hence can also be done in $O(\lg n)$ time. In the case of parsing these grammars conditions a and b turn out to be satisfied because the algebraic expressions in question live in a mathematical

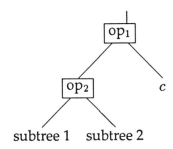

FIGURE VI.28. Before pruning

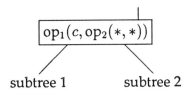

FIGURE VI.29. After pruning

system with only a finite number of values — i.e. it is as if there were only a finite number of numbers upon which to do arithmetic operations.

The idea of this algorithm is to take a syntax tree and prune it in a certain way so that it is reduced to a single node in $O(\lg n)$ steps. This is nothing but the Parallel Tree Contraction method, described in § 2.2 (page 320). Each step of the pruning operation involves a partial computation of the value of the expression and in the final step the entire expression is computed. Figures VI.28 and VI.29 illustrate this pruning operation.

Here $c$ is a *constant* and $op_1$ and $op_2$ represent operations or formulas of the form $f(x; a, b, c, d)$ in remark 3.4.4 above. Note that every node of the syntax tree has two children — it is not hard to see that in such a tree at least half of all of the nodes are leaves. This pruning operation removes nodes with the property that one of their children is a leaf. The total number of leaves is halved by this operation when it is properly applied — we will discuss what this means shortly. Since the property of having two children is preserved by this operation, the tree will be reduced to a single node (the root) in $O(\lg n)$ steps.

The phrase "properly applied" in the sentence above means that the operation cannot be applied to two adjacent nodes in the same step. We solve this problem by numbering all of:

1. the leaf nodes that are right children; and
2. the leaf nodes that are left children.

In each step of the computation we perform the pruning operation of all odd-numbered right-child leaf nodes, then on all odd numbered left-child leaf nodes. The net result is that the number of leaf-nodes is at least halved in a step. We re-number the leaf-nodes of the new graph by halving the numbers of the remaining nodes (all of which will have even numbers).

The numbers are initially assigned to the leaves by an algorithm called the *Euler-Tour Algorithm*. This is a simple and ingenious algorithm due to Tarjan and Vishkin, that is described in § 2.1 in chapter VI of this book. The main thing that is relevant to the present topic is that the execution-time of this algorithm is $O(\lg n)$.

Now put two numbers, **Right-count**, and **Left-count**, on each node of this path:

1. Non leaf nodes get 0 for both values;
2. A right-child leaf node gets 1 for **Right-count** and 0 for **Left-count**;
3. A left-child leaf node gets 0 for **Right-count** and 1 for **Left-count**;

Now form cumulative totals of the values of **Right-count** and **Left-count** over this whole Euler Tour. This can be done in $O(\lg n)$ time and results in a numbering of the leaf-nodes. This numbering scheme is arbitrary except that adjacent leaf-nodes (that are both right-children or left-children) will have consecutive numerical values.

The whole algorithm for evaluating the expression is:

1. Form the syntax tree of the expression.
2. Number the leaf-nodes of this syntax tree via the Euler-Tour algorithm (2.2).
3. Associate an identity matrix with every non leaf node of the tree. Following remark 3.4, the *identity matrix,* $\begin{pmatrix} 1 & 0 \\ 0 & 1 \end{pmatrix}$, is associated with the function $f(x; 1, 0, 0, 1) = x$ — the *identity function*. Interior nodes of the tree will, consequently, have the following data associated with them:

    1. an operation;
    2. two children;
    3. a $2 \times 2$ matrix;

The meaning of this data is: calculate the values of the two child-nodes; operate upon them using the operation, and plug the result (as $x$) into the equation $(ax + b)/(cx + d)$, where is the matrix associated with the node.

4. Perform $\lg n$ times:

    1. the pruning operation on odd-numbered right-child leaf-nodes. We refer to diagrams VI.28 and VI.29. The matrix, $M'$, on the node marked $op_1(op_2(*, *), c)$ is computed using the matrix, $M$, on the node marked $op_2$ and the operation coded for $op_1$ via:
        • If the corresponding *left* sibling is *also* a constant, then just perform the indicated operation and compute the result; *otherwise*

- a. If $op_1 = $ '+' then $M' = \begin{pmatrix} 1 & c \\ 0 & 1 \end{pmatrix} \cdot M$;

  b. If $op_1 = $ '*' then $M' = \begin{pmatrix} c & 0 \\ 0 & 1 \end{pmatrix} \cdot M$;

  c. If $op_1 = $ '$-$' then $M' = \begin{pmatrix} 1 & -c \\ 0 & 1 \end{pmatrix} \cdot M$;

  d. If $op_1 = $ '/' then $M' = \begin{pmatrix} 1 & 0 \\ 0 & c \end{pmatrix} \cdot M$;

Perform a pruning operation on odd-numbered left-child leaf nodes, using a diagram that is the mirror image of VI.28 and:

- If the corresponding *right* sibling is *also* a constant, then just perform the indicated operation and compute the result; *otherwise*

- a. If $op_1 = $ '+' then $M' = \begin{pmatrix} 1 & c \\ 0 & 1 \end{pmatrix} \cdot M$;

  b. If $op_1 = $ '*' then $M' = \begin{pmatrix} c & 0 \\ 0 & 1 \end{pmatrix} \cdot M$;

  c. If $op_1 = $ '$-$' then $M' = \begin{pmatrix} -1 & c \\ 0 & 1 \end{pmatrix} \cdot M$;

  d. If $op_1 = $ '/' then $M' = \begin{pmatrix} 0 & c \\ 1 & 0 \end{pmatrix} \cdot M$;

In every case the old matrix is left-multiplied by a new matrix that represents the effect of the operation with the value of the subtree labeled $op_2$ (see figures VI.28 and VI.29) plugged into an equation of the form $(ax + b)/(cx + d)$ (as $x$). For instance, consider the case of a reduction of a *left* leaf-node where the operation is *division*. In this case the value of the whole tree is $c/x$, where $x = $ the value of the subtree labeled $op_2$. But $c/x = (0 \cdot x + c)/(1 \cdot x + 0)$, so its matrix representation is $\begin{pmatrix} 0 & c \\ 1 & 0 \end{pmatrix}$, and the value computed for the subtree labeled $op_2$ is to be plugged into this equation. Remark 3.4 implies that this is equivalent to multiplying the matrix, $M$, associated with this node $(op_2)$ by $\begin{pmatrix} 0 & c \\ 1 & 0 \end{pmatrix}$.

Divide the numbers (i.e. the numbers computed using the Euler Tour algorithm — 2.2) of the remaining leaves (all of which will be even), by 2. This is done to number the vertices of the new graph — i.e. we *avoid* having to perform the Euler Tour Algorithm all over again.

When this is finished, we will have a matrix, $\begin{pmatrix} a & b \\ c & d \end{pmatrix}$, at the root node and one child with a value '$t$' stored in it. Plug this value into the formula corresponding to the

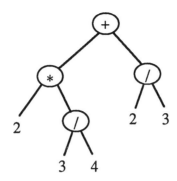

FIGURE VI.30. Original parse tree

matrix to get $(at + b)/(ct + d)$ as the value of the whole expression.

We conclude this section with an example. Suppose our initial expression is $(2 + 3/4) * 5 + 2/3$.

1. We fully-parenthesize this expression, using algorithm 3.1, to get: $(((((2)) + ((3)/(4)))) * (5)) + ((2)/(3)))$.

2. We carry out the parenthesis-matching algorithm, 3.2 (or the somewhat more efficient algorithm described in the appendix), to construct the parse-tree depicted in figure VI.30.

3. We "decorate" this tree with suitable matrices in order to carry out the pruning operation described above. We get the parse-tree in figure VI.31.

4. We order the leaf-nodes of the syntax tree, using the Euler Tour algorithm described in § 2.2 on page 315. We get the digram in figure VI.32.

5. We carry out the pruning algorithm described above:
   a. Prune odd-numbered, right-child leaf nodes. There is only 1 such node — the child numbered 3. Since the left sibling is also a number, we compute the value at the parent — and we get figure VI.33.
   b. Prune odd-numbered, left-child leaf nodes. There are two of these — nodes 1 and 5. We get the parse-tree in figure VI.34.
   We renumber the leaf-nodes. All of the present node-numbers are *even* since we have pruned all odd-numbered nodes. We just divide these indices by 2 to get the tree in figure VI.35.
   c. Again, prune odd-numbered right-child leaf-nodes — there is only node 3. In this step we actually carry out one of the matrix-modification steps listed on page 377. The result is figure VI.36.
   Now prune the odd-numbered left-child leaf node — this is node 1. The result of this step is the value of the original expression.

**3.4. Discussion and Further Reading.** There is an extensive literature on the problem of parsing and evaluating arithmetic expressions, and related problems.

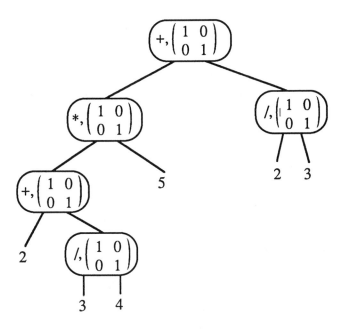

FIGURE VI.31.  Labeled parse tree

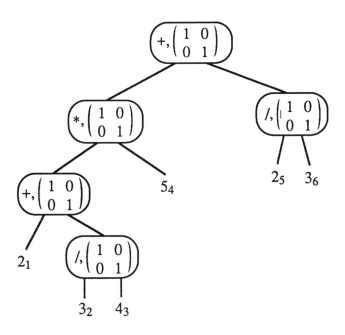

FIGURE VI.32.  Inorder traversal of the labeled parse tree

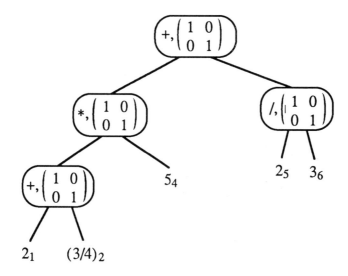

FIGURE VI.33. Result of pruning odd-numbered, right-child leaf nodes

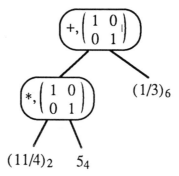

FIGURE VI.34. Result of pruning odd-numbered, left-child leaf nodes

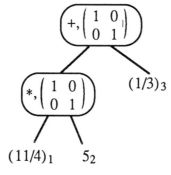

FIGURE VI.35. Result of renumbering leaf-nodes

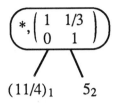

$$\left( *, \begin{pmatrix} 1 & 1/3 \\ 0 & 1 \end{pmatrix} \right)$$

$(11/4)_1 \qquad 5_2$

FIGURE VI.36. Second pruning operation

We have already discussed Brent's Theorem (§ 7.1 on page 48) — see [20]. Also see [21] and [93]. This work considered how one could evaluate computational networks for arithmetic expressions (and how one could find computational networks for an expression that facilitated this evaluation).

In [91] (1975), Kosoraju showed that the Cocke-Younger-Kasami algorithm for parsing context-free grammars lent itself to parallelization on an array processor. An improved version of this algorithm appeared in [24] (1987) by Chang, Ibarra, and Palis.

An **NC** algorithm for parsing general context-free grammars was published by Rytter in 1985 — see [129]. This algorithm executes in $O(\lg^2 n)$ time using $O(n^6)$ processors on a PRAM computer. This algorithm can be implemented on cube-connected cycles and perfect shuffle computers with no asymptotic time degradation — see [130].

Also see [73], by Ibarra Sohn, and Pong for implementations of parallel parsing algorithms on the hypercube. In [131] Rytter showed that *unambiguous* grammars can be parsed in $O(\lg n)$ time on a PRAM computer, but the number of processors required is still very large in general (although it is polynomial in the complexity of the input). If we restrict our attention to *bracket languages* the number of processors can be reduced to $O(n/\lg n)$, as remarked above — see [132].

The Japanese $5^{th}$ Generation Computer Project has motivated some work on parallel parsing algorithms — see [109] by Y. Matsumoto.

In 1982 Dekel and Sahni published an algorithm for putting arithmetic expressions in postfix form — see [43].

Gary L. Miller, Vijaya Ramachandran, and Erich Kaltofen have developed a fast parallel[12] algorithm for evaluating certain types of computational networks — see [115]. Their algorithm involved performing matrix operations with $n \times n$ matrices to evaluate computational networks that have $n$ vertices. Their algorithm has a faster execution-time than what one gets by a straightforward application of Brent's Theorem (the time is $O(\lg n(\lg nd))$ for a computational network of total size $n$ and depth $d$), but uses many more processors ($O(n^{2.376})$, where $n^{2.376}$ is the number of processors needed to multiply two $n \times n$ matrices).

---

[12]In general it is much faster than using Brent's Theorem, as presented on page 49 in the present book.

In [10], Baccelli and Fleury, and in [11], Baccelli and Mussi consider problems that arise in evaluating arithmetic expressions *asynchronously*.

Richard Karp and Vijaya Ramachandran have written a survey of parallel parsing algorithms — see [84].

Besides the obvious applications of parsing algorithms to compiling computer languages, there are interesting applications to *image processing*. See [25] by Chang and Fu for a discussion of parallel algorithms for parsing pattern and tree grammars that occurred in the analysis of Landsat images.

**Exercises:**

1. Consider the language consisting of *boolean expressions* — that is, expressions of the form $a \lor (b \land c)$, where $\lor$ represents OR and $\land$ represents AND. This is clearly an operator-precedence grammar (in fact, it is the same grammar as that containing arithmetic expressions, if we require all variables to take on the values 0 and 1). Recall the Circuit Value Problem on page 44. Does the algorithm in the present section provide a solution to this problem[13]?

2. Modify the algorithms of this section to *compile* an expression. Assume that we have a stack-oriented machine. This does all of its computations on a computation stack. It has the following operations:

   LOAD op — pushes its operand onto the stack

   **ADD** — pops the top two elements of the stack, adds them, and pushes the sum.

   **MULT** — pops the top two elements of the stack, multiplies them, and pushes the product.

   **SUB** — pops the top two elements of the stack, subtracts the second element from the top element them, and pushes the difference.

   **DIV** — pops the top two elements of the stack, divides the second element by the top element them, and pushes the quotient.

   **EXCHA** — exchanges the top two elements of the stack.

   The idea is to write a program using these instructions that leaves its result on the top of the stack. Incidentally, one advantage of this type of assembly language is that you rarely have to be concerned about addresses. There exist some actual machines with this type of assembly language. If Ex is an expression and C(Ex) represents the *code* required to *compute* the expression, you can generate code to compute general expressions by following the rules:

   a. C(Ex1 '+' Ex2) = C(Ex1)| |C(Ex2)| |'ADD' ;

   b. C(Ex1 '*' Ex2) = C(Ex1)| | C(Ex2)| |'MULT' ;

---

[13] And, therefore, a proof that *no* inherently sequential problems exist!

    c. C(Ex1 '−' Ex2) = C(Ex2)||C(Ex1)||'SUB';

    d. C(Ex1 '/' Ex2) = C(Ex2)||C(Ex1)||'DIV';

Here '||' represents *concatenation* of code.

The problem is to modify the algorithm of §3 for evaluating an expression so that it produces *assembly code* that will *compute* the value of the expression when run on a computer. This algorithm should execute in time $O(\lg n)$ using $n$ processors. The main problem is to replace the $2 \times 2$ matrices used in the algorithm (actually, the formula $(ax + b)/(cx + d)$ that the matrices represent) by a data structure that expresses a list of assembly language instructions with an unknown sublist — this unknown sublist takes the place of the variable $x$ in the formula $(ax + b)/(cx + d)$. This data structure should designed in conjunction with an algorithm to insert another such list into the position of the unknown sublist in unit time.

**3.5. Appendix: Parenthesis-matching algorithm.** We perform the following additional steps at the beginning of the parentheses-matching algorithm:

1. Partition the array (of characters) into $n/\lg n$ segments of length $\lg n$ each. Assign one processor to each of these segments.

2. Each processor scans its segment for matching pairs of parentheses within its segment. This uses a simple *sequential* algorithm (using a stack, for instance). The parentheses found in this step are marked and not further considered in this matching-algorithm. The remaining unmatched parentheses in each segment form a sequence of right parentheses followed by left parentheses: ')) (('. We assume that the each processor forms a data-structure listing these unmatched parentheses in its segment — this might be a linked list of their positions in the character array. Since each segment is $\lg n$ characters long, this can be done in $O(\lg n)$ time.

Next each processor matches its *leftmost* unmatched *left* parenthesis and its *rightmost* unmatched *right* parenthesis — we will call these the *extreme parentheses* — with corresponding parentheses in the whole *original* sequence. This is done by a kind of binary search algorithm that we will discuss shortly.

Note that this operation of matching these extreme parentheses in each sequence basically solves the entire matching-problem. Essentially, if we match the rightmost right parenthesis in the sequence: ')))) ' with the leftmost left parenthesis in another sequence '((((((( ', the *remaining* parentheses can be matched up by simply *scanning* both sequences. Here we are making *crucial* use of the fact that these sequences are *short* ($2 \lg n$ characters) — i.e. we can get away with using a straightforward sequential algorithm for scanning the sequences:

Now we use the binary search algorithm in § 3, 3.2, to match up the extreme parentheses in each sequence, and then use the sequential scan in each sequence to match up the remaining parentheses.

*Date*: 92/07/17 13:29:30

## 4. Searching and Sorting

In this section we will discuss some important parallel sorting algorithm. We have already seen one sorting algorithm at the end of §2 in chapter II. That algorithm sorted $n$ numbers in time $O(\lg^2 n)$ using $O(n)$ processors. The theoretical lower bound in time for sorting $n$ numbers is $O(\lg n)$ (since this many comparisons must be performed). The algorithms presented here take advantage of the full power of SIMD computers.

**4.1. Parallel searching.** As is usually the case, before we can discuss sorting, we must discuss the related operation of *searching*. It turns out that, on a SIMD computer we can considerably improve upon the well-known *binary search* algorithm. We will follows Kruskal's treatment of this in [92].

PROPOSITION 4.1. It is possible to search a sorted list of $n$ items using $p$ processors in time $\lceil \log(n+1)/\log(p+1) \rceil$.

This algorithm will be called **Search**$_p(n)$.

PROOF. We will prove by induction that $k$ comparison steps serve to search a sorted list of size $(p+1)^k - 1$. In actually implementing a search algorithm the inductive description of the process translates into a recursive program. The result is certainly true for $k = 0$. Assume that it holds for $k - 1$. Then to search a sorted list of size $(p+1)^k - 1$ we can compare the element being searched for to the elements in the sorted list subscripted by $i(p+1)^{k-1} - 1$ for $i = 1, 2, \ldots$. There are no more than $p$ such elements (since $(p+1)(p+1)^k - 1 > (p+1)^k - 1$). Thus the comparisons can be performed in one step and the problem is reduced to searching a list of size $(p+1)^{k-1} - 1$. $\square$

Exercises:

Write a C* program to implement this search algorithm.

**4.2. Sorting Algorithms for a PRAM computer.** We have already seen the Batcher sorting algorithm. It was one of the first parallel sorting algorithms developed. In 1978 Preparata published several SIMD-parallel algorithms for sorting $n$ numbers in $O(\lg n)$ time — see [123]. The simplest of these sorting algorithms required $O(n^2)$ processors. Essentially this algorithm used the processors to compare every key value with every other and then, for each key, counted the number of key values less than that key. When the proper key sequence was computed one additional step sufficed to move the data into sequence. This is fairly characteristic of unbounded parallel sorting algorithms — they are usually *enumeration sorts* and most of the work involves *count* *acquisition* (i.e. determining where a given key value belongs in the overall set of key values).

A more sophisticated algorithm in the same paper required only $O(n \lg n)$ processors to sort $n$ numbers in $O(\lg n)$ time. That algorithm uses a parallel *merging* algorithm due to Valiant (see [156]) to facilitate count acquisition.

In 1983, Ajtai, Komlós, Szemerédi published (see [3] and [4]) a description of a sorting network that has $n$ nodes and sorts $n$ numbers in $O(\lg n)$ time. The original description of the algorithm had a very large constant of proportionality ($2^{100}$), but recent work by Lubotzky, Phillips and Sarnal implies that this constant may be more manageable — see [105].

### 4.2.1. *The Cole Sorting Algorithm — CREW version.*

This section will discuss a sorting algorithm due to Cole in 1988 (see [28]) that is the fastest parallel sort to data. Unlike the algorithm of Ajtai, Komlós, Szemerédi, it is an *enumeration-sort* rather than a network. There are versions of this algorithm for a CREW-PRAM computer and an EREW-PRAM computer. Both versions of the algorithm use $n$ processors and run in $O(\lg n)$ time although the version for the EREW computer has a somewhat larger constant of proportionality. We will follow Cole's original treatment of the algorithm in [28] very closely.

The high-level description of the Cole sorting algorithm is fairly simple. We will assume that all inputs to the algorithm are *distinct*. If this *is not* true, we can modify the comparison-step so that a comparison of two elements always finds one strictly larger than the other.

Suppose we want to sort $n = 2^k$ data-items. We begin with a binary-tree with $n$ leaves (and $k$ levels), and each node of the tree has three *lists* or *arrays* attached to it (actual implementations of the algorithm do not require this but the description of the algorithm is somewhat simpler if we assume the lists are present). The leaves of the tree start out with the data-items to be sorted (one per leaf). Let $p$ be a node of the tree.

> Throughout this discussion, this tree will be called the **sorting tree** of the algorithm.

DEFINITION 4.2. We call the lists stored at each node:

1. $L(p)$ — this is a list that contains the result of sorting all of the data-items that are stored in leaf-nodes that are descendants of $p$. The whole algorithm is essentially an effort to compute $L$(the root). Note: this list would not be defined in the actual algorithm — we are only using it as a kind of descriptive device.

2. $\text{UP}(p)$ — this contains a sorted list of *some* of the data-items that started out at the leaves of the tree that were descendants of $p$. $\text{UP}(p) \subseteq L(p)$. In the beginning of the algorithm, all nodes of the tree have $\text{UP}(p) = \{\}$ — the empty list, except for the leaf-nodes: they contain the *input data* to the algorithm.

3. $\mathrm{SUP}(p)$ — this is a subsequence of $\mathrm{UP}(p)$ that is computed during the execution of the algorithm in a way that will be described below. $\mathrm{SUP}(p) \subseteq \mathrm{UP}(p)$.

Although we use the term *list* to describe these sequences of data-items, we will often want to refer to data within such a list by its index-position. Consequently, these lists have many of the properties of *variable-sized arrays*. Actual programs implementing this algorithm will have to define them in this way.

This algorithm is based upon the concept of *pipelining* of operations. It is essentially a *merge-sort* in which the recursive merge-operations are *pipelined* together, in order the reduce the overall execution-time. In fact, Cole describes it as sorting algorithm that uses an $O(\lg n)$-time merging algorithm. A *straightforward* implementation of a merge-sort would require an execution-time of $O(\lg^2 n)$ time. We pipeline the merging operations so that they take place concurrently, and the execution-time of the *entire* algorithm is $O(\lg n)$. The list SUP is an abbreviation for "sample-UP". They are *subsets* of the UP-list and are the *means* by which the pipelining is accomplished — the algorithm merges small subsets of the whole lists to be merged, and use them in order to compute the full merge operations. The *partial* merges execute in *constant* time.

We will classify nodes of the tree as *external* or *internal*:

DEFINITION 4.3. A node, $p$, of the tree will be called *external* if $\mathrm{UP}(p) = L(p)$ — otherwise it will be called *internal*.

1. Note that we don't have to know what $L(p)$ is, in order to determine whether $\mathrm{UP}(p) = L(p)$ — we know the size of the list $L(p)$ (which is $2^k$, if $p$ is in the $k^{\mathrm{th}}$ level from the leaves of the tree), and we just compare sizes.

2. The status of a node will change as the algorithm executes. At the beginning of the algorithm all nodes will be internal except the leaf-nodes. The interior-nodes will be internal because they contain *no data* so that $\mathrm{UP}(p)$ is *certainly* $\neq L(p)$. The leaf-nodes will be external because they start out each containing a single item of the *input-data* — and a single data-item is trivially sorted.

3. All nodes of the tree eventually become external (in this *technical* sense of the word). The algorithm terminates when the root of the tree becomes external.

ALGORITHM 4.4. **Cole Sorting Algorithm — CREW version** With the definitions of 4.2 in mind, the execution of the algorithm consists in propagating the data up the tree in a certain fashion, while carrying out parallel merge-operations. In each phase we perform operations on the lists defined above that depend upon whether a node is *internal* or *external*. In addition the operations that we carry out on the *external* nodes depend on *how long* the node has been external. We define a variable e_age to represent this piece of data: when a node becomes external e_age= 1. This variable is incremented each phase of the algorithm thereafter. With all of this in mind, the operations at a node $p$ are:

1. **$p$ internal:**
   a. Set the list $SUP(p)$ to every fourth item of $UP(p)$, measured from the right. In other words, if $|UP(p)| = t$, then $SUP(p)$ contains items of rank $t - 3 - 4i$, for $0 \leq i < \lfloor t/4 \rfloor$.
   b. If $u$ and $v$ are $p$'s immediate children, set $UP(p)$ to the result of merging $SUP(u)$ and $SUP(v)$. The merging operation is described in 4.6, to be described later.
2. **$p$ external with e_age= 1:** Same as step 1 above.
3. **$p$ external with e_age= 2:** Set the list $SUP(p)$ to every second item of $UP(p)$, measured from the right. In other words, if $|UP(p)| = t$, then $SUP(p)$ contains items of rank $t - 1 - 2i$, for $0 \leq i < \lfloor t/2 \rfloor$.
4. **$p$ external with e_age= 3:** Set $SUP(p) = UP(p)$.
5. **$p$ external with e_age> 3:** Do nothing. This node no longer actively participates in the algorithm.

Several aspects of this algorithm are readily apparent:

1. The activity of the algorithm "moves up" the tree from the leaves to the root. Initially all nodes except the leaves are internal, but they also have no data in them. Consequently, in the *beginning*, the actions of the algorithm have no real effect upon any nodes more than one level above the leaf-nodes. As nodes become external, and *age* in this state, the algorithm ceases to perform any activity with them.
2. Three phases after a given node, $p$, becomes external, its *parent* also becomes external. This is due to:
   a. $UP(p) = L(p)$ — the sorted list of *all* data-items that were input at leaf-nodes that were descendants of $p$.
   b. the rule that says that $SUP(p) = UP(p)$ in the third phase after node $p$ has become external.
   c. step 2 of the algorithm for internal nodes (the parent, $p'$, of $p$ in this case) that says that $UP(p')$ to the result of merging $SUP(p)$ its the data in its sibling $SUP(p'')$.

   It follows that the execution time of the algorithm will be $\leq 2\lg n \cdot K$, where $K$ is $>$ the time required to perform the merge-operations described above for internal nodes. The *interesting* aspect of this algorithm is that $K$ is bounded by a constant, so the overall execution-time is $O(\lg n)$.
3. The algorithm correctly sorts the input-data. This is because the sorted sequences of data are *merged* with each other as they are copied up the tree to the root.

The remainder of this section will be spent proving that the algorithm executes in the stated time. The only thing that must be proved is that the merge-operations can be carried out in constant time. We will need several definitions:

DEFINITION 4.5. **Rank-terminology.**

1. Let $e$, $f$, and $g$ be three data-items, with $e < g$. $f$ is *between* $e$ and $g$ if $e \leq f < g$. In this case $e$ and $g$ *straddle* $f$.
2. Let $L$ and $J$ be sorted lists. Let $f$ be an item in $J$, and let $e$ and $g$ be two adjacent items in $L$ that straddle $f$ (in some cases, we assume $e = -\infty$ or $g = \infty$). With this in mind, the *rank* of $f$ in $L$ is defined to be the rank of $e$ in $L$. If $e = -\infty$, the rank of $f$ is defined to be 0.
3. If $e \in L$, and $g$ is the next larger item then we define $[e, g)$ to be the *interval induced by* $e$. It is possible for $e = -\infty$ and $g = \infty$.
4. If $c$ is a positive integer, $L$ is a *c-cover* of $J$, if each interval induced by an item in $L$ contains at most $c$ items from $J$. This implies that, if we merge $L$ and $J$, at most $c$ items from $J$ will ever get merged between any two items of $L$.
5. $L$ is defined to be *ranked in* $J$, denoted $L \to J$, if for each item in $L$, we know its rank in $J$. This basically means that we know where each item in $L$ would go if we decided to merge $L$ and $J$.
6. Two lists $L$ and $J$ are defined to be *cross ranked*, denoted $L \leftrightarrow J$, if $L \to J$ and $J \to L$. When two lists are cross-ranked it is very easy to merge them in constant time: the rank of an element in the result of merging the two lists is the *sum* of its rank in *each* of the lists.

In order to prove that the merge can be performed in constant time, we will need to keep track of $\mathrm{OLDSUP}(v)$ — this is the $\mathrm{SUP}(v)$ in the *previous phase* of the algorithm. $\mathrm{NEWUP}(v)$ is the value of $\mathrm{UP}(v)$ in the *next* phase of the algorithm. The main statement that will imply that the merge-operations can be performed in constant time is:

> In each phase of the algorithm, and for each vertex, $u$ of the graph $\mathrm{OLDSUP}(u)$ is a 3-cover of $\mathrm{SUP}(u)$. This and the fact that $\mathrm{UP}(u)$ is the result of merging $\mathrm{OLDSUP}(v)$ and $\mathrm{OLDSUP}(w)$ (where $v$ and $w$ are the children of $u$) will imply that $\mathrm{UP}(u)$ is a 3-cover of $\mathrm{SUP}(v)$ and $\mathrm{SUP}(w)$.

Now we describe the merging-operations. We assume that:

1. $\mathrm{UP}(u) \to \mathrm{SUP}(v)$ and $\mathrm{UP}(u) \to \mathrm{SUP}(w)$ **are given at the start of the merge-operation.**
2. **The notation** $a \cup b$ **means "the result of merging lists $a$ and $b$".**

ALGORITHM 4.6. **Cole Merge Algorithm — CREW version** We perform the merge in two phases:

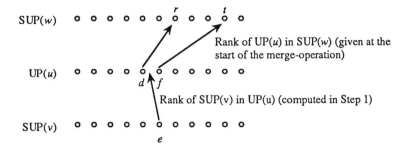

FIGURE VI.37. Ranking SUP($v$) in SUP($w$)

**Phase 1.** In this phase we compute $\mathrm{NEWUP}(u)$. Let $e$ be an item in $\mathrm{SUP}(v)$; the rank of $e$ in $\mathrm{NEWUP}(u) = \mathrm{SUP}(v) \cup \mathrm{SUP}(w)$ is equal to the sum of its ranks in $\mathrm{SUP}(v)$ and $\mathrm{SUP}(w)$. We must, consequently, *cross-rank* $\mathrm{SUP}(v)$ and $\mathrm{SUP}(w)$, by a procedure described below. Having done that, for each item $e \in \mathrm{SUP}(v)$, we know:

1. Its rank in $\mathrm{NEWUP}(u)$;
2. The two items $d, f \in \mathrm{SUP}(w)$ that straddle $e$.
3. We know the ranks of $d$ and $f$ in $\mathrm{NEWUP}(u)$.

For each item in $\mathrm{NEWUP}(u)$ we record:

- whether it came from $\mathrm{SUP}(v)$ or $\mathrm{SUP}(w)$;
- the ranks of the straddling items from the *other* set.

This completes the description of phase 1. We must still describe how to cross-rank $\mathrm{SUP}(v)$ and $\mathrm{SUP}(w)$:

**Step 1** For each item in $\mathrm{SUP}(v)$, we compute its rank in $\mathrm{UP}(u)$. This is performed by the processors associated with the items in $\mathrm{UP}(u)$ as follows:

If $y \in \mathrm{UP}(u)$, let $I(y)$ be the interval induced by $y$ in $\mathrm{UP}(u)$ — recall that this is the interval from $y$ to the next higher element of $\mathrm{UP}(u)$. Next consider the items of $\mathrm{SUP}(v)$ contained in $I(y)$ — there are at most 3 such items, by the 3-cover property. The processor associated with $y$ assigns ranks to each of these 3 elements — this step, consequently, requires constant time (3 units of the time required to assign ranks).

**Step 2** For each item $e \in \mathrm{SUP}(v)$, we compute its rank on $\mathrm{SUP}(w)$. This is half of the effort in cross-ranking $\mathrm{SUP}(v)$ and $\mathrm{SUP}(w)$. We determine the two items $d, f \in \mathrm{UP}(u)$ that straddle $e$, using the rank computed in the step immediately above. Suppose that $d$ and $f$ have ranks $r$ and $t$, respectively, in $\mathrm{SUP}(u)$ — we can determine this by the information that was *given* at the start of the merge-operation. See figure VI.37. Then:

- all items of rank $\leq r$ are *smaller* than item $e$ — since all inputs are *distinct* (one of the hypotheses of this sorting algorithm).

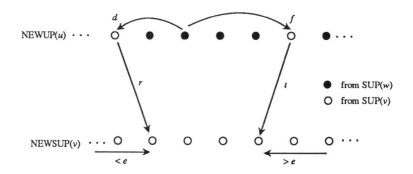

FIGURE VI.38. Finding the rank of items from SUP($w$) in NEWSUP($v$)

- all items of rank $> t$ are *larger* than $e$.

The only items about which there is any question are those with ranks between $r$ and $t$. The 3-cover property implies that there are at most 3 such items. Computing the rank of $e$ and these ($\leq 3$) other items can be computed with at most 2 comparisons.

The cross-ranking of $\mathrm{SUP}(v)$ and $\mathrm{SUP}(w)$ is completed by running these last two steps with $v$ and $w$ interchanged. Once we know these ranks, we can perform a parallel move-operation to perform the merge (remember that we are using a PRAM computer).

We have made *essential use* of rankings $\mathrm{UP}(u) \to \mathrm{SUP}(v)$ and $\mathrm{UP}(u) \to \mathrm{SUP}(w)$. In order to be able to *continue* the sorting algorithm in later stages, we must be able to provide this kind of information for *later* merge-operations. This is phase 2 of the merge:

**Phase 2.** We will compute rankings $\mathrm{NEWUP}(U) \to \mathrm{NEWSUP}(v)$ and $\mathrm{NEWUP}(u) \to \mathrm{NEWSUP}(w)$. For each item $e \in \mathrm{NEWUP}(U)$ we will compute its rank in $\mathrm{NEWSUP}(v)$ — the corresponding computation for $\mathrm{NEWSUP}(w)$ is entirely analogous. We start by noting:

- Given the ranks for an item from $\mathrm{UP}(u)$ in both $\mathrm{SUP}(v)$ and $\mathrm{SUP}(w)$, we can deduce the rank of this item in $\mathrm{NEWUP}(u) = \mathrm{SUP}(v) \cup \mathrm{SUP}(w)$ — this new rank is just the sum of the old ranks.
- Similarly, we obtain the ranks for items from $\mathrm{UP}(v)$ in $\mathrm{NEWUP}(v)$.
- This yields the ranks of items from $\mathrm{SUP}(v)$ in $\mathrm{NEWSUP}(v)$ — since each item in $\mathrm{SUP}(v)$ came from $\mathrm{UP}(v)$, and $\mathrm{NEWSUP}(v)$ comprises every fourth item in $\mathrm{NEWUP}(v)$.

It follows that, for every item $e \in \mathrm{NEWUP}(u)$ that came from $\mathrm{SUP}(v)$, we know its rank in $\mathrm{NEWSUP}(v)$. It remains to compute this rank for items that came from $\mathrm{SUP}(w)$. Recall that for each item $e$ from $\mathrm{SUP}(w)$ we computed the straddling items $d$ and $f$ from $\mathrm{SUP}(v)$ (in phase 1 above) — see figure VI.38.

We know the ranks $r$ and $t$ of $d$ and $f$, respectively, in $\mathrm{NEWSUP}(v)$. Every item of

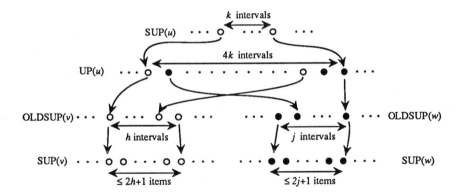

FIGURE VI.39

rank $\leq r$ in $\text{NEWSUP}(v)$ is smaller than $e$, while every item of rank $> t$ is larger than $e$. Thus the only items about which there is any doubt concerning their size relative to $e$ are items with rank between $r$ and $t$. But the 3-cover property implies that there are at most 3 such items. The relative order of $e$ and these (at most) three items can be determined be means of at most two comparisons. This ranking step can, consequently, be done in constant time.

We still haven't proved that it works. We have to prove the 3-cover property that was used throughout the algorithm.

LEMMA 4.7. Let $k \geq 1$. In each iteration, any $k$ adjacent intervals in $\text{SUP}(u)$ contain at most $2k + 1$ items from $\text{NEWSUP}(u)$.

PROOF. We prove this by induction on the number of an iteration. The statement of the lemma is true initially because:

1. When $\text{SUP}(u)$ is *empty*, $\text{NEWSUP}(u)$ contains at most one item.
2. The first time $\text{SUP}(u)$ is nonempty, it contains one item and $\text{NEWSUP}(u)$ contains at most two items.

Now we give the *induction* step: We want to prove that $k$ adjacent intervals in $\text{SUP}(u)$ contain at most $2k + 1$ items from $\text{NEWSUP}(u)$, assuming that the result is true in the previous iteration — i.e., for all nodes $u'$ and for all $k' \geq 1$, $k'$ intervals in $\text{OLDSUP}(u')$ contain at most $2k' + 1$ items from $\text{SUP}(u')$.

We first suppose that $u$ is not external at the start of the current iteration — see figure VI.39.

Consider a sequence of $k$ adjacent intervals in $\text{SUP}(u)$ — they cover the same range as some sequence of $4k$ adjacent intervals in $\text{UP}(u)$. Recall that $\text{UP}(u) = \text{OLDSUP}(v) \cup \text{OLDSUP}(w)$. The $4k$ intervals in $\text{UP}(u)$ overlap some $k \geq 1$ adjacent intervals in $\text{OLDSUP}(v)$ and some $k \geq 1$ adjacent intervals in $\text{OLDSUP}(w)$, with $h + j = 4k + 1$. The $h$ intervals in $\text{OLDSUP}(v)$ contain at most $2h + 1$ items in $\text{SUP}(v)$, by the inductive hypothesis, and similarly, the $j$ intervals in $\text{OLDSUP}(w)$ contain at most $2j + 1$ items from $\text{SUP}(w)$. Recall that $\text{NEWUP}(u) = \text{SUP}(v) \cup \text{SUP}(w)$. It follows that the $4k$ intervals in $\text{UP}(u)$ contain at most $2h + 2j + 2 = 8k + 4$ items from $\text{NEWUP}(u)$. But $\text{NEWSUP}(u)$ is formed by selecting every fourth item in $\text{NEWUP}(u)$ — so that the $k$ adjacent intervals in $\text{SUP}(u)$ contain at most $2k + 1$ items from $\text{NEWSUP}(u)$.

At this point we are almost done. We must still prove the lemma for the first and second iterations in which $u$ is external. In the third iteration after $u$ becomes external, there are no $\text{NEWUP}(u)$ and $\text{NEWSUP}(u)$ arrays.

Here we can make the following *stronger* claim involving the relationship between $\text{SUP}(u)$ and $\text{NEWSUP}(u)$:

> $k$ adjacent intervals in $\text{SUP}(u)$ contain exactly $2k$ items from $\text{NEWSUP}(u)$ and every item in $\text{SUP}(u)$ occurs in $\text{NEWSUP}(u)$.

Proof of claim: Consider the first iteration in which $u$ is external. $\text{SUP}(u)$ is made up of every fourth item in $\text{UP}(u) = L(u)$, and $\text{NEWSUP}(u)$ contains every second item in $\text{UP}(u)$. The claim is clearly true in this iteration. A similar argument proves the claim in the following iteration. $\square$

COROLLARY 4.8. For all vertices $u$ in the sorting-tree, $\text{SUP}(u)$ is a 3-cover of $\text{NEWSUP}(u)$.

PROOF. Set $k = 1$ in 4.7 above. $\square$

We have seen that the algorithm executes in $O(\lg n)$ time. A detailed analysis of the algorithm shows that it performs $(15/2)n \lg n$ comparisons. This analysis also shows that the number of active elements in all of the lists in the tree in which the sorting takes place is bounded by $O(n)$, so that the algorithm requires $O(n)$ processors.

4.2.2. *Example.* We will conclude this section with an example. Suppose our initial input-data is the 8 numbers $\{6, 1, 5, 3, 2, 0, 7, 4\}$. We put these numbers at the leaves of a complete binary tree — see figure VI.40.

In the first step, the leaf-vertices are external and all other nodes are internal, in the terminology of 4.3 on page 386, and UP will equal the data stored there.

In the notation of 4.4 on page 386, the leaf-vertices have e_age equal to 1. In the first step of computation, SUP of the leaf vertices are set to every $4^{\text{th}}$ value in UP of that vertex — this clearly has no effect. Furthermore, nothing significant happens at higher levels of the tree.

In step 2 of the algorithm, nothing happens either, since the SUP-lists of the leaf vertices are set to every second element of the UP-lists.

In step 3 of the algorithm e_age is 3 and SUP of each leaf vertex is set to the corresponding UP set. Nothing happens at higher levels of the tree. Since the e_age

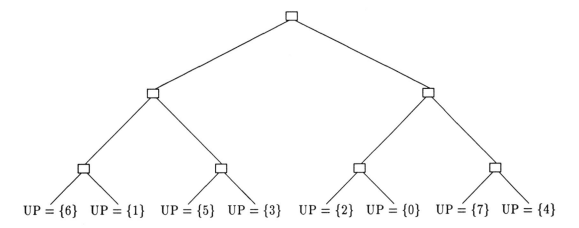

UP = {6}   UP = {1}   UP = {5}   UP = {3}   UP = {2}   UP = {0}   UP = {7}   UP = {4}

FIGURE VI.40. Sort-tree with the input-data.

of leaf vertices is $\geq 4$ in the remaining steps, no further activity takes place at the leaf vertices after this step.

In step 4 the UP lists of the vertices one level above the leaf-vertices are given non-null values. These vertices will be external in the next step with e_age equal to 1. Our tree is changed to the one in figure VI.41.

In step 5 the only vertices in which any significant activity takes place are those one level above the leaf-nodes. these vertices are external with e_age equal to 1. No assignment to the SUP lists occurs in this step, due to the rules in 4.4 on page 386.

In step 6 the active vertices from step 5 have assignments made to the SUP lists, the resulting tree appears in figure fig:colesortexampd3.

The leaf-vertices have significant data stored in them, but do not participate in future phases of the algorithm in any significant way, so we mark them "Inactive".

In step 8 we assign data to the UP lists one level higher, and we expand the SUP lists in the next lower level to get the sort-tree in figure VI.43.

In step 9, we expand the UP lists of the vertices one level below the top. We are only merging a few more data-items into lists that are already sorted. In addition, the data we are merging into the UP lists are "well-ranked" with respect to the data already present. Here the term "well-ranked" means that UP is a 3-cover of the SUP lists of the child vertices — see line 3 of 4.5 on page 388. These facts imply that the merging operation can be carried out in constant time. We get the sort-tree in figure VI.44.

In step 10, we expand the SUP lists (in constant time, making use of the fact that the UP lists are a 3-cover of them), and put some elements into the UP lists of the root-vertex. The result is the sort-tree in figure VI.45.

In step 11, we:

- Put elements into the SUP list of the root.

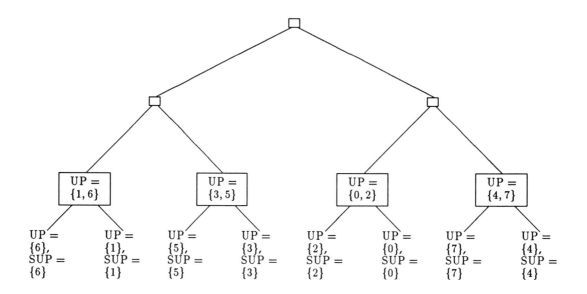

FIGURE VI.41. The sort-tree in step 4 of the Cole sorting algorithm.

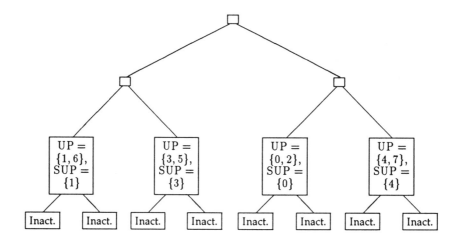

FIGURE VI.42. The sort-tree at the end of step 6.

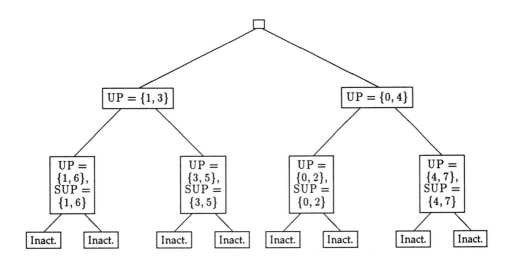

FIGURE VI.43. Sort-tree at the end of step 8.

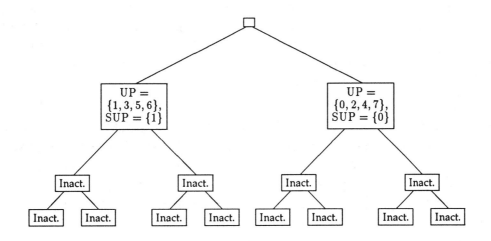

FIGURE VI.44. The sort-tree at the end of step 9.

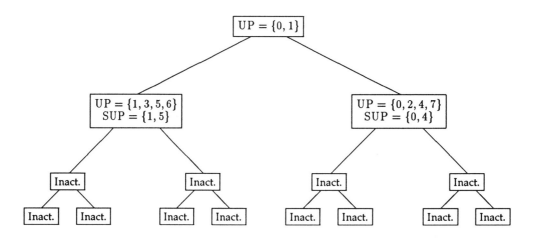

FIGURE VI.45. The sort-tree at the end of step 10.

- Merge more elements into the UP list of the root.
- Expand the SUP lists of the vertices below the root.

The result is the sort-tree in figure VI.46.

In step 13, we essentially complete the sorting operation — the resulting sort-tree is shown in figure VI.47.

The next step would be to update the SUP list of the root. It is superfluous.

4.2.3. *The Cole Sorting Algorithm — EREW version.* Now we will consider a version of the Cole sorting algorithm that runs on an EREW computer in the same *asymptotic* time as the one above (the constant of proportionality is larger, however). The basic sorting algorithm is almost the same as in the CREW case. The only part of the algorithm that is not EREW (and must be radically modified) is the *merging operations.* We will present an EREW version of the merging algorithm, described in 4.6 on page 388.

We will need to store some additional information in each node of the sorting-tree:

ALGORITHM 4.9. Cole Sorting Algorithm — EREW version This is the same as the CREW version 4.4 (on page 386) except that we maintain lists (or variable-sized arrays) $UP(v)$, and $DOWN(v)$, $SUP(v)$, and $SDOWN(v)$ at each node, $v$, of the sorting tree. The $SDOWN(v)$-list is composed of every fourth item of the $DOWN(v)$-list.

Consider a small portion of the sorting-tree, as depicted in figure VI.48.

At node $v$, in each step of the sorting algorithm, we:

1. Form the arrays $SUP(v)$ and $SDOWN(v)$;
2. Compute $NEWUP(v) = SUP(x) \cup SUP(y)$. Use the merge algorithm 4.10 described below.
3. Compute $NEWDOWN(v) = SUP(w) \cup SDOWN(u)$. Use the merge algorithm 4.10 described below.

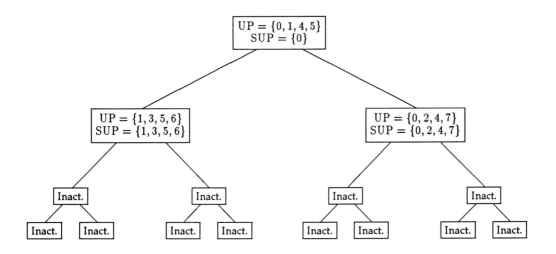

FIGURE VI.46. The sort-tree at the end of step 11.

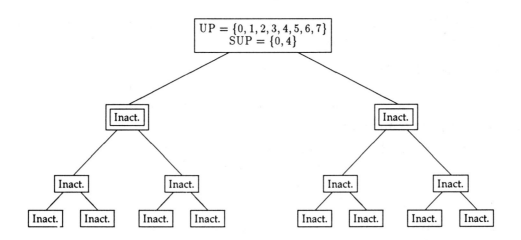

FIGURE VI.47. Sort-tree at the end of the Cole sorting algorithm.

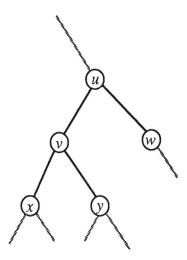

FIGURE VI.48. A small portion of the sorting-tree

In addition, we maintain the following arrays in order to perform the merge-operations in constant time:

- $UP(v) \cup SDOWN(v)$;
- $SUP(v) \cup SDOWN(v)$;

We have omitted *many* details in this description — the EREW version of the merge-operation requires many more cross-rankings of lists than are depicted here. The remaining details of the algorithm will be given below, on page 399.

Here $SDOWN(v)$ is a 3-cover of $NEWSDOWN(v)$ — the proof of this is identical to the proof of the 3-cover property of the $SUP$ arrays in 4.8 on page 392.

ALGORITHM 4.10. **Cole Merging Algorithm — EREW Case** Assume that $J$ and $K$ are two sorted arrays of distinct items and $J$ and $K$ have no items in common. It is possible to compute $J \leftrightarrow K$ in constant time (and, therefore, also $L = J \cup K$), given the following arrays and rankings:

1. Arrays $SK$ and $SJ$ that are 3-covers of $J$ and $K$, respectively;
2. $SJ \leftrightarrow SK$ — this amounts to knowing $SL = SJ \cup SK$;
3. $SK \to J$ and $SJ \to K$;
4. $SJ \to J$ and $SK \to K$.

These input rankings and arrays are depicted in figure VI.49.

This merge algorithm will also compute $SL \to L$, where $L = J \cup K$.

The algorithm is based upon the observation that the interval $I$ between two adjacent items $e$ and $f$, from $SL = SJ \cup SK$, contains at most three items from each of $J$ and $K$. In order to cross-rank $J$ and $K$, it suffices, for each such interval, to determine the relative order of the (at most) six items it contains. To carry out this procedure, we

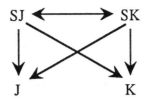

FIGURE VI.49. Input-rankings for the EREW form of the Cole Merge Algorithm

associate one processor with each interval in the array $SL$. The number of intervals is one larger than the number of items in the array $SL$. The ranking takes place in two steps:

1. We identify the two sets of (at most) three items contained in $I$. These are the items straddled by $e$ and $f$. If $e$ is in $\left\{\begin{array}{c} SJ \\ SK \end{array}\right\}$, we determine the leftmost item of these (at most) three items using $\left\{\begin{array}{c} SJ \to J \\ SK \to J \end{array}\right\}$; the rightmost item is obtained in the same way. The (at most) three items from $K$ are computed analogously.

2. For each interval in $SL$, we perform at most five comparisons to compute the rankings of the at most three items from each of $J$ and $K$.

We compute $SL \to L$ as follows:

For each item $e \in SL$, we simply add its ranks in $J$ and $K$, which wields its rank in $L$. These ranks are obtained from

$$\begin{cases} SJ \to J \text{ and } SJ \to K & \text{if } e \text{ is from } SJ \\ SK \to J \text{ and } SK \to K & \text{if } e \text{ is from } SK \end{cases}$$

This completes the description of the merging procedure. It is clear, from its *hypotheses*, that a considerable amount of information is required in each step of 4.9 in order to *carry out* the merges described there. In fact, at each step of 4.9, we will need the following rankings:

### 4.11. Input Rankings

1. $\mathrm{OLDSUP}(x) \leftrightarrow \mathrm{OLDSUP}(y)$
2. $\mathrm{OLDSUP}(v) \to \mathrm{SUP}(v)$;
3. $\mathrm{OLDSUP}(w) \leftrightarrow \mathrm{OLDSDOWN}(u)$;
4. $\mathrm{OLDSDOWN}(v) \to \mathrm{SDOWN}(v)$;
5. $\mathrm{SUP}(v) \leftrightarrow \mathrm{SDOWN}(v)$;
6. $\mathrm{UP}(v) \leftrightarrow \mathrm{SDOWN}(v)$;
7. $\mathrm{SUP}(v) \leftrightarrow \mathrm{DOWN}(v)$;
8. Since $\mathrm{DOWN}(v) = \mathrm{OLDSUP}(w) \cup \mathrm{OLDSDOWN}(u)$, and since we have the cross-ranking in line 7, we get $\mathrm{OLDSUP}(w) \to \mathrm{SUP}(v)$, and

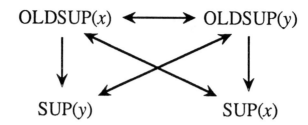

FIGURE VI.50. Input template for the merge in step 1

9. $\text{OLDSDOWN}(u) \to \text{SUP}(v)$.
10. Since $\text{UP}(v) = \text{OLDSUP}(x) \cup \text{OLDSUP}(y)$, and we have the cross-ranking in line 7, we get the rankings $\text{OLDSUP}(x) \to \text{SDOWN}(v)$ and $\text{OLDSUP}(y) \to \text{SDOWN}(v)$.

The remainder of the sorting algorithm involves five more steps, that each apply the merge-algorithm 4.10 to compute the information in the list above. The following facts should be kept in mind during these five merge-operations:

> The node $v$ is the *current node* of the graph. All of the merge-operations are being carried out in order to compute information for node $v$. Node $v$ is surrounded by *other* nodes that contribute information for the merge-operations — see figure VI.48 on page 398.

Each merge-operation requires four lists with five-rankings between them as input — we represent these by a "template", as in figure VI.49 on page 399.

**Step 1:** We compute the rankings in lines 1 and 2 in the list above. We begin by computing $\text{SUP}(x) \leftrightarrow \text{SUP}(y)$. This computation also gives the rankings $\text{UP}(v) \to \text{NEWUP}(v)$ and $\text{SUP}(v) \to \text{NEWSUP}(v)$. This is a straightforward application of the merge-algorithm 4.10, using the following input-information (that we *already have*) and template VI.50:

- $\text{OLDSUP}(x) \leftrightarrow \text{OLDSUP}(y)$, from line 1 in the list above, *at* node $v$.
- $\text{OLDSUP}(x) \leftrightarrow \text{SUP}(y)$, from line 8 *at* node $y$.
- $\text{OLDSUP}(y) \to \text{SUP}(x)$, from line 8 *at* node $x$.
- $\text{OLDSUP}(x) \to \text{SUP}(x)$, from line 2 *at* node $x$.
- $\text{OLDSUP}(y) \to \text{SUP}(y)$, from line 2 *at* node $y$.

Here, we have made essential use of knowledge of the rankings in 4.11 at nodes *other than* the current node.

**Step 2:** Compute $\text{SUP}(w) \leftrightarrow \text{SDOWN}(u)$, giving rise to $\text{NEWDOWN}(v)$, $\text{DOWN}(v) \to \text{NEWDOWN}(v)$, and $\text{DOWN}(v) \to \text{NEWSDOWN}(v)$. Again, we perform the merge-algorithm 4.10 using the (known) data and template VI.51:

- $\text{OLDSUP}(w) \leftrightarrow \text{OLDSDOWN}(u)$, from line 3 of statement 4.11 *at* node $v$.
- $\text{OLDSUP}(w) \to \text{SDOWN}(u)$, from line 10 of statement 4.11 *at* node $u$.

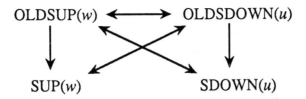

FIGURE VI.51. Input template for the merge in step 2

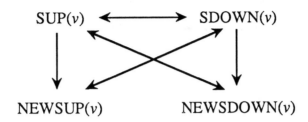

FIGURE VI.52. Input template for the merge in step 3

- $\text{OLDSDOWN}(u) \rightarrow \text{SUP}(w)$, from line 9 of statement 4.11 *at* node $w$.
- $\text{OLDSUP}(w) \rightarrow \text{SUP}(w)$, from line 2 of statement 4.11 *at* node $w$.
- $\text{OLDSDOWN}(u) \rightarrow \text{SDOWN}(u)$, from line 4 of statement 4.11 *at* node $u$.

**Step 3:** Compute $\text{NEWSUP}(v) \leftrightarrow \text{NEWSDOWN}(v)$. As before, we have the following input-information for algorithm 4.10 on page 398 and template VI.52:

1. $\text{SUP}(v) \leftrightarrow \text{SDOWN}(v)$, from line 5 of statement 4.11 *at* node $v$.
2. $\text{SUP}(v) \leftrightarrow \text{NEWDOWN}(v)$, and, therefore, $\text{SUP}(v) \rightarrow \text{NEWSDOWN}(v)$. This is computed from
   a. $\text{SUP}(v) \leftrightarrow \text{SUP}(w)$, at step 1 and node $u$, and
   b. $\text{SUP}(v) \leftrightarrow \text{SDOWN}(u)$, from step 2 at node $w$.
   These rankings give rise to $\text{SUP}(v) \leftrightarrow [\text{SUP}(w) \cup \text{SDOWN}(u)] = \text{SUP}(v) \leftrightarrow \text{SUP}(w)$.
3. $\text{NEWUP}(v) \leftrightarrow \text{SDOWN}(v)$, and, therefore, $\text{SDOWN}(v) \rightarrow \text{NEWSUP}(v)$. This is computed from
   a. $\text{SUP}(x) \leftrightarrow \text{SDOWN}(v)$, at step 2 and node $y$, and
   b. $\text{SUP}(y) \leftrightarrow \text{SDOWN}(v)$, from step 2 at node $x$.
   These rankings give rise to $[\text{SUP}(x) \cup \text{SUP}(y)] \leftrightarrow \text{SDOWN}(v) = \text{NEWUP}(v) \leftrightarrow \text{SDOWN}(v)$.
4. $\text{SUP}(v) \rightarrow \text{NEWSUP}(v)$, from step 1 at node $v$.
5. $\text{SDOWN}(v) \rightarrow \text{NEWSDOWN}(v)$ from step 3 at node $v$.

**Step 4:** Compute $\text{NEWUP}(v) \leftrightarrow \text{NEWSDOWN}(v)$. This is an application of the merge-algorithm 4.10 on page 398 using the (known) input-data and template VI.53:

- $\text{NEWSUP}(v) \leftrightarrow \text{SDOWN}(v)$, from step 3.3 above, at node $v$.

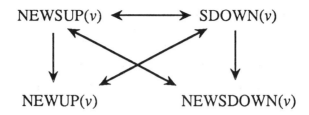

FIGURE VI.53. Input template for the merge in step 4

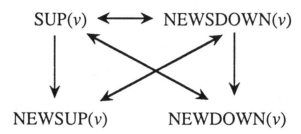

FIGURE VI.54. Input template for the merge in step 5

- $SDOWN(v) \rightarrow NEWUP(v)$, from step 3.3 above, at node $v$.
- $NEWSUP(v) \rightarrow NEWSDOWN(v)$, form step 3 at node $v$.
- $NEWSUP(v) \rightarrow NEWUP(v)$.
- $SDOWN(v) \rightarrow NEWSDOWN(v)$, from step 2 at node $v$.

**Step 5:** Compute $NEWSUP(v) \leftrightarrow NEWDOWN(v)$. We have the input information to the merge algorithm and template VI.54:

- $SUP(v) \leftrightarrow NEWDOWN(v)$, from step 3.2 above, applied to node $v$.
- $SUP(v) \rightarrow NEWDOWN(v)$, from step 3.2 above, applied to node $v$.
- $NEWSDOWN(v) \rightarrow NEWSUP(v)$, from step 3 at node $v$.
- $SUP(v) \rightarrow NEWSUP(v)$, from step 1 at node $v$.
- $NEWSDOWN(v) \rightarrow NEWDOWN(v)$.

**4.3. The Ajtai, Komlós, Szemerédi Sorting Network.** In this section we will present an asymptotically optimal sorting algorithm developed by Ajtai, Komlós, Szemerédi. It differs from the Cole sorting algorithm of the previous section in that:

- it is a sorting *network*. Consequently, it could (at least *in principle*) be used as a *substitute* for the Batcher sort in the simulation-algorithms in chapter II.
- it is *not uniform*. This means that, *given* a value of $n$, we can construct (with a *great deal* of effort) a sorting network with $O(n \lg n)$ comparators and with depth $O(\lg n)$. Nevertheless, we don't have an $O(\lg n)$-time sorting *algorithm* in the sense of the Cole sorting algorithm. The complexity-parameter, $n$, cannot be regarded as one of the *inputs* to the sorting algorithm. In other words, we have a *different algorithm* for each value of $n$.

As remarked above, this algorithm uses $O(n)$ processors and executes in $O(\lg n)$ time. Unfortunately, the constant factor in this $O(\lg n)$ may well turn out to be *very* large. It turns out that this constant depends upon knowledge of a certain combinatorial construct known as an *expander graph*. The only known *explicit constructions* of these graphs give rise to very large graphs (i.e., $\approx 2^{100}$ vertices), which turn out to imply a very large constant factor in the algorithm. Recent results suggest ways of reducing this constant considerably — see [105].

On the other hand, there is a probabilistic argument to indicate that these known constructions are extraordinarily *bad*, in the sense that there are many known "small" expander graphs. This is an area in which a great deal more research needs to be done. See §4.5 for a discussion of these issues.

With all of these facts in mind, we will regard this algorithm as essentially a *theoretical* result — it proves that a sorting network with the stated properties *exists*.

DEFINITION 4.12. 1. Given a graph $G = (V, E)$, and a set, $S$, of vertices $\Gamma(S)$ is defined to be the set of *neighbors* of $S$ — i.e., it is the set of vertices defined by:

$$z \in \Gamma(S) \Leftrightarrow \exists x \in S \text{ such that } (z, e) \in E$$

2. A bipartite graph (see definition 8.8 on page 104 and figure III.9 on page 104) $G(V_1, V_2, E)$, with $|V_1| = |V_2| = n$ is called an *expander graph* with parameters $(\lambda, \alpha, \mu)$ if

- for any set $A \subset V_1$ such that $|A| \leq \alpha n$ we have $|\Gamma(A)| \geq \lambda |A|$
- The maximum number of edges incident upon any vertex is $\leq \mu$

In § 8 in chapter III, expander graphs were used with $\alpha = n/(2c-1)$, $\lambda = (2c-1)/b$, $\mu = 2c - 1$, in the notation of lemma 8.9 on page 105. That result also gives a probabilistic proof of the *existence* of expander graphs, since it shows that sufficiently large *random graphs* have a nonvanishing probability of being expander graphs.

Our algorithm requires that we have expander-graphs with certain parameters to be given at the outset. We will also need the concept of a *1-factor* of a graph:

DEFINITION 4.13. Let $G = (V, E)$ be a graph. A *1-factor* of $G$ is a set $S = \{e_1, \ldots, e_k\}$ of disjoint edges that *span* the graph (regarded as a subgraph).

Recall that a subgraph of a graph *spans* it if all the vertices of the containing graph are also in the subgraph. It is not hard to see that a graph that *has* a 1-factor must have an *even* number of vertices (since each edge in the 1-factor has two end-vertices). A 1-factor of a graph can also be called a *perfect matching* of the graph. Here is an example. Figure VI.55 shows a graph with a few of its 1-factors.

We will need an expander graph with *many* 1-factors. The question of whether a graph has even one 1-factor is a nontrivial one — for instance it is clear that any graph

Original graph

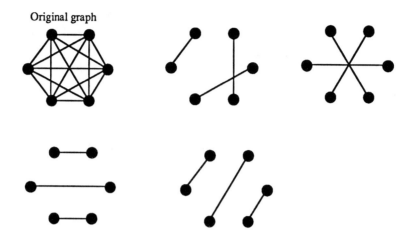

FIGURE VI.55. A graph with some of its 1-factors

with an odd number of vertices has *no* 1-factor[14]. We can ensure that this property exists *by construction* — we do this by forming the union of many 1-factors on the same set of vertices.

Section 4.5 discusses some of the probabilistic arguments that can be used to study these random bipartite graphs. Counting arguments like that used in lemma 8.9 on page 105 show that there exists an expander graph with parameters $(2, 1/3, 8)$. In fact, counting arguments show that a bipartite graph that is the union of $\mu$ *random* 1-factors is *probably* an expander graph with parameters $(\lambda, 1/(\lambda + 1), \mu)$ if only

$$\mu >= 2\lambda(\ln(\lambda) + 1) + 1 - \frac{2 + \ln(\lambda)}{3\lambda} + O(\lambda^{-3})$$

— see 4.34 on page 421 for the precise statement. Such a graph has precisely $\mu$ edges incident upon each vertex. A random bipartite 1-factor on the vertex-sets $V_1$ and $V_2$, with $|V_1| = |V_2| = m$ is easily specified by giving a random *permutation* on $m$ objects — say, $\sigma$. We simply connect the $i^{\text{th}}$ vertex in $V_1$ to the $\sigma(i)^{\text{th}}$ vertex of $V_2$.

We will, consequently, assume that our expander-graphs with parameters $(\lambda, \alpha, \mu)$ has $\mu$ distinct 1-factors.

The *idea* of this algorithm (in its crudest form) is as follows:

Assume that we have $n$ numbers stored in $n$ storage locations and $cn$ processors, where each processor is associated with 2 of the storage locations, and $c$ is a constant. Now in $c \lg n$ parallel steps, each involving $n$ of the processors the following operation is carried out:

Each processor compares the numbers in the two storage locations it can access and *interchanges them* if they are out of sequence.

---

[14]Since each edge has two ends.

It turns out that *different* sets of $n$ processors must be used in different phases of the algorithm so that we ultimately need $cn$ processors. If we think of these processors as being connected together in a network-computer (i.e., like the butterfly computer, or the shuffle-exchange computer), we get a computer with $cn$ processors and $n$ distinct blocks of RAM.

Each block of RAM must be shared by $c$ processors. Such a network computer can, using the Ajtai, Komlós, Szemerédi sorting algorithm, simulate a PRAM computer with a time-degradation factor of $c \lg n$ rather than the factor of $O(\lg^2 n)$ that one gets using the Batcher Sorting algorithm (see chapter II, § 2).

Throughout this algorithm we will assume that $\epsilon', \epsilon$, $A$, and $c$ are three numbers, such that

$$\epsilon' \ll \epsilon \ll 1 \ll c$$

and $1 \ll A$. Here $\ll$ means "sufficiently smaller than". It is known that

- Given any sufficiently small value of $\eta$, there exists some value of $\epsilon$ that makes the algorithm work, and the algorithm works for any smaller values of $\epsilon$
- Given any sufficiently small values of $\eta$ and $\epsilon$, there exists some value of $\epsilon' <$
  $\dfrac{\epsilon}{\lg(1/\epsilon)}$ that makes the algorithm work, and the algorithm works for any smaller values of $\epsilon$.
- $c$ must be large enough that there exists expander-graphs of any size with the parameters $((1 - \epsilon')/\epsilon', \epsilon', c)$.
- $A$ must be $> \epsilon^{-1/4}$.

In [3], Ajtai, Komlós, Szemerédi suggest the value of $10^{-15}$ for $\epsilon'$, $10^{-9}$ for $\epsilon$, and $10^{-6}$ for $\eta$. We will also assume given an expander graph with parameters $((1-\epsilon')/\epsilon', \epsilon', c)$.

Now we will present the algorithm in somewhat more detail. We divide it into three *sub-algorithms*:

ALGORITHM 4.14. $\epsilon'$-**halving** In a fixed finite time an $\epsilon'$-halver on $m$ registers puts the lower half of the $m$ numbers in the lower half of the registers with at most $\epsilon m$ errors. In fact, for all $k$, $1 \leq k \leq m/2$, the first $k$ numbers are in the lower half of the registers with at most $\epsilon k$ errors (and similarly for the top $k$ numbers). This is the step that makes use of Expander Graphs. Our expander graph, $G(V_1, V_2, E)$, has a total of $m$ vertices and parameters $((1 - \epsilon')/\epsilon', \epsilon', c)$. The sets $V_1$ and $V_2$ are each of size $m/2$ and we associate $V_1$ with the lower half of the $m$ registers, and $V_2$ with the upper half. See figure VI.56. We decompose $G(V_1, V_2, E)$ into $c$ different 1-factors: $\{F_1, \ldots, F_c\}$

for $i = 1, \ldots, c$,
    for all edges $e \in F_i$, do in parallel
        Compare numbers at the ends of $e$
            and interchange them if they are

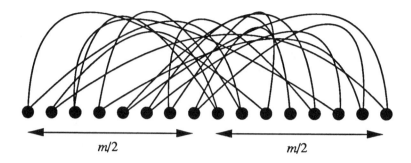

FIGURE VI.56. Expander graphs and approximate halving operations

        out of sequence
endfor

This requires at most $c$ parallel operations.

Note that it is possible to do all of these parallel comparison- and interchange-operations because the edges of a 1-factor are *disjoint*.

This algorithm bears a striking resemblance to the Halving operation in the Quicksort Algorithm. Unlike that algorithm, however, we do not scan the upper and lower halves of our set of $m$ numbers and make decisions based upon the values we encounter. Instead, we perform a kind of "sloppy" halving operation: we compare (and sort) *fixed* sequences of elements (determined by 1-factors of the expander-graph)*regardless* of the values we encounter. Because we have been so sloppy, some large elements may end up in the lower half of the set, and small elements may end up in the upper set. Nevertheless, the combinatorial property of an expander graph guarantees that the number of "errors" in our halving operation will be limited. The following result gives a precise statement of this fact:

PROPOSITION 4.15. Suppose the set $S = \{a_1, \dots, a_n\}$ of numbers is the result of performing the $\epsilon'$-halving algorithm described above on some set of $n$ numbers. Then:

- for all $k \le \lfloor n/2 \rfloor$, $a_i > a_{\lfloor n/2 \rfloor}$ for at most $\epsilon' k$ values of $i \le k$.
- for all $k > \lfloor n/2 \rfloor$, $a_i \le a_{\lfloor n/2 \rfloor}$ for at most $\epsilon' k$ values of $i > k$.

In general, when we prove that this algorithm works, we will assume that the input consists of some permutation of the sequence of numbers $S = \{1, 2, \dots, n\}$. Since the algorithm as a whole is a *sorting network*, it suffices to prove that it works for all sequences of 0's and 1's (see the discussion of the 0-1 principle on page 22). Clearly, if we prove the algorithm for all permutations of the set $S$, it will work for all sequences of 0's and 1's, and, therefore, for all possible sequences of numbers.

PROOF. This follows directly from the combinatorial definition of an expander graph. Suppose that $Z = \{1, \dots, k\}$ is a set of numbers embedded in the input somewhere, with the property that *more than* $\epsilon'k$ elements of $Z$ end up in the upper half of the output of the $\epsilon'$-halving algorithm.

But, if $> \epsilon'k$ elements of $Z$ end up in the upper half of the output, then it follows that $< (1 - \epsilon')k$ elements of $Z$ are left to end up in the lower half of the output.

Then these $\epsilon'k$ elements must be incident upon *at least* $(1 - \epsilon')k$ elements on the lower half of the output, by the defining property of the $(\epsilon', (1 - \epsilon')/\epsilon', c)$-expander graph used to implement the $\epsilon'$-halving algorithm. Since the number of elements of $Z$ than lies in the lower half of the output is strictly less than $(1 - \epsilon')k$, it follows that one of the elements of $Z$ in the upper-half of the output must be incident upon an element of the lower half of the output that *is not* in the set $Z$. But this is a contradiction, since elements of the upper half of the output must be $>$ any element of the lower half, *that they are incident upon*[15]. It is impossible for any element of $Z$ to be larger than any element of the complement of $Z$ — $Z$ was chosen to consist of the smallest numbers. $\square$

This result implies that the $\epsilon'$-halving algorithm works with reasonably good accuracy at the *ends* of the sequence $\{1, \ldots, n\}$. Most of the errors of the algorithm are concentrated in the middle.

DEFINITION 4.16. Let $\pi$ be a permutation of the sequence $(1, \ldots, m)$. Let $S \subseteq (1, \ldots, m)$ be a set of integers. Define:

1. $\pi S = \{\pi i | i \in S\}$;
2. Given $\epsilon > 0$
$$S^\epsilon = \{1 \le j \le m | \, |j - i| \le \epsilon m\}$$
3. A permutation $\pi$ is $\epsilon$-*nearsorted* if
$$|S - \pi S^\epsilon| < \epsilon|S|$$
holds for all initial segments $S = (1, \ldots, k)$ and endsegments $S = (k, \ldots, m)$, where $1 \le k \le m$.

PROPOSITION 4.17. Suppose $\pi$ is an $\epsilon$-nearsorted permutation. Statement 3 in definition 4.16 implies that:
$$|S - \pi S^\epsilon| < 3\epsilon m$$
holds for all sequences $S = (a, \ldots, b)$.

ALGORITHM 4.18. $\epsilon$-**nearsort** This is a kind of *approximate* sorting algorithm based upon the $\epsilon'$-halving algorithm mentioned above. It also executes in a fixed finite amount of time that is independent of $n$. For a given value of $\epsilon$, it sorts the $n$ input values into $n$ storage locations with at most $\epsilon n$ mistakes. In fact, for every $k$ such that $1 \le k \le \epsilon n$, the first $k$ numbers are in the lower $\epsilon'n$ storage locations with at most $\epsilon'k$

---

[15]By the way the $\epsilon'$-halving algorithm works — it has a comparator on each of the edges of the graph.

mistakes. To carry out the nearsort apply an $\epsilon'$-halver to the whole set of $n$ numbers, then apply $\epsilon'$-halvers to the top and bottom half of the result, then to each quarter, eighth, etc, until the pieces each have size $< n\epsilon'$. (The algorithm executes in constant time because we stop it when the pieces become a *fixed fraction* of the size of the original set of numbers). It is not hard to see that each piece of size $w = n\epsilon$ has at most $\epsilon w$ errors.

The *upshot* of all of this is that we have an algorithm for approximately sorting $n$ numbers in *constant* time. The sorting operation is approximate in the sense that a small fraction of the $n$ numbers may end up being *out of sequence* when the algorithm is over.

Note the similarities between the $\epsilon$-nearsort algorithm and the *quicksort* algorithm. The $\epsilon$-nearsort algorithm executes in a constant time. Intuition says that, if we can reduce the number of "errors" in the sorting process by a constant factor (of $\epsilon$) each time we perform an $\epsilon$-nearsort, we may be able to get an *exact* sorting algorithm by performing an $\epsilon$-nearsort $O(\lg n)$ times. This is the basic idea of the Ajtai, Komlós, Szemerédi sorting algorithm. One potential problem arises in following our intuitive analysis:

> Simply applying the $\epsilon$-nearsort to the same set of numbers $O(\lg n)$ times might not do any good. It may make the "same mistakes" each time we run it.

Something like this turns out to be true — we must use some finesse in repeating the $\epsilon$-nearsort algorithm. We use the fact that the $\epsilon'$-halving algorithm make fewer mistakes at the ends of the range of numbers it is sorting to get some idea of where the errors will occur in the $\epsilon$-nearsort.

The remainder of the Ajtai, Komlós, Szemerédi sorting algorithm consists in applying the $\epsilon$-nearsort to the $n$ numbers in such a way that the "errors" in the sorting operation get *corrected*. It *turns out* that this requires that the $\epsilon'$-nearsort operation be carried out $O(\lg n)$ times.

Let $\{m_1, \ldots, m_n\}$ be the memory locations containing the numbers to be sorted. Now the sorting algorithm is divided into $\lg n$ *phases*, each of which is composed of three smaller steps.

In each of these steps we perform an $\epsilon$-nearsort separately and simultaneously on each set in a partition of the memory-locations.

The procedure for partitioning the registers is somewhat arcane. Although we will give *formulas* describing these partitions later, it is important to get a somewhat more conceptual mental image of them. We will use a binary tree of depth $\lg n$ as a *descriptive device* for these partitions. In other words, we don't actually use a binary tree in the algorithm — we merely perform $\epsilon$-nearsorts on sets of memory-locations $\{m_{i_1}, \ldots, m_{i_s}\}$. *Describing* these sets, however, is made a *little* easier if we first consider

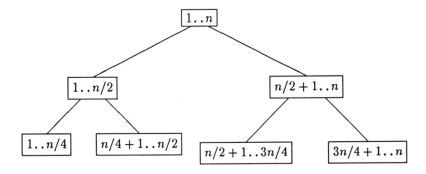

FIGURE VI.57. A complete binary tree of depth 3, with natural intervals

a binary tree[16].

DEFINITION 4.19. Consider a complete binary tree of depth $\lg n$. To each vertex, $v$, there corresponds a *natural interval*, $I(v)$, of memory locations — regard the vertices at each level as getting *equal subdivisions* of the $n$ registers. This means that the natural interval of the root is *all* of the locations from 1 to $n$. The natural interval of the *two children* of the root is 1 to $n/2$ for the left child, and $n/2 + 1$ to $n$ for the right child, respectively. See figure VI.57.

In the following description we will assume given a *parameter* $A$ that satisfies the condition that $A \ll 1/\epsilon$ — $A = 100$, for instance. In general $A$ must satisfy the condition that $\epsilon < A^{-4}$ — see [3], section 8. Although the authors do not state it, it turns out that their term $\alpha$ must be equal to $1/A$.

We will define a *set* of memory-locations associated to each vertex in the binary tree in phase $i$ — recall that $0 \le i \le \lg n$:

DEFINITION 4.20. The following two steps describe how the memory-locations are distributed among the vertices of the binary tree in phase $i$ of the Ajtai, Komlós, Szemerédi sorting algorithm

1. **Initial Assignment Step:**
   - Vertices of depth $>$ have an empty set of memory-locations associated with them.
   - Vertices of depth $i$ have their natural interval of memory-locations associated with them — see 4.19 and figure VI.57.
   - Vertices of depth $j < i$ have a *subset* of their natural interval associated with them — namely the lower $A^{-(i-j)}$ and upper $A^{-(i-j)}$ portions. In other words, if the natural interval would have had $k$ elements, this vertex

---

[16]The original paper of Ajtai, Komlós, Szemerédi ([4]) simply gave mathematical formulas for the partitions, but people found the formulas hard to understand.

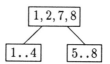

FIGURE VI.58

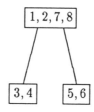

FIGURE VI.59. Phase 2

has a subset of that natural interval associated with it, composed of terms
1 through $\lfloor A^{-(i-j)}k \rfloor$ and terms $k - \lfloor A^{-(i-j)}k \rfloor$ through $k$.

2. **Sifting Step:** The partitioning-scheme described above has each memory location associated with *several* vertices of the binary-tree. Now we cause each
memory-location to be associated with a *unique* vertex via the following rule:

Each memory-location only remains associated with the *highest* (i.e.,
lowest depth) vertex that step 1 assigned it to. In other words, higher
vertices have higher priority in getting memory-locations assigned
to them.

Basically the algorithm distributes the memory locations among the vertices of a
depth-$t$ subtree in phase $t$. Now we are in a position to describe the sorting-steps of
the algorithm.

EXAMPLE 4.21. Suppose $n = 8$ and $A = 4$. In phase 1, the tree is of depth 0 and we
only consider the root, with its natural interval of memory locations $1 . . 8$.

In phase 2, the initial assignment step gives rise to the arrangement depicted in
figure VI.58. The sifting step modifies this to get what is depicted in figure VI.59.

In phase 3, the assignment of memory locations filters down to the leaves of the tree.
We ultimately get the result that appears in figure VI.60.

DEFINITION 4.22. Define a *triangle* of the tree to be a parent vertex and its two children.
Define the set of memory-locations associated with the triangle of vertices to be the
*union* of the sets of memory-locations associated with its vertices (as described in 4.20
above).

**Zig-step** Partition the tree into triangles with apexes at *even* levels, and perform
independent $\epsilon$-nearsort operations on the sets of vertices associated with each

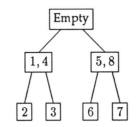

FIGURE VI.60. Phase 3

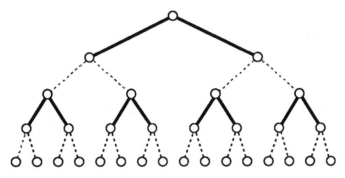

FIGURE VI.61. Partition in the **Zig** phase

of these triangles. Each triangle of vertices defines one *set* in the partition of the memory locations. See figure VI.61.

**Zag step** Partition the binary tree into triangles with apexes at *odd* levels and perform the $\epsilon$-nearsort to every triangle, as described above. See figure VI.62.

Now we can describe the entire Ajtai, Komlós, and Szemerédi sorting algorithm:

ALGORITHM 4.23. The Ajtai, Komlós, and Szemerédi sorting algorithm consists in performing the following steps:

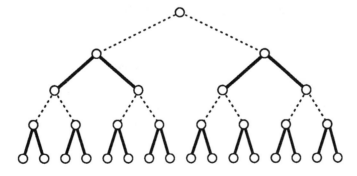

FIGURE VI.62. Partition in the **Zag** phase

for all $i \leftarrow 1$ to $\lg n$ do
   {
   Compute the association of memory-locations with vertices
      for phase $i$ in/ 4.20
   Perform:
        **Zig**
        **Zag**
        **Zig**
   }
endfor

At the end of this procedure, the number in the original memory-locations will be correctly sorted.

Note that this is a sorting *network* because:

- Each $\epsilon'$-halving operation is given as a network of comparators;
- Each $\epsilon$-nearsort can be described in terms of a sorting network, using the description of $\epsilon'$-halving operations above.
- Whenever we must perform an $\epsilon$-nearsort on disjoint sets of memory-locations, we merely *splice* suitably-sized copies of the networks described in the previous line, into our whole sorting network.

Also note that this algorithm is *far* from being uniform. Although the execution of the algorithm doesn't depend upon the data-values being sorted, it does depend upon the number $n$ of data-items. We basically have a *set* of algorithms parameterized by $n$. The amount of work needed to compute the partitions in 4.20 and 4.22, and the expander-graphs used in the $\epsilon'$-halving operations can be very significant.

*If* we perform all of these preliminary computations[17], *set* up the algorithm, and simply *feed data* to the resulting sorting network, then the time required for this data to *pass through* the network (and become sorted) is $O(\lg n)$.

It is clear that the algorithm executes in $O(\lg n)$-time (if we assume that all partitions, expander-graphs, etc. have been pre-computed). We must still prove that it *works*.

The pattern used to associate the memory-locations with the vertices of the tree is interesting. *Most* of the memory-locations in phase $t$ are associated with the vertices of depth $t$ in the tree, but a *few* memory-locations remain associated with the higher vertices — namely the lower $1/A^{\text{th}}$ and upper $1/A^{\text{th}}$ (in depth $t - 1$). The purpose of this construction becomes clear now — these small intervals of registers serve to *catch* the data that belongs *outside* a given triangle — the $\epsilon$-nearsort is sufficiently accurate that there is very little of this data.

---

[17]I.e., computation of the partitions in 4.20 and 4.22, and the expander-graphs used in the $\epsilon'$-halving operations.

FIGURE VI.63. A triangle of vertices

Note that, if these small intervals (of size $1/A$) did not exist, the *activity* of the algorithm in phase $t$ would be *entirely concentrated* in the vertices of depth $t$ in the tree. The partitions of the data would be disjoint and remain so throughout the algorithm.

DEFINITION 4.24. We define the *wrongness* of a memory location $M$ at time $t$. Suppose it is associated to a vertex $v$. If the data $x \in M$ lies in the natural interval $I(v)$ (as defined in 4.19 on page 409) the *wrongness* $w(R)$ is defined to be 0. If $x$ lies in the parent vertex of $v$ the wrongness is defined to be 1, and so on. Since the natural interval of the root vertex is the entire range of numbers, wrongness is always well-defined.

We will prove that the following *inductive hypothesis* is satisfied in every phase of the algorithm:

> At every vertex, $v$, of the tree, the fraction of memory-locations associated with $v$ that have wrongness $\geq r$ is $\leq (8A)^{-3r}$.

Note that wrongness starts out being 0 for all memory locations, because they are all initially assigned to the root of the tree. If the inductive hypothesis is still satisfied at the end of the algorithm, all of the data-items will have been correctly sorted, since each vertex will have only a single data-item in it, so none of the memory-locations associated with a vertex will have any wrongness.

As the time-step advances from $t$ to $t + 1$ the wrongness of most of the memory-locations is increased by 1. This is due to the fact that we have redefined the partitions of the memory-locations, and have *refined* them.

Consider the triangle in figure VI.63, and the memory-locations with wrongness 1:

- If they are associated with vertex Y, they should be associated with vertex Z. This is because the natural interval of vertex X is equal to the unions of the natural intervals of Y and Z. The **ZigZagZig** operation will move most of them[18] into memory-locations associated with vertices X or Z, and decrease wrongness.
- A corresponding argument applies to memory-locations with wrongness of 1 that are associated with Z.

Now consider memory-locations with wrongness $r > 1$ that really belong to *lower numbered* vertices than X. *In* the data associated with the triangle X-Y-Z (called the *cherry* of X, by Ajtai, Komlós, Szemerédi) these memory-locations will form an initial

---

[18]All but $3\epsilon$ of them, by 4.17 on page 407.

segment, hence will be $\epsilon$-nearsorted more accurately than data in the center of the interval — see line 3 of 4.16 on page 407. They will, consequently, be sorted into one of the *initial segments* of vertex X — recall how the memory-locations are associated with X, Y, and Z in this step.

Although most of the memory-locations in this phase are associated with Y and Z, $2/A$ of the natural interval of X *remains associated* with X — the lower $1/A$ of this interval and the upper $1/A$. After it is sorted into the initial segment of vertex X, its wrongness has been decreased by 1.

**4.4. Detailed proof of the correctness of the algorithm.** Now we will make the heuristic argument at the end of the previous section *precise*. We will assume that we are in the beginning of phase $t$ of the algorithm. *As before*, we will assume that the original set of numbers being sorted by the algorithm was initially some random permutation of the numbers $\{1, \ldots, n\}$.

DEFINITION 4.25. We have a partition of $\{1, \ldots, n\}$ into intervals $\{J_1, \ldots, J_m\}$, as defined in 4.20 on page 409. Each such interval is associated with a vertex of the tree and consists of a consecutive sequence of memory-locations $\{x, x+1, \ldots, y-1, y\}$. Note that we can order the intervals and write $J_i < J_k$, if every element of $J_i$ is less than every element of $J_k$.

1. If $i$ is a memory-location, $R(i)$ denotes its *contents*. In like manner, if $J$ is some *interval* of memory-locations, $R(J)$ is defined to be the *set* of values that occur in the memory-locations of $J$.

2. If $v$ is a vertex of the tree, the set of memory-locations assigned to $v$ and its two child-vertices, is called the *cherry* associated to $v$.

3. If $J$ is an interval, the *lower section* $L(J)$ is the union of all intervals $\leq J$ and the *upper section* $U(J)$ is the union of all intervals $\geq J$.

4. Let $J$ and $K$ be two intervals with $K < J$ that are *not neighbors* in the partition $\{J_1, \ldots, J_m\}$ (i.e. if $a$ is the highest element of $K$, then $a+1$ is strictly less than every element of $J$). Then $d(J, K)$ is defined to be the distance between the vertices of the tree associated with these intervals.

5. Given an interval $J$ and an integer $r \geq 0$, set $S_1 = \max |R(J) \cap L(K)|$, where the maximum is taken over all intervals $K$, $K < J$, $K$ not adjacent to $J$ (in the partition of all of the processors, $\{J_1, \ldots, J_m\}$), such that $d(J, K) \geq r$. Set $S_2 = \max |R(J) \cap U(K)|$, where the maximum is taken over all intervals $K$, $K > J$, $K$ not adjacent to $J$, such that $d(J, K) \geq r$. Given these definitions, define

$$\Delta_r(J) = \frac{\max(J_1, J_2)}{|J|}$$

This is essentially the proportion of elements of $J$ whose "wrongness" (as defined in 4.24 on page 413) is $\geq r$.

6. If $r \geq 0$, $\Delta_r = \max_J \Delta_r(J)$

7. Given an interval $J$, define $\delta(J) = \dfrac{|R(J) \setminus J|}{|J|}$. Here $\setminus$ denotes *set difference*.

Define $\delta = \max_J \delta(J)$.

This measures the proportion of element that are mis-sorted in any given step of the algorithm.

At the beginning of the first phase of the algorithm, $\delta = 0$ and $\Delta_r = 0$ for all $r \geq 0$. The main result of Ajtai, Komlós, Szemerédi is:

THEOREM 4.26. After each phase of the algorithm:

1. $\Delta_r < A^{-(3r+40)}$, for $r \geq 1$.
2. $\delta < A^{-30}$.

This result implies that the sorting algorithm works because, in the final phase of the algorithm, each interval has a size of 1. The remainder of this section will be spent proving this result.

LEMMA 4.27. Suppose $\Delta_r$ and $\delta$ are the values of these quantities at the end of a given phase of the algorithm, and $\Delta'_r$ and $\delta'$ are the values at the beginning of the next phase *after* the new partition of the memory-locations is computed. Then

- $\Delta'_r < 6A\Delta_{r-4}$, $r \geq 6$.
- $\delta' < 6A(\delta + \epsilon)$

This shows that the initial refinement of the partition of the processors at the beginning of a phase of the algorithm, wrongness is generally *increased*.

PROOF. A new interval $J'$ is the union of at most three subintervals, each one contained in an old interval $J$, and $|J'| > |J|/(2A)$ for any one of them. Each such subinterval is at most two levels away (in the tree) from the old interval, $J$. Similarly, a new lower-section $L(K')$ is contained in an old lower-section $L(K)$, and $K'$ is at most two levels away from $K$ in the tree. This implies that the total *distance* in the tree is $\leq 4$. $\square$

LEMMA 4.28. Let $\Delta_r$ and $\delta$ be the error-measures before a **Zig** (or a **Zag**)-step, and let $\Delta'_r$ and $\delta'$ be the values after it. If $\delta < 1/A^2$, then

- $\Delta'_r < 8A(\Delta_r + \epsilon\Delta_{r-2})$, for $r \geq 3$;
- $\Delta'_r < 8A(\Delta_r + \epsilon)$, for $r = 1, 2$;
- $\delta' < 4A(\delta + \epsilon)$

This implies that the **Zig** or **Zag**-steps compensate for the increases in the errors that took place in 4.27.

PROOF. Each *cherry* of the tree (see line 2 of 4.25, on page 414 for the definition) has $\leq 6$ intervals associated with it. For any interval $K$, and any cherry of the tree, the closest (in the sequence of intervals) interval of the cherry that is outside of a lower-section $L(K)$ is either:

- the closest (on the *tree*); or
  adjacent to $K$ (*in the list of intervals*).

Most elements of $L(K)$ will be sorted to the left, or to this closest interval. The proportion of elements that are *not* (the *exceptional* elements) will be $< \epsilon$, which is $< 8A\epsilon \times$ (the size of any interval in the cherry). Since this closest interval (in the list of intervals) is also the closest in the tree, the size of the errors cannot increase except for the exceptional elements — and, for these elements, the errors (measured by levels of the tree) can increase by at most 2.

We assume that $\delta < 1/A^2$ to ensure that the extreme interval of the cherry (which represents about $1/(4A)$ of the entire cherry), can accommodate all of the foreign elements. This extreme interval might be empty. In this case, however, the total number of memory-locations associated with the cherry is $< 4A$, and $4A\delta < 1$ — all memory-locations associated with the cherry contain the proper (sorted) elements so $R(i) = i$. $\square$

LEMMA 4.29. If $\delta < 1/A^4$, then a **Zig-Zag** step will change the errors as follows:

- $\Delta'_r < 64A^2(\Delta_{r+1} + 3\epsilon\Delta_{r-4}), r \geq 5;$
- $\Delta'_r < 64A^2(\Delta_{r+1} + 3\epsilon), r = 1, 2, 3, 4;$
- $\delta' < 16A^2(\delta + 2\epsilon)$

PROOF. This is essentially the same as the proof of the previous lemma. We only make one additional remark:

Given any intervals $J$ and $L$ with $d(J, L) \geq 1$, if $J$ was closest to $L$ (in the sequence of intervals) in the **Zig** step, then it won't be the closest (in the tree) in the succeeding **Zag**-step. This implies that the errors won't increase (as a result of composing **Zig** and **Zag** steps). $\square$

We finally have:

LEMMA 4.30. After a completed phase of the algorithm, we have:

$$(102) \qquad \delta < 10 \left( A\epsilon + \sum_{r \geq 1}(4A)^r \Delta_r \right) < \alpha^{30}$$

PROOF. We consider an interval $J$ and estimate the number $x = |R(J) \cap U(J')|$, where $J'$ is the interval adjacent to $J$ on its left (i.e., $J' < J$). This number is certainly bounded by the number $y = |R(L(J)) \cap U(J')|$, which is equal to $z = |R(U(J')) \cap L(J)|$. The equality $y = z$ implies the identity:

$$
\begin{aligned}
y_1 - x_1 =&|R(J) \cap (U(J') \setminus J')| - |R(J') \cap (L(J) \setminus J)| \\
&+ |R(L(J) \setminus J) \cap U(J')| - |R(U(J') \setminus J') \cap L(J)| \\
=&x_2 - y_2 + x_3 - y_3
\end{aligned}
$$

where $x_1 = |R(J) \cap J'|$ and $y_1 = |R(J') \cap J|$.

Now we estimate the terms on the right-hand side; for reasons of symmetry, it is enough to work with $x_2$ and $x_3$. Clearly, $x_2 \leq \Delta_1 \cdot |J|$, and the hard part is estimating $x_3$.

We partition $L(J) \setminus J$ into intervals. We may have an interval $J_0$ among them which is as a distance 0 from $J'$. For this $J_0$

$$
|R(J_0) \cap U(J')| \leq |R(J_0) \cap U(J)| < \Delta_1 \cdot |J_0| < \Delta_1 \cdot |J|
$$

The number of intervals of this partition that are at a distance $r \geq 1$ from $J'$ is at most $2^{r+1}$, and their size is at most $(2A)^r |J|$. Thus

$$
(103) \qquad x_3 < |J| \left( \Delta_1 + \sum_{r \geq 1} 2^{2r+1} A^r \Delta_r \right)
$$

We have shown that $|y_1 - x_1|$ is small. Assume that the intervals $J$ and $J'$ are in the same cherry for a **Zig**-step (if they are not in the same cherry in either a **Zig** or a **Zag** step, then they belong to different components, and in this case $\delta(J) = 0$). If the bound on the right side of equation (103) is less than $A^{-35}$ after a **Zag**-step, then the next **Zig**-step will exchange all foreign elements $J$ and $J'$ except for at most:

$$
|x_1 - y_1| + 8A(\Delta_1' + \epsilon)|J| < (8A + 20A\Delta_1' + \sum_{r \geq 2} 2^{2r+1} A^r \Delta_r')|J| < A^{-30}|J|
$$

$\square$

We have tacitly use the following fact:

For any $k$, $1 \leq k \leq n$ the numbers

$$
|R(\{1, \ldots, k\}) \cap \{k+1, \ldots, n\}|
$$

and

$$
|R(\{k+1, \ldots, n\}) \cap \{1, \ldots, k\}|
$$

are monotone decreasing throughout the algorithm.

**4.5. Expander Graphs.** We will compute the probability that the union of a set of random 1-factors is an expander-graph with parameters $(\lambda, 1/(\lambda+1), \mu)$. It turns out that it will be easier to compute the probability that such a graph *is not* an expander graph.

PROPOSITION 4.31. Let $n$ be an integer $> 1$ and let $G(V_1, V_2)$ be the union of $\mu$ random bipartite 1-factors on $2n$ elements. Let $S$ be a given set of vertices in $V_1$ with $|S| = g$. The probability that $|\Gamma(S)|$ is contained in some given set of size $\beta$ is $\leq$

$$(104) \qquad \left( \frac{\beta!(n-g)!}{n!(\beta-g)!} \right)^{\mu}$$

PROOF. We assume that the sets $V_1$ and $V_2$ are ordered. Since the 1-factors are *random* we can assume, without loss of generality, that

- The set $S$ consists of the first $g$ elements of $V_1$;
- The target-set (of size $\beta$) is the first $\beta$ elements of $V-2$ — if this isn't true, we can compose all of the 1-factors with the permutation that maps $S$ into the first $g$ elements of $V_1$ and the permutation that maps the target-set into the first $\beta$ elements of $V_2$.

Now we consider the probability that a random 1-factor maps $S$ into the first $\beta$ element of $V_2$. The number of ways that it can map $S$ into $V_2$ (respecting ordering) is

$$\frac{n!}{(n-g)!}$$

Similarly, the number of ways it can map $S$ into the first $\beta$ elements of $V_2$ is

$$\frac{\beta!}{(\beta-g)!}$$

It follows that the probability that it maps $S$ into the first $\beta$ elements of $V_2$ is

$$\frac{\beta!/(\beta-g)!}{n!/(n-g)!} = \frac{\beta!(n-g)!}{n!(\beta-g)!}$$

□

COROLLARY 4.32. Under the hypotheses of the previous result, the probability that there exists a set $S$ of size $g$ such that $|\Gamma(S)| \leq \beta$ is

$$(105) \qquad \binom{n}{g}\binom{n}{\beta}\left( \frac{\beta!(n-g)!}{n!(\beta-g)!} \right)^{\mu}$$

PROOF. We have simply multiplied equation (104) by the number of ways of choosing sets of size $g$ and $\beta$. □

COROLLARY 4.33. Let $\lambda, n, \mu$ be integers $> 1$. If $G(V_1, V_2)$ is a bipartite graph composed of $\mu$ random 1-factors, and $S \subset V_1$ is of size $n/(\lambda + 1)$, the probability that $|\Gamma(S)| \leq \lambda|S|$ is asymptotic to

$$\left\{ \frac{\lambda^{2\lambda}}{(\lambda - 1)^{\lambda - 1}(\lambda + 1)^{\lambda + 1}} \right\}^{\frac{\mu(n-1/2)}{\lambda+1}} \left( \frac{\lambda - 1}{\lambda} \right)^{\frac{\mu}{\lambda+1}}$$

as $n \to \infty$.

Here, the term "asymptotic" means that the ratio of the two quantities approaches 1 as $n \to \infty$.

PROOF. If we plug $g = n/(\lambda + 1)$ and $\beta = \lambda n/(\lambda + 1)$ into equation (104), we get

$$\left\{ \frac{\left( \dfrac{\lambda n}{\lambda + 1} \right)! \left( \dfrac{\lambda n}{\lambda + 1} \right)!}{n! \left( n \dfrac{\lambda - 1}{\lambda + 1} \right)!} \right\}^{\mu}$$

Now we use Stirling's Formula for the factorial function — see page 106. It states that $k!$ is asymptotic to $k^{k-.5}e^{-k}$. The factors of $e$ cancel out and we get:

$$\left( \left( \frac{n\lambda}{\lambda + 1} \right)^{2\frac{n\lambda}{\lambda+1} - 1} e^n e^{\frac{n(\lambda-1)}{\lambda+1}} e^{-2\frac{n\lambda}{\lambda+1}} n^{-n+1/2} \left( \frac{n(\lambda - 1)}{\lambda + 1} \right)^{-\frac{n(\lambda-1)}{\lambda+1}+1/2} \right)^{\mu}$$

At this point, all of the factors of $n$ inside the large brackets *cancel out* and we get:

$$= \left( \lambda^{2\lambda-1/(n-1/2)}(\lambda - 1)^{-\lambda+1+1/(n-1/2)}(\lambda + 1)^{-\lambda-1} \right)^{\frac{\mu(n-1/2)}{\lambda+1}}$$

$$= \left\{ \frac{\lambda^{2\lambda}}{(\lambda - 1)^{\lambda - 1}(\lambda + 1)^{\lambda + 1}} \right\}^{\frac{\mu(n-1/2)}{\lambda+1}} \left( \frac{\lambda - 1}{\lambda} \right)^{\frac{\mu}{\lambda+1}}$$

This proves the result. □

We will use this to get a crude estimate of the probability that a graph composed of random 1-factors is an expanding graph.

The result above can be used to estimate the probability that a graph *is not* an expander. We want to bound from above by a simpler expression. We get the following formula, which is *larger* than the probability computed in 4.33:

(106)
$$\left\{ \frac{\lambda^{2\lambda}}{(\lambda - 1)^{\lambda - 1}(\lambda + 1)^{\lambda + 1}} \right\}^{\frac{\mu n}{\lambda+1}}$$

This is an *estimate* for the probability that a single randomly chosen set has a neighbor-set that is not $\lambda$ times larger than itself. In order to compute the probability that such a set *exists* we must form a kind of *sum* probabilities over all possible subsets of $V_1$ satisfying the size constraint. Strictly speaking, this is not a sum, because we also have to take into account *intersections* of these sets — to do this computation *correctly* we would have to use the Möbius Inversion Formula. This leads to a fairly complex expression. Since we are only making a crude estimate of the probability of the existence of such a set, we *will* simply sum the probabilities. Moreover, we will assume that all of these probabilities are the *same*. The result will be a quantity that is strictly *greater* than the true value of the probability.

Now we apply 4.32 and Stirling's formula to this to get:

$$\binom{n}{\frac{n}{\lambda+1}}\binom{n}{\frac{\lambda n}{\lambda+1}} = n\lambda^{-\frac{2\lambda n - \lambda - 1}{\lambda+1}}(\lambda+1)^{2n-2}$$

$$= n\frac{\lambda}{\lambda+1}\frac{(\lambda+1)^{2n}}{\lambda^{\frac{2\lambda n}{\lambda+1}}}$$

$$= n\frac{\lambda}{\lambda+1}\left\{\frac{(\lambda+1)^{2(\lambda+1)}}{\lambda^{2\lambda}}\right\}^{\frac{n}{\lambda+1}}$$

Since we want to get an upper bound on the probability, we can ignore the factor of $\frac{\lambda}{\lambda+1}$ so we get:

(107) $$\binom{n}{\frac{n}{\lambda+1}}\binom{n}{\frac{\lambda n}{\lambda+1}} \geq n\left\{\frac{(\lambda+1)^{2(\lambda+1)}}{\lambda^{2\lambda}}\right\}^{\frac{n}{\lambda+1}}$$

Our estimate for the probability that any set of size $n/(\lambda+1)$ has a neighbor-set of size $n\lambda/(\lambda+1)$ is the product of this with equation (106):

$$\left\{\frac{\lambda^{2\lambda}}{(\lambda-1)^{\lambda-1}(\lambda+1)^{\lambda+1}}\right\}^{\frac{\mu n}{\lambda+1}}\left\{n\left\{\frac{(\lambda+1)^{2(\lambda+1)}}{\lambda^{2\lambda}}\right\}^{\frac{n}{\lambda+1}}\right\}$$

If we combine terms with exponent $n/(\lambda+1)$, we get

$$n\left\{\left(\frac{\lambda^{2\lambda}}{(\lambda-1)^{\lambda-1}(\lambda+1)^{\lambda+1}}\right)^{\mu}\frac{(\lambda+1)^{2(\lambda+1)}}{\lambda^{2\lambda}}\right\}^{\frac{n}{\lambda+1}}$$

Now the quantity raised to the power of $n/(\lambda+1)$ effectively overwhelms all of the other terms, so we can restrict our attention to:

$$\left\{ \left( \frac{\lambda^{2\lambda}}{(\lambda - 1)^{\lambda-1}(\lambda + 1)^{\lambda+1}} \right)^{\mu} \frac{(\lambda + 1)^{2(\lambda+1)}}{\lambda^{2\lambda}} \right\}^{\frac{n}{\lambda+1}}$$

and this approaches 0 as $n \to \infty$ if and only if

$$\left( \frac{\lambda^{2\lambda}}{(\lambda - 1)^{\lambda-1}(\lambda + 1)^{\lambda+1}} \right)^{\mu} \frac{(\lambda + 1)^{2(\lambda+1)}}{\lambda^{2\lambda}} < 1$$

Our conclusion is:

PROPOSITION 4.34. Let $\lambda, n, \mu$ be integers $> 1$. If $G(V_1, V_2)$ is a bipartite graph composed of $\mu$ random 1-factors. The probability that there exists a set $S \subset V_1$ of size $n/(\lambda + 1)$ such that $|\Gamma(S)| \leq \lambda|S|$, approaches 0 as $n \to \infty$ if and only if

(108)
$$\left\{ \frac{\lambda^{2\lambda}}{(\lambda - 1)^{\lambda-1}(\lambda + 1)^{\lambda+1}} \right\}^{\mu} \frac{(\lambda + 1)^{2(\lambda+1)}}{\lambda^{2\lambda}} < 1$$

or

$$\mu > \frac{2\ln(\lambda + 1)(\lambda + 1) - 2\lambda\ln(\lambda)}{\ln(\lambda - 1)(\lambda - 1) + \ln(\lambda + 1)(\lambda + 1) - 2\lambda\ln(\lambda)}$$

(109)
$$= 2\lambda(\ln(\lambda) + 1) + 1 - \frac{2 + \ln(\lambda)}{3\lambda} + O(\lambda^{-3})$$

PROOF. We simply took the logarithm of equation (106) and solved for $\mu$. After that, we obtained an asymptotic expansion of the value of $\mu$.  □

## 5. Computer Algebra

**5.1. Introduction.** In this section we will discuss applications of parallel processing to symbolic computation, or computer algebra. We will basically give efficient parallel algorithms for performing operations on polynomials on a SIMD-parallel computer.

At first glance, it might seem that the SIMD model of computation wouldn't lend itself to symbolic computation. It turns out that performing many algebraic operations with polynomials can be easily accomplished by using the Fourier Transform. See § 2.2 in chapter V, particularly the discussion on page 202.

Suppose we are given two polynomials:

(110)
$$p(x) = \sum_{i=1}^{n} a_i x^i$$

(111)
$$q(x) = \sum_{j=1}^{m} b_j x^j$$

It is not difficult to write down a formula for the coefficients of the product of these polynomials:

(112)
$$(p \cdot q)(x) = \sum_{k=1}^{n+m} c_k x^k$$

where

(113)
$$c_k = \sum_{i+j=k} a_i \cdot b_j$$

The discussion in that section show that we can get the following algorithm for computing the $\{c_k\}$:

ALGORITHM 5.1. We can compute the coefficients, $\{c_k\}$, of the product of two polynomials by performing the following sequence of operations:

1. Form the Fourier Transforms of the sequences $A = \{a_i\}$ and $B = \{b_j\}$, giving sequences $\{\mathcal{F}_\omega(A)\}$ and $\{\mathcal{F}_\omega(B)\}$;
2. Form the *element-wise* product of these (Fourier transformed) sequences — this is clearly very easy to do in parallel;
3. Form the inverse Fourier transform of the product-sequence $\{\mathcal{F}_\omega(A) \cdot \mathcal{F}_\omega(B)\}$. The result is the sequence $\{c_k\}$.

The procedure described above turns out to be asymptotically optimal, *even* in the *sequential* case — in this case it leads to an $O(n \lg n)$-time algorithm for computing the $\{c_k\}$, rather than the execution-time of $O(n^2)$[19]. In the parallel case, the advantages of this algorithm become even greater. Step 2 is easily suited to implementation on a SIMD machine. In effect, the procedure of taking the Fourier Transform has converted multiplication of polynomials into ordinary *numerical* multiplication. The same is true for addition and subtraction.

Division presents additional complexities: even when it is possible to form the termwise quotient of two Fourier Transforms, the result may not be meaningful. It will only be meaningful if is somehow (magically?) *known* beforehand, that the denominator *exactly divides* the numerator.

**Exercises:** Modify the program on page 213 to compute the product of two polynomials, by performing steps like those described in algorithm 5.1. The program should prompt the user for coefficients of the two polynomials and print out the coefficients of the product.

---

[19]This is what one gets by using the naive algorithm

**5.2. Number-Theoretic Considerations.** When one does the exercise 5.1, one often gets results that are somewhat mysterious. In many cases it is possible to plug polynomials into the programs that have a known product, and the values the program prints out don't resemble this known answer. This is a case of a very common phenomena in numerical analysis known as *round-off error*. It is a strong possibility whenever many cascaded floating-point calculations are done. There are many techniques [20] for minimizing such errors, and one technique for totally eliminating it: perform *only* fixed-point calculations.

The reader will probably think that this suggestion is not particularly relevant:

- Our calculations with Fourier Transforms involve irrational, and even complex numbers. How could they possibly be rephrased in terms of the integers?
- The polynomials that are input to the algorithm might have rational numbers as coefficients.
- When we use integers (assuming the the two problems above can be solved) we run into the problem of fixed point overflow.

The second objection is not too serious — we can just clear out the denominator of all of the rational number that occur in expressing a polynomial (in fact we could use a data-structure for storing the polynomials that has them in this form).

It turns out that there are several solutions to these problems. The important point is that Fourier Transforms (and their inverses) really only depend upon the existence of a "quantity" $s$ that has the following two *structural properties*

1. $s^n=1$; (in some suitable sense).
2.

$$(114) \qquad \sum_{i=0}^{n-1} s^{ik} = 0 \text{ for all } 0 < k < n-1$$

3. the number $n$ has a multiplicative inverse.

The structural conditions can be fulfilled in the integers if we work with numbers *modulo* a prime number[21]. In greater generality, we could work modulo any positive integer that had suitable properties. In order to explore these issues we need a few basic results in number theory:

PROPOSITION 5.2. Suppose $n$ and $m$ are two integers that are relatively prime. Then there exist integers $P$ and $Q$ such that

$$Pn + Qm = 1$$

---

[20]In fact an entire theory
[21]Recall that a prime number is not divisible by any other number except 1. Examples: $2, 3, 5, 7\ldots$

Recall that the term "relatively prime" just means that the only number that exactly divides both $n$ and $m$ is 1. For instance, 20 and 99 are relatively prime. Numbers are relatively prime if and only if their greatest common divisor is 1.

PROOF. Consider all of the values taken on by the linear combinations $Pn + Qm$ as $P$ and $Q$ run over all of the integers, and let $Z$ be the smallest *positive* value that occurs in this way.

Claim 1: $Z < n, m$. If not, we could subtract a copy of $n$ or $m$ from $Z$ (by altering $P$ or $Q$) and make it smaller. This would contradict the fact that $Z$ is the *smallest* positive value $Pn + Qm$ takes on.

Claim: $n$ and $m$ are both divisible by $Z$. This is a little like the last claim. Suppose $n$ is not divisible by $Z$. Then we can write $n = tZ + r$, where $t$ is some integer and $r$ is the remainder that results from dividing $n$ by $Z - r < Z$. We plug $Z = Pn + Qm$ into this equation to get

$$n = t(Pn + Qm) + r$$

or

$$(1 - tP)n - Qm = r$$

where $r < Z$. The existence of this value of $r$ contradicts the assumption that $Z$ was the smallest positive value taken on by $Pn + Qm$.

Consequently, $Z$ divides $n$ and $m$. But the only positive integer that divides both of these numbers is 1, since they are relatively prime.  □

This implies:

COROLLARY 5.3. Let $m$ be a number $> 1$ and let $a$ be a number such that $a$ is relatively prime to $m$. Then there exists a number $b$ such that

$$ab \equiv 1 \pmod{m}$$

PROOF. Proposition 5.2 implies that we can find integers $x$ and $y$ such that $ax + my = 1$. When we reduce this modulo $m$ we get the conclusion.  □

COROLLARY 5.4. Let $p$ be a prime number and let $a$ and $b$ be two integers such that $0 \leq a, b < p$. Then $ab \equiv 0 \pmod{p}$ implies that either $a \equiv 0 \pmod{p}$ or $b \equiv 0 \pmod{p}$

We will also need to define:

DEFINITION 5.5. **The Euler $\phi$-function.**

1. If $n$ and $m$ are integers, the notation $n|m$ — stated as "$n$ divides $m$"— means that $m$ is exactly divisible by $n$.
2. Let $m$ be a positive integer. Then the Euler $\phi$-function, $\phi(m)$ is defined to be equal to the number of integers $k$ such that:
   a. $0 < k < m$;

   b. $k$ is relatively prime to $m$. This means that the only integer $t$ such that
   $t|m$ and $t|k$, is $t = 1$.

The Euler $\phi$ function has many applications in number theory. There is a simple
formula (due to Euler) for calculating it:

(115) $$\phi(m) = m \prod_{p|m} \left(1 - \frac{1}{p}\right)$$

where $\prod_{p|m}$ means "form the product with $p$ running over all primes such that $m$ is
divisible by $p$". It is not hard to calculate: $\phi(36) = 12$ and $\phi(1000) = 400$. It is also
clear that, if $p$ is a prime number, then $\phi(p) = p - 1$.

   Many computer-algebra systems perform symbolic computations by considering
numbers modulo a prime number. See [76] and [77] for more information on this
approach. The following theorem (known as Euler's Theorem) gives one important
property of the Euler $\phi$-function:

THEOREM 5.6. Let $m$ be any positive integer and let $a$ be any nonzero integer that is
relatively prime to $m$. Then

$$a^{\phi(m)} \equiv 1 \pmod{m}$$

PROOF.     CLAIM 5.7. There exists some number $k$ such that $a^k \bmod m = 1$.

   First, consider the result of forming higher and higher powers of the number $a$, and
reducing the results **mod** $m$. Since there are only a finite number of possibilities, we
have to get $a^u \bmod m = a^v \bmod m$, for some $u$ and $v$, with $u \neq v$ — suppose $u < v$.
Corollary 5.3 implies that we can find a value $b$ such that $ab \bmod m = 1$, and we can
use this cancel out $a^u$:

$$a^u b^u \bmod m = a^v b^u \bmod m$$
$$1 = a^{v-u} \bmod m$$

so $k = v - u$. Now consider the set, $S$, of all numbers $i$ such that $1 \leq i < m$ and
$i$ is relatively prime to $m$. There are $\phi(m)$ such numbers, by definition of the Euler
$\phi$-function. If $i \in S$, define $Z_i$ to be the the set of numbers $\{i, ia, ia^2, \ldots, ia^{k-1}\}$, all
reduced **mod** m. For instance $Z_a$ is $\{a, a^2, \ldots, a^{k-1}, a^k\} = \{a, a^2, \ldots, a^{k-1}, 1\} = Z_1$.

   CLAIM 5.8. If $i, j \in S$, and there exists any number $t$ such that $t \in Z_i \cap Z_j$, then
$Z_i = Z_j$.

   This means that the sets $Z_i$ are either equal, or entirely disjoint. If $t \in Z_i$, then
$t \bmod m = ia^u \bmod m$, for some integer $u$. Similarly, $t \in Z_j$ implies that $t \bmod m =
ja^v \bmod m$, for some integer $v$. If we multiply the first equation by $b^u \bmod m$ (where
$ab \bmod m = 1$, we get $i = tb^u \bmod m$ and this implies that $i = ja^v b^u \bmod m$, so

$i \in Z_j$. Since all multiples of $i$ by powers of $a$ are also in $Z_j$, it follows that $Z_i \subseteq Z_j$. Since these sets are of the same size, they must be equal.

All of this implies that the set $S$ is the union of a number of disjoint sets $Z_i$, each of which has the same size. This implies that the size of $S$ (i.e., $\phi(m)$) is *divisible* by the size of each of the sets $Z_i$. These sets each have $k$ elements, where $k$ is the smallest integer $> 1$ such that $a^k \bmod m = 1$. Since $\phi(m)$ is divisible by $k$ it follows that $a^{\phi(m)} \bmod m = 1$. $\square$

This leads to the corollary, known as Fermat's Little Theorem:

THEOREM 5.9. Let $p$ be a prime number and let $a$ be any nonzero integer. Then

$$a^{p-1} \equiv 1 \quad (\bmod \ p)$$

Some of these numbers $a$ have the property that they are *principal* $p - 1^{\text{th}}$ roots of 1 modulo $p$ i.e., $a^i \not\equiv 1 \ (\bmod \ p)$ for $0 < i < p - 1$. In this case it turns out that the property expressed by equation (114) is also true for such numbers. For instance, suppose $p = 5$ and $n = 2$. Then 2 is a principal $4^{\text{th}}$ root of 1 modulo 5, since $2^2 \equiv 4$ $(\bmod \ 5)$, $2^3 = 8 \equiv 3 \ (\bmod \ 5)$, and $2^4 = 16 \equiv 1 \ (\bmod \ 5)$.

PROPOSITION 5.10. Let $p$ be a prime number, and let $a$ be a principal $p - 1^{\text{th}}$ root of 1 modulo $p$. Then

(116)
$$\sum_{i=0}^{n-1} a^{ik} = 0$$

for all $0 < k < p$.

The proof is essentially the same as that of equation 23. We have to show that $(a - 1) \sum_{i=0}^{n-1} a^{ik} = 0$ implies that $\sum_{i=0}^{n-1} a^{ik} = 0$.

Since principal roots modulo a prime have the two required structural properties, we can use them to compute Fourier Transforms.

The advantages to this will be:

- there is be no round-off error in the computations because we are working over the integers, and
- there is be no problem of integer overflow, because all numbers will bounded by $p$

We also encounter the "catch" of using these number-theoretic techniques:

**The results of using mod-$p$ Fourier Transforms to do Computer Algebra will only be correct mod $p$.**

In many cases this is good enough. For instance:

PROPOSITION 5.11. If we know a number $n$ satisfies the condition $-N/2 < n < N/2$, where $N$ is some big number, then $n$ is *uniquely determined* by its mod $N$ reduction. If we pick a very big value for $p$, we may be able to use the mod $p$ reductions of the coefficients of the result of our calculations to compute the results themselves.

Suppose $z$ is the reduction of $n$ modulo $N$. If $z > N/2$, then $n$ must be equal to $-(N-z)$.

We need one more condition to be satisfied in order to use the Fast Fourier Transform algorithm, described in § 2.2 of chapter V:

**The number of elements in the sequence to be transformed must be an exact power of 2.**

This imposes a very significant restriction on how we can implement the Fast Fourier Transform since Fermat's Theorem (5.9) implies that the principal roots of 1 are the $p-1^{\text{th}}$-ones. In general, the main constraint in this problem is the size of $n = 2^k$. This must be a number $>$ the maximum exponent that will occur in the computations, and it determines the number of processors that will be used.

> Claim: Suppose $p$ is a prime with the property that $p = tn + 1$, for some value of $t$, and suppose $\ell$ is a principal $p-1^{\text{th}}$ root of 1 modulo $p$. Then $\ell^t$ will be a principal $n^{\text{th}}$ root of 1.

We must, consequently, begin with $n = 2^k$ and find a multiple $tn$ of $n$ with the property that $tn + 1$ is a prime number.

This turns out to be fairly easy to do. In order to see why, we must refer to two famous theorems of number theory: the Prime Number Theorem and the Dirichlet Density theorem:

THEOREM 5.12. Let the function $\pi(x)$ be defined to be equal to the number of primes $\le x$. Then $\pi(x) \sim x/\log(x)$ as $x \to \infty$.

The statement that $\pi(x) \sim x/\log(x)$ as $x \to \infty$ means that $\lim_{x\to\infty} \dfrac{\pi(x)}{x/\log(x)} = 1$. This was conjectured by Gauss and was proved almost simultaneously by Hadamard and C. J. de La Vallée Poussin in 1896. See [138] and [139].

THEOREM 5.13. Let $m$ be a number $\ge 2$ and let $M > m$ be an integer. For all $0 < k < m$ that are relatively prime to $m$, define $z(k, M)$ to be the proportion of primes $< M$ and $\equiv k \pmod m$. Then $\lim z(k, M) = 1/\phi(m)$ as $M \to \infty$.

Note that $\lim_{M\to\infty} z(k, M) = 0$, if $k$ is *not* relatively prime to $m$, since all sufficiently large primes are relatively prime to $m$. The number of numbers $< m$ and relatively prime to $m$ is equal to $\phi(m)$, so that the Dirichlet Density Theorem says that the primes tend to be evenly distributed among the numbers $\pmod m$ that can *possibly* be $\equiv$ to

| $k$ | $n = 2^k$ | $t$ | $p = tn + 1$ | $\omega$ | $\omega^{-1} \pmod{p}$ | $n^{-1} \pmod{p}$ |
|-----|-----------|-----|--------------|----------|------------------------|-------------------|
| 7   | 128       | 2   | 257          | 9        | 200                    | 255               |
| 8   | 256       | 1   | 257          | 3        | 86                     | 256               |
| 9   | 512       | 15  | 7681         | 7146     | 7480                   | 7681              |
| 10  | 1024      | 12  | 12289        | 10302    | 8974                   | 12277             |
| 11  | 2048      | 6   | 12289        | 1945     | 4050                   | 12283             |
| 12  | 4096      | 3   | 12289        | 1331     | 7968                   | 12286             |
| 13  | 8192      | 5   | 40961        | 243      | 15845                  | 40956             |
| 14  | 16384     | 4   | 65537        | 81       | 8091                   | 65533             |
| 15  | 32768     | 2   | 65537        | 9        | 7282                   | 65535             |
| 16  | 65536     | 1   | 65537        | 3        | 21846                  | 65536             |
| 16  | 131072    | 6   | 786433       | 213567   | 430889                 | 786427            |
| 18  | 262144    | 3   | 786433       | 1000     | 710149                 | 786430            |
| 19  | 524288    | 11  | 5767169      | 177147   | 5087924                | 5767158           |

TABLE VI.1. Indices for performing the FFT modulo a prime number

primes. In a manner of speaking this theorem basically says that primes behave like *random numbers*[22] when we consider their reductions modulo a fixed number $m$.

This implies that *on the average* the smallest value of $t$ with the property that $tn + 1$ is a prime is $\leq \log(n)$. This is because the Prime number Theorem (5.12) implies that there are approximately $n$ primes $< \log(n)n + 1$, and the Dirichlet Density theorem (5.13) implies that *on the average* one of these primes will be $\equiv 1 \pmod{n}$. Table VI.1 lists primes and primitive roots of 1.

We can now carry out the Fast Fourier Transform Algorithm 2.7 on page 212 using this table:

1. We start with a value of $k$ such that we have $n = 2^k$ processors available for computations.
2. We perform all computations modulo the prime $p$ that appears in the same row of the table.
3. We use the corresponding principal $n^{\text{th}}$ root of 1 in order to perform the Fourier Transform.

EXAMPLE 5.14. Suppose we are performing the computations on a Connection Machine, and we have $2^{13} = 8192$ processors available. When we execute the Fast Fourier Transform algorithm, it is advantageous to have one processor per data element. We perform the calculations modulo the prime 40961 and set $\omega = 243$ and $\omega^{-1} = 15845$. We will also use $40956 \equiv 8192^{-1} \pmod{40961}$.

---

[22]*Truly* random numbers, of course, would also be $\equiv \pmod{m}$ to numbers that are *not* relatively prime to $m$

Here is a program that implements this version of the Fast Fourier Transform:

```
#include <stdio.h>
#include <math.h>
shape [8192]linear;
unsigned int MODULUS;

unsigned int n;          /* Number of data points. */
int k;           /* log of number of
                 * data-points. */
unsigned int inv_n;

unsigned int:linear temp;
int j;
void fft_comp(unsigned int, unsigned int:current, unsigned int:current *);

int clean_val(unsigned int, unsigned int);

void fft_comp(unsigned int omega, unsigned int:current in_seq,
        unsigned int:current * out_seq)
{
    /* Basic structure to hold the data-items. */
    int:linear e_vals;   /* Parallel array to hold
                     * the values of the e(r,j) */
    unsigned int:linear omega_powers[13];    /* Parallel array to
                         * hold the values of
                         * omega^e(r,j). */
    unsigned int:linear work_seq;   /* Temporary variables,
                     * and */
    unsigned int:linear upper, lower;

    /*
     * This block of code sets up the e_vals and the
     * omega_powers arrays.
     */

    with (linear)
    where (pcoord(0) >= n)
    {
        in_seq = 0;
```

```
    *out_seq = 0;
  }
  with (linear)
  {
    int i;
    int:linear pr_number = pcoord(0);
    int:linear sp;
      e_vals = 0;
    for (i = 0; i < k; i++)
    {
      e_vals <<= 1;
      e_vals += pr_number % 2;
      pr_number >>= 1;
    }
    /*
     * Raise omega to a power given by
     * e_vals[k−1]. We do this be repeated
     * squaring, and multiplying omega^(2^i),
     * for i corresponding to a 1−bit in the
     * binary representation of e_vals[k−1].
     */

      temp = omega;

  omega_powers[k − 1]= 1;
  sp = e_vals;
  for (i = 0; i < 31; i++)
  {
      where (sp % 2 == 1)
      omega_powers[k − 1]
      = (omega_powers[k − 1]* temp) % MODULUS;

      sp = sp >> 1;
      temp = (temp * temp) % MODULUS;
  }

  for (i = 1; i < k; i++)
  {
      omega_powers[k − 1 − i]= (omega_powers[k − i]*
        omega_powers[k − i]) % MODULUS;
```

```
      }
      work_seq = in_seq;
      pr_number = pcoord(0);
      for (i = 0; i < k; i++)
      {
         int:linear save;

         save = work_seq;
         lower = pr_number & (~(1 << (k − i − 1)));
         upper = lower | (1 << (k − i − 1));
         where (pr_number == lower)
         {
            [lower]work_seq = ([lower]save
                    +[lower]omega_powers[i]*[upper]save
               + MODULUS) % MODULUS;
            [upper]work_seq = ([lower]save +
                    [upper]omega_powers[i]*[upper]save
               + MODULUS) % MODULUS;
         }
      }
   }
   with (linear)
   where (pcoord(0) < n)
      [e_vals]* out_seq = work_seq;
}


/*
 * This routine just maps large values to negative numbers.
 * We are implicitly assuming that the numbers that
 * actually occur in the course of the computations will
 * never exceed MODULUS/2.
 */
int clean_val(unsigned int val, unsigned int modulus)
{
   if (val < modulus / 2)
      return val;
   else
      return val − modulus;
}
```

```
void main()
{
    unsigned int:linear in_seq;
    unsigned int:linear out_seq;
    int i, j;
    unsigned int primroot = 243;
    unsigned int invprimroot = 15845;

    MODULUS = 40961;
    k = 13;
    n = 8912;        /* Number of data-points. */
    inv_n = 40956;
    with (linear) in_seq = 0;

    [0]in_seq = (MODULUS - 1);
    [1]in_seq = 1;
    [2]in_seq = 1;
    [3]in_seq = 2;
    fft_comp(primroot, in_seq, &out_seq);

    /*
     * Now we cube the elements of the Fourier
     * Transform of the coefficients of the polynomial.
     * After taking the inverse Fourier Transform of
     * the result, we will get the coefficients of the
     * cube of the original polynomial.
     */
    with (linear)
    {
        in_seq = out_seq * out_seq % MODULUS;
        in_seq = in_seq * out_seq % MODULUS;
    }
        fft_comp(invprimroot, in_seq, &out_seq);
    with (linear)
    where (pcoord(0) < n)
        out_seq = (inv_n * out_seq) % MODULUS;

    for (i = 0; i < 20; i++)
        printf("i=%d, coefficient is %d\n", i,
        clean_val([i]out_seq, MODULUS));
```

}

Here we have written a subroutine to compute the Fourier Transform with respect to a given root of unity. Using theorem 2.2 in § 2.2 of chapter V, we compute the *inverse* Fourier transform the of the result by:

- taking the Fourier Transform with respect to a root of unity that is a multiplicative inverse of the original root of unity.
- multiplying by the multiplicative inverse of the number of data-points — this is 40956, in the present case

In the program above, the original input was a sequence $\{-1, 1, 1, 2, \ldots, 0\}$. result of taking the Fourier Transform and the inverse is the sequence $\{40960, 1, 1, 2, \ldots, 0\}$, so that the $-1$ in the original sequence has been turned into 256. Since we know that the original input data was inside the range $-40961/2, +40961/2$, we subtract 40961 from any term $> 40961/2$ to get the correct result.

This program illustrates the applicability of Fast Fourier Transforms to symbolic computation. In this case, we compute the Fourier Transform and *cube* the resulting values of the Fourier Transform. When we take the Inverse Transform, we will get the result of forming a convolution of the original sequence with itself 3 times. If the original sequence represented the polynomial $1 + x + x^2 + 2x^3$, the final result of this whole procedure will be the coefficients of $(1 + x + x^2 + 2x^3)^3$. We will have used 8192 processors to compute the cube of a cubic polynomial! Although this sounds more than a little ridiculous, the simple fact is that the step in which the cubing was carried out is *entirely* parallel, and we might have carried out much more complex operations than cubing.

Suppose the coefficients of the problem that we want to study are *too large* — i.e., they exceed $p/2$ in absolute value. In this case we can perform the calculations with *several* different primes and use the Chinese Remainder Theorem. First, recall that two integers $n$ and $m$ are called *relatively prime* if the only number that exactly divides both of them is 1. For instance 7 and 9 are relatively prime, but 6 and 9 aren't (since they are both divisible by 3).

THEOREM 5.15. Suppose $n_1, \ldots, n_k$ are a sequence of numbers that are pairwise relatively prime (i.e. they could be distinct primes), and suppose we have congruences:

$$M \equiv z_1 \pmod{n_1}$$
$$\cdots$$
$$M \equiv z_k \pmod{n_k}$$

Then the value of $M$ is uniquely determined modulo $\prod_{i=1}^{k} n_i$.

The term "pairwise relatively prime" means that *every pair* of numbers $n_i$ and $n_j$ are relatively prime. If a given prime is too small for the calculations modulo that prime to determine the answer, we can perform the calculations modulo many different primes and use the Chinese Remainder Theorem. Table VI.2 lists several primes that could be used for the problem under considerations.

We will now prove the Chinese Remainder theorem, and in the process, give an algorithm for computing the value $M$ modulo the product of the primes. We need the following basic result from Number Theory:

ALGORITHM 5.16. Given relatively prime integers $m$ and $n$ such that $0 < m < n$, we can compute a number $z$ such that $zm \equiv 1 \pmod{n}$ by computing $m^{\phi(n)-1} \mod n$.

This follows immediately from Euler's theorem, 5.6, on page 425. The running time of this algorithm is clearly $O(\lg n)$ because $\phi(n) < n$ (which follows from formula (115) on page 425), and we can compute $m^{\phi(n)-1} \mod n$ by *repeated squaring*.

Now we are in a position to prove the Chinese Remainder theorem. Suppose we have congruences:

$$M \equiv z_1 \pmod{n_1}$$

$$\cdots$$

$$M \equiv z_k \pmod{n_k}$$

Now set $P = \prod_{i=1}^{k} n_i$ and multiply the $i^{\text{th}}$ equation by $P/n_i = \prod_{j \neq i} n_j$. We get:

$$M \left( \frac{P}{n_1} + \cdots + \frac{P}{n_k} \right) = z_1 \frac{P}{n_1} + \cdots + z_k \frac{P}{n_k} \pmod{P}$$

The fact that the $\{n_i\}$ are relatively prime implies that $\left( \dfrac{P}{n_1} + \cdots + \dfrac{P}{n_k} \right)$ and $P$ are relatively prime. We can use 5.16 to compute a multiplicative inverse $J$ to $\left( \dfrac{P}{n_1} + \cdots + \dfrac{P}{n_k} \right)$. We compute

$$M = J \cdot \left( z_1 \frac{P}{n_1} + \cdots + z_k \frac{P}{n_k} \right) \pmod{P}$$

Table VI.2 gives a list of primes we can use in an application of the Chinese Remainder Theorem.

We will conclude this section with an example. We will do some algebraic calculations over two different primes, $p_1 = 40961$ and $p_2 = 65537$, and use the Chinese Remainder Theorem to patch the results together. The final results will be correct modulo $2^{31} < p_1 p_2 = 2684461057 < 2^{32}$. Since we will be using a machine with a word-size[23] of 32 bits, we will have to use *unsigned integers* for all calculations. In addition, we will need special routines for addition and multiplication, so that the

---

[23]We won't use any of the *unique* features of the CM-2, such as the availability of large (or variable) words.

| $p$ | $8192^{-1} \pmod{p}$ | $\omega =$Principal $n^{\text{th}}$ root of 1 | $\omega^{-1}$ |
|---|---|---|---|
| 40961 | 40956 | 243 | 15845 |
| 65537 | 65529 | 6561 | 58355 |
| 114689 | 114675 | 80720 | 7887 |
| 147457 | 147439 | 62093 | 26569 |
| 163841 | 163821 | 84080 | 15743 |
| 188417 | 188394 | 59526 | 383 |

TABLE VI.2. Distinct primes for the FFT with 8912 processors

calculations don't produce an overflow-condition. Since the primes $p_1$ and $p_2$ are both $< \sqrt{2^{32}}$, we will only need these special routines in the step that uses the Chinese Remainder Theorem. We will also need:

- $a_1 = p_2/(p_1 + p_2) \equiv 894711124 \pmod{p_1 p_2}$
- $a_2 = p_1/(p_1 + p_2) \equiv 1789749934 \pmod{p_1 p_2}$

```
#include <stdio.h>
#include <math.h>
shape [8192]linear;
unsigned int p1=40961;
unsigned int p2=65537;
unsigned int MODULUS=40961*65537;
/* (40961+65537)&~(-1) mod 40961*65537 */
unsigned int invpsum=597020227;

unsigned int n;        /* Number of data points. */
int k;            /* log of number of
                * data-points. */
unsigned int inv_n;

unsigned int:linear temp;
int j;
void fft_comp(unsigned int, unsigned int:current, unsigned int:current *);

int clean_val(unsigned int, unsigned int);

void fft_comp(unsigned int omega, unsigned int:current in_seq,
        unsigned int:current * out_seq)
{
  /* Basic structure to hold the data-items. */
```

```
int:linear e_vals;    /* Parallel array to hold
                 * the values of the e(r,j) */
unsigned int:linear omega_powers[13];    /* Parallel array to
                           * hold the values of
                           * omega^e(r,j). */
unsigned int:linear work_seq;    /* Temporary variables,
                 * and */
unsigned int:linear upper, lower;


/*
 * This block of code sets up the e_vals and the
 * omega_powers arrays.
 */

with (linear)
where (pcoord(0) >= n)
{
    in_seq = 0;
    *out_seq = 0;
}
with (linear)
{
    int i;
    int:linear pr_number = pcoord(0);
    int:linear sp;
        e_vals = 0;
    for (i = 0; i < k; i++)
    {
        e_vals <<= 1;
        e_vals += pr_number % 2;
        pr_number >>= 1;
    }
    /*
     * Raise omega to a power given by
     * e_vals[k-1]. We do this be repeated
     * squaring, and multiplying omega^(2^i),
     * for i corresponding to a 1-bit in the
     * binary representation of e_vals[k-1].
     */
```

```
      temp = omega;

   omega_powers[k − 1]= 1;
   sp = e_vals;
   for (i = 0; i < 31; i++)
   {
      where (sp % 2 == 1)
      omega_powers[k − 1]
      = (omega_powers[k − 1]* temp) % MODULUS;

      sp = sp >> 1;
      temp = (temp * temp) % MODULUS;
   }

   for (i = 1; i < k; i++)
   {
      omega_powers[k − 1 − i]= (omega_powers[k − i]*
         omega_powers[k − i]) % MODULUS;
   }
   work_seq = in_seq;
   pr_number = pcoord(0);
   for (i = 0; i < k; i++)
   {
      int:linear save;

      save = work_seq;
      lower = pr_number & (~(1 << (k − i − 1)));
      upper = lower | (1 << (k − i − 1));
      where (pr_number == lower)
      {
         [lower]work_seq = ([lower]save
                 +[lower]omega_powers[i]*[upper]save
            + MODULUS) % MODULUS;
         [upper]work_seq = ([lower]save +
                 [upper]omega_powers[i]*[upper]save
            + MODULUS) % MODULUS;
      }
   }
}
with (linear)
```

```
    where (pcoord(0) < n)
        [e_vals]* out_seq = work_seq;
}


/*
 * This routine just maps large values to negative numbers.
 * We are implicitly assuming that the numbers that
 * actually occur in the course of the computations will
 * never exceed MODULUS/2.
 */
int clean_val(unsigned int val, unsigned int modulus)
{
    if (val < modulus / 2)
        return val;
    else
        return val − modulus;
}


void main()
{
    unsigned int:linear in_seq;
    unsigned int:linear out_seq;
    int i, j;
    unsigned int primroot = 243;
    unsigned int invprimroot = 15845;

    MODULUS = 40961;
    k = 13;
    n = 8912;       /* Number of data−points. */
    inv_n = 40956;
    with (linear) in_seq = 0;

    [0]in_seq = (MODULUS − 1);
    [1]in_seq = 1;
    [2]in_seq = 1;
    [3]in_seq = 2;
    fft_comp(primroot, in_seq, &out_seq);

    /*
     * Now we cube the elements of the Fourier
```

```
* Transform of the coefficients of the polynomial.
* After taking the inverse Fourier Transform of
* the result, we will get the coefficients of the
* cube of the original polynomial.
*/
with (linear)
{
    in_seq = out_seq * out_seq % MODULUS;
    in_seq = in_seq * out_seq % MODULUS;
}
    fft_comp(invprimroot, in_seq, &out_seq);
with (linear)
where (pcoord(0) < n)
    out_seq = (inv_n * out_seq) % MODULUS;

for (i = 0; i < 20; i++)
    printf("i=%d, coefficient is %d\n", i,
        clean_val([i]out_seq, MODULUS));
}
```

The procedure in this program only works when the coefficients of the result lie in the range $-2684461057/2, +2684461057/2$. If the coefficients of the result do not meet this requirement, we must perform the calculations over several *different* primes (like the ones in table VI.2 on page 435) and use the Chinese Remainder Theorem on page 433 to patch up the results.

### Exercises:

1. Given a prime $p$ and a positive number $n$, give an algorithm for computing the inverse of $n$ modulo $p$. (Hint: use Fermat's theorem — 5.9.

2. Suppose $m$ is a large number[24]. Give algorithms for performing computations modulo $m$. Note:

   a. A number modulo $m$ can be represented as a linked list of words, or an array.

   b. Multiplication can be carried out using an algorithm involving a Fast Fourier Transform[25]. The hard part is reducing a number $> m$ that

---

[24]In other words, it is large enough that it is impossible to represent this number in a word on the computer system to be used for the computations

[25]Where have we heard of that before! See the discussion on using Fourier Transforms to perform

represented as an array or linked list of words, modulo $m$.

3. Most computer-algebra systems (i.e., Maple, Reduce, Macsyma, etc.) have some number-theoretic capabilities and have an associated programming language. If you have access to such a system, write a program in the associated programming language to:

    a. Write a function that takes an exponent $k$ as its input and:

        i. Finds the smallest prime $p$ of the form $t2^k + 1$;

        ii. Finds a principal $p - 1^{\text{th}}$ root of 1 modulo $p$

        iii. Raises that principal root of 1 to the $t^{\text{th}}$ power modulo $p$ in order to get a principal $n^{\text{th}}$ root of 1 modulo $p$, where $n = 2^k$.

    For instance, the author wrote such a function in the programming language bundles with Maple on a Macintosh Plus computer to compute table VI.1 (in about 20 minutes).

    b. Write a function to take two exponents $k$ and $k'$ as parameters and find the largest prime $p < 2^k$ such that $p$ is or the form $t2^{k'} + 1$.

---

multiplication of binary numbers on page 203.

# VII
# Probabilistic Algorithms

## 1. Introduction and basic definitions

In this chapter we will discuss a topic whose importance has grown considerably in recent years. Several breakthroughs in sequential algorithms have been made, in such diverse areas as computational number theory (with applications to factoring large numbers and breaking Public-Key Encryption schemes, etc.), and artificial intelligence (with the development of Simulated Annealing techniques). A complete treatment of this field is beyond the scope of the present text. Nevertheless, there are a number of simple areas that we can touch upon.

Perhaps the first probabilistic algorithm ever developed predated the first computers by 200 years. In 1733 Buffon published a description of the so-called *Buffon Needle Algorithm* for computing $\pi$ — see [22]. Although it is of little practical value, many modern techniques of numerical integration can be regarded as direct descendants of it. This algorithm requires a a needle of a precisely-known length and a floor marked with parallel lines with the property that the distance between every pair of neighboring lines is exactly double the length of the needle. It turns out that if the needle is dropped on this floor, the probability that it will touch one of the lines is equal to $1/\pi$. It follows that if a person randomly drops this needle onto the floor and keeps a record of the number of times it hits one of the lines, he or she can calculate an approximation of $\pi$ by dividing the number of times the needle hits a line of the floor into the total number of trials. This "algorithm" is not practical because it converges fairly slowly, and because there are much better ways to compute $\pi$.

There are several varieties of probabilistic algorithms:

**1.1. Numerical algorithms.** The Buffon Needle algorithm falls into this category. These algorithms involve performing a large number of independent trials and produce an answer that converges to the correct answer as the number of trials increase. They are *ideally suited* to parallelization, since the trials are completely independent — *no*

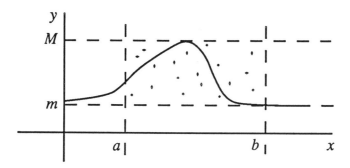

FIGURE VII.1. Monte-Carlo Integration

*communication* between processors is required (or even wanted). In fact the distinction between SIMD and MIMD algorithms essentially disappears when one considers these algorithms.

1.1.1. *Monte Carlo Integration.* This is a very unfortunate name since Monte Carlo Integration is *not* a Monte Carlo algorithm, as defined above. It is based upon the principle that is used in the *rain gauge* — the amount of rain hitting a given surface is proportional to the *rate* of rainfall and the *area* of the surface. We can also regard it as a direct generalization of the Buffon needle algorithm. We will begin by giving a very simple example of this type of algorithm. We want to solve the basic problem of computing a definite integral like

$$A = \int_a^b f(x)\, dx$$

Let $M = \max_{a \leq x \leq b} f(x)$ and $m = \min_{a \leq x \leq b} f(x)$. Now we pick a large number of points in the rectangle $a \leq x \leq b$, $m \leq y \leq M$ at *random* and count the number of points that lie below the curve $y = f(x)$. If the randomly-chosen points are *uniformly distributed* the expected count of the points below the curve $y = f(x)$ is proportional to the integral $A$ — see figure VII.1.

The algorithm for computing integrals is extremely simple:

ALGORITHM 1.1. All processors allocate two counters $i$ and $t$. Perform the following steps as until a desired degree of accuracy is achieved:

In parallel all processors perform the following sequence of steps:

1. Randomly generate a point $(x, y)$, of the domain of integration — this is given by $a \leq x \leq b$, $m \leq y \leq M$ in the example above. Increment the $t$ counter.
2. Determines whether its generated point lies under the curve $y = f(x)$. This is just a matter of deciding whether the inequality $y \leq f(x)$ is satisfied. It, therefore, involves a computation of $f(x)$.
3. If the inequality above is satisfied, increment the $i$ counter.

Form the totals of all of the $t$, and $i$ counters and call them $T$, and $R$, respectively. The estimate of the value of

$$A = \int_a^b f(x)\,dx$$

$$\frac{(M-m)(b-a)T}{R}$$

The nice features of this algorithm include:

- There is no communication whatsoever needed between processors in this algorithm. It is, consequently, ideally suited for parallelization.
- This algorithm has essentially complete utilization of the processors. In other words, the parallel version of the algorithm it almost exactly $n$ times faster than the sequential version, where $n$ is the number of processors involved.

Monte Carlo integration is of most interest when we want to compute a multiple integral. Deterministic algorithms using systematic methods to sample the values of the function to be integrated generally require a sample size that grows exponentially with the dimension, in order to achieve a given degree of accuracy. In Monte Carlo integration, the dimension of the problem has little effect upon the accuracy of the result. See [144] for a general survey of Monte Carlo integration.

**Exercises:**

1. Write a C* program to perform Monte Carlo Integration to evaluate the integral

$$\int_0^2 \frac{1}{\sqrt{1+x^3+x^7}}\,dx$$

(Use the **prand**-function to generate random numbers in parallel. It is declared as **int:current prand(void)**.)

**1.2. Monte Carlo algorithms.** Some authors use this term for *all* probabilistic algorithms. Today the term Monte Carlo algorithms usually refers to algorithms that make random choices that cause the algorithm to either produce the correct answer, or a completely wrong answer. In this case, the wrong answers are *not* approximations to the correct answer. These algorithms are equipped with procedures for comparing answers — so that the probability of having a recognizably-correct answer increases with the number of trials. There is always a finite probability that the answer produced by a Monte Carlo algorithm is wrong.

DEFINITION 1.2. We distinguish certain types of Monte Carlo algorithms:

1. A Monte Carlo algorithm is called *consistent* if it never produces two *distinct* correct answers.
2. If the probability that a Monte Carlo algorithm returns a correct answer in one trial is $p$, where $p$ is a real number between $1/2$ and $1$, then the algorithm is called *p-correct*. The value $p - 1/2$ is called the *advantage* of the algorithm.
3. Suppose $y$ is some possible result returned by a consistent Monte Carlo algorithm, $A$. The $A$ is called *y-biased* if there exists a subset $X$ of the problem-instances such that:
    a. the solution returned by $A$ is always correct whenever the instance to be solved in not in $X$.
    b. the correct answer to all instances that belong to $X$ is always $y$.
   We do not require the existence of a procedure for testing membership in $X$.

We will be interested in analyzing $y$-biased Monte Carlo algorithms. It turns out that such algorithms occur in many interesting situations (involving parallel algorithms), and there are good criteria for the correctness of the answers that these algorithms return.

PROPOSITION 1.3. If a $y$-biased Monte Carlo algorithm returns $y$ as its answer, the it is correct.

PROOF. If a problem-instance is in $X$, and the algorithm returns $y$, the answer is correct. If the instance is not in $X$ it always returns correct answers.  □

PROPOSITION 1.4. Suppose $A$ is a $y$-biased Monte Carlo algorithm, and we call $A$ $k$ times and receive the answers $\{y_1, \ldots, y_k\}$. In addition, suppose that $A$ is $p$-correct. Then:

1. If for some $i$, $y_i = y$, then this is the correct answer.
2. If $y_i \neq y_j$ for some values of $i$ and $j$, then $y$ is the correct answer. This is due to the fact that the algorithm is consistent. If two different answers are received, they cannot be the correct answers. Consequently, the problem-instances must have been in $X$, in which case it is known that the correct answer is $y$.
3. If all of the $y_i = \bar{y}$, for all $1 \leq i \leq k$, and $\bar{y} \neq y$, then the probability that this is a wrong answer is $(1 - p)^k$.

As an example of a $y$-biased, we will examine the following simple algorithm for testing whether a polynomial is identically zero:

PROPOSITION 1.5. Let $p(x)$ be a polynomial, and suppose we have a "black box" that returns the value of $p(x)$, given a value of $x$. Our Monte Carlo algorithm plugs sets $x$ to a random number and computes $p(x)$ (using the "black box"). The algorithm returns $p(x) \neq 0$ if this result is nonzero and reports that $p(x) = 0$ if this computation is zero.

Clearly, this is $y$-biased, where $y$ is the answer that says that $p(x) \neq 0$, since a polynomial, $p(x)$, that is nonzero for any value of $x$ cannot be identically zero.

**1.3. Las Vegas algorithms.** These are algorithms that, unlike Monte Carlo algorithm, never produce incorrect answers. These algorithms make random choices that sometimes prevent them from producing an answer at all. These algorithms generally do not lend themselves to SIMD implementation. The random choices they make alter the flow of control, so they need independent processes. The main result involving these algorithms that is of interest to us is:

PROPOSITION 1.6. If a Las Vegas algorithm has a probability of $p$ of producing an answer, then the probability of getting an answer from it in $n$ trials is $1 - (1 - p)^n$.

PROOF. If the probability of producing an answer in one trial is $p$, then the probability of not producing an answer in one trial is $1 - p$. The probability of this outcome occurring repeatedly in $n$ trials (assuming all trials are independent) is $(1 - p)^n$. Consequently, the probability of getting a answer within $n$ trials is $1 - (1 - p)^n$.  $\square$

If we define the *expected number of repetitions until success* as the average number of trials that are necessary until we achieve as success. This is a weighted average of the number of trials, weighted by the probability that a given number of trials is necessary.

PROPOSITION 1.7. Given a Las Vegas algorithm with a probability of success in one trial equal to $p$, the expected number of trials required for success is:

$$\frac{1}{p}$$

PROOF. The probability of no success within $k$ trials is $q^k$, where $q = 1 - p$. The probability of achieving success at precisely the $k + 1^{\text{st}}$ trial is $pq^k$. Consequently, the weighted average of the number of trials, weighted by the probability of success in a given trial is

$$S = \sum_{n=1}^{\infty} npq^{n-1}$$

We can evaluate this quantity exactly. If we multiply $S$ by $q$, we get

(117)
$$qS = \sum_{n=1}^{\infty} npq^n$$

Consider the infinite series $T = \sum_{n=1}^{\infty} q^n$. If we multiply $T$ by $q$, we get $qT = \sum_{n=2}^{\infty} q^n = T - q$, so $T$ satisfies the equation $qT = T - q$ and $0 = (1 - q)T - q$ or $T = q/(1 - q)$. Now, if we add $T$ to $qS$ in equation (117), we get

$$qS + T = \sum_{n=1}^{\infty} (n + 1)pq^n$$
$$= S - p$$

so $qS + T = S - p$ and $T = (1 - q)S - p$ and $T + p = (1 - q)S$ and $S = (T + p)/(1 - q)$. Now we use the fact that $T = q/(1 - q)$ to get $S = (p + q)/(1 - q) = 1/p$. $\square$

## 2. The class RNC

These are algorithms that always work, and always produce correct answers, but the *running time* is indeterminate, as a result of random choices they make. In this case, the probable running time may be very fast, but the *worst case* running time (as a result of unlucky choices) may be bad. This type of algorithm is of particular interest in Parallel Processing. We define the class **RNC** to denote the class of problems that can be solved by probabilistic parallel algorithms in probable time that is poly-logarithmic, using a polynomial number of processors (*in* the complexity-parameter — see 35).

The precise definition is

DEFINITION 2.1. Let A be a probabilistic parallel algorithm. Then:

1. the *time-distribution* of this algorithm is a function $p(t)$ with the property that the probability of the execution-time of the algorithm being between $t_0$ and $t_1$ is

$$\int_{t_0}^{t_1} p(t)\, dt$$

2. the *expected execution-time* of the algorithm is

$$\mu_1(A) = \int_0^{\infty} tp(t)\, dt$$

3. a problem with complexity-parameter $n$ is in the class **RNC** if there exists a probabilistic parallel algorithm for it that uses a number of processors that is a polynomial in $n$ and which has expected execution-time that is $O(\lg^k n)$ for some value of $k$.

We will also distinguish a type of algorithm that is a *Monte Carlo* **RNC** algorithm. This is a Monte Carlo algorithm that executes in expected time $O(\lg^k n)$ with a polynomial number of processors. This class of such problems will be called **m-RNC**.

The expected execution is a weighted average of all of the possible execution-times. It is weighted by the probabilities that given execution-times actually occur. It follows that, if we run an **RNC** algorithm many times with the same input and compute the average running time, we will get something that is bounded by a power of the logarithm of the input-size.

We must distinguish the class m-**RNC**, because these algorithms do not necessarily produce the correct answer in expected poly-logarithmic time — they only produce an answer with some degree of confidence. We can repeat such an algorithm many times to increase this level of confidence, but the algorithm will still be in m-**RNC**.

**Exercises:**

1. Suppose we define a class of algorithms called lv-**RNC** composed of Las Vegas algorithms whose expected execution-time is poly-logarithmic, using a poly-nomial number of processors. Is this a new class of algorithms?

**2.1. Work-efficient parallel prefix computation.** This section discusses a simple application of randomization in a parallel algorithm. Recall the discussion at the end of § 7.2.2 regarding the use of the Brent Scheduling Principle to perform parallel-prefix operations on $n$ items stored in an array, using $O(n/\lg n)$ processors. For many years there was no known deterministic algorithm for the corresponding problem in which the input data is in a linked list. In 1984 Uzi Vishkin discovered a probabilistic algorithm for this problem in 1984 — see [161]. In 1988 Anderson and Miller developed an simplified version of this in [8]. We will discuss the simplified algorithm here.

ALGORITHM 2.2. Let $\{a_0, \ldots, a_{n-1}\}$ be data-structures stored in an array, that define a linked list, and let $\star$ be some associative operation. Suppose that the data of element $a_i$ is $d_i$. Then there exists an **RNC** algorithm for computing the quantity $d_0 \star \cdots \star d_{n-1}$. The expected execution-time is $O(\lg n)$ using $O(n/\lg n)$ processors (on an EREW-PRAM computer).

Although the input-data is stored in an array, this problem is not equivalent to the original problem that was considered in connection with the Brent Scheduling Principle: in the present problem the $a_i$ are not stored in *consecutive* memory locations. Each of the $a_i$ has a **next**-pointer that indicates the location of the next term.

We begin the algorithm by assigning $O(\lg n)$ elements of the linked list to each processor in an arbitrary way. Note that if we could know that each processor had a range of consecutive elements assigned to it, we could easily carry out the *deterministic* algorithm for computing $d_0 \star \cdots \star d_{n-1}$ — see 1.5 on page 307.

The algorithm proceeds in phases. Each phase selects certain elements of the list, deletes them, and splices the pieces of the list together. After carrying out these two steps, we recursively call the algorithm on the new (shorter) list. When we delete an entry from the linked list, we modify the following entry in such a way that the data contained in the deleted entry is not lost. In fact, if we delete the $i^{th}$ entry of the linked list (which, incidentally, is probably *not* the $i^{th}$ entry in the array used to store the data) we perform the operation

$$d_{i+1} \leftarrow d_i \star d_{i+1}$$

In order for this statement to make any sense, our deletions must satisfy the condition that

**No two adjacent elements of the list are ever deleted in the same step.**

In addition, in order to maximize parallelism and make the algorithm easier to implement, we also require that:

**At most one of the elements of the linked list that are assigned to a given processor is deleted in any phase of the algorithm.**

We will show that it is possible to carry out the deletions in such a way that the two conditions above are satisfied, deletions require constant parallel time, *and* that, on average, $O(\lg n)$ phases of the algorithm are need to delete all of the entries of the linked list.

The deletions are selected by the following procedure:

### 2.3. Selection Procedure:

1. Each processor selects one of its associated list-elements still present in the linked list. We will call the list-element selected by processor $i$, $e_i$.
2. Each processor "flips" a coin — i. e., generates a random variable that can equal 0 or 1 with equal probability. Call the $i^{th}$ "coin value" $c_i$ (1=heads).
3. If $c_i = 1$ and $c_{i+1} \neq 1$, then $e_i$ is selected for deletion.

Note that this procedure selects elements that satisfy the two conditions listed above, and that the selection can be done in constant time. We must analyze the behavior of this algorithm. The probability that a processor will delete its chosen element in a given step is $1/4$ since

1. The probability that it will get $c_i = 1$ is $1/2$.
2. The probability that $c_{i+1} = 0$ is also $1/2$.

This heuristic argument implies that the algorithm completes its execution in $O(\lg n)$ phases. In order to prove this rigorously, we must bound the probability that very few list-elements are eliminated in a given step. The argument above implies that, on the average, $O(n/\lg n)$ list elements are eliminated in a given step. The problem is that this average might be achieved by having a few processors eliminate all of their list-elements rapidly, and having the others hold onto theirs. The mere statement that each phase of the algorithm has the expected average behavior doesn't prove that the expected execution time is what we want. We must also show that the *worst case* behavior is very unlikely.

We will concentrate upon the behavior of the list-entries selected by a single processor in multiple phases of the algorithm. In each phase, the processor has a probability of $1/4$ of deleting an element from the list. We imagine the entries selected in step 1 of the selection procedure above as corresponding to "coins" being flipped, where the probability of "heads" (i.e., the chosen element being deleted from the list) is $1/4$. In $c \lg n$ trials, the probability of $\leq \lg n$ "heads" occurring is

$$\text{Prob}\left[\{H \leq \lg k\}\right] \leq \binom{c \lg n}{\lg n} \left(\frac{3}{4}\right)^{c \lg n - \lg n}$$

Here, we simply compute the probability that $c \lg n - \lg n$ *tails* occur and we ignore all of the other factors. The factor $\left(\frac{3}{4}\right)^{c \lg n - \lg n}$ is the probability the first $c \lg n - \lg n$ trials result in tails, and the factor of $\binom{c \lg n}{\lg n}$ is the number of ways of distributing these results among all of the trials. We use Stirling's formula to estimate the binomial coefficient:

$$\binom{c \lg n}{\lg n} \left(\frac{3}{4}\right)^{c \lg n - \lg n} = \binom{c \lg n}{\lg n} \left(\frac{3}{4}\right)^{(c-1) \lg n}$$

$$\leq \left(\frac{ec \lg n}{\lg n}\right)^{\lg n} \left(\frac{3}{4}\right)^{(c-1) \lg n}$$

$$= \left(ec \left(\frac{3}{4}\right)^{c-1}\right)^{\lg n}$$

$$\leq \left(\frac{1}{4}\right)^{\lg n}$$

$$= \frac{1}{n^2}$$

as long as $c \geq 20$ (this is in order to guarantee that Stirling's formula is sufficiently accurate). This is the probability that *one* processor will still have list-elements left over at the end of $c \lg n$ phases of the algorithm. The probability that any of the $n/\lg n$ processors will have such list elements left over is

$$\leq \frac{n}{\lg n} \cdot \frac{1}{n^2} \leq \frac{1}{n}$$

and this shows that the expected execution-time is $O(\lg n)$.

**Exercises:**

1. What is the worst-case running time of the randomized parallel prefix algorithm?
2. How would the algorithm have to be modified if we wanted to compute all of the values $\{d_0, d_0 \star d_1, \ldots, d_0 \star \cdots \star d_{n-1}\}$?

**2.2. The Valiant and Brebner Sorting Algorithm.** This algorithm is due to Valiant and Brebner (see [157]). It was one of the first **RNC** algorithms to be developed and illustrates some of the basic ideas involved in all **RNC** algorithms. It is also interesting because an incomplete implementation of it is built into a physical piece of hardware — the Connection Machine. The implementation on the Connection Machine is incomplete because it doesn't use the *randomized* aspects of the Valiant-Brebner algorithm. As a result this implementation occasionally suffers from the problems that randomization was intended to correct — bottlenecks in the data-flow.

Suppose we have an $n$-dimensional hypercube computer. Then the vertices will be denoted by $n$-bit binary numbers and adjacent vertices will be characterized by the fact that their binary numbers differ in at most one bit. This implies that we can route data on a hypercube by the following algorithm:

LEMMA 2.4. Suppose we want to move data from vertex $a_1 \ldots a_n$ to vertex $b_1 \ldots b_n$, where these are the binary representations of the vertex numbers. This movement can be accomplished in $2n$ steps by:

1. scanning the numbers from left to right and;
2. whenever the numbers differ (in a bit position) moving the data along the corresponding communication line.

Suppose we have a hypercube computer with information packets at each vertex. The packets look like <data,target_vertex>, and we send them to their destinations in two steps as follows:

Phase 1. Generate temporary destinations for each packet and send the packets to these temporary destinations. These are *random* $n$-bit binary numbers (each bit has a $1/2$ probability of being 1),

Phase 2. Send the packets from their temporary destinations to their true final destinations.

In carrying out these data movements we must:

1. use the left to right algorithm for routing data, and;

2. whenever multiple packets from phase 1 or phase 2 appear at a vertex they are queued and sequentially sent out;

3. whenever packets from both phases appear at a vertex the packets from phase 1 have priority over those from phase 2.

The idea here is that bottlenecks occur in the routing of data to its final destination because of patterns in the numbers of the destination-vertices. This problem is solved by adding the intermediate step of sending data to random destinations. These random temporary destinations act like hashing — they destroy regularity of the data so that the number of collisions of packets of data is minimized.

The main result is:

THEOREM 2.5. The probability is $(.74)^d$ that this algorithm, run on a hypercube of $d$ dimensions, takes more than $8d$ steps to complete.

Since the number of vertices of a hypercube of dimension $d$ is $2^d$, we can regard $n = 2^d$ as the number of input data values. The expected execution-time of this algorithm is, consequently, $O(\lg n)$.

This result has been generalized to computers on many bounded-degree networks by Upfal in [154].

**2.3. Maximal Matchings in Graphs.** In this section we will discuss a probabilistic algorithm for performing an important graph-theoretic computation.

DEFINITION 2.6. Let $G = (V, E)$ denote an undirected graph.

1. A *maximal matching* of $G$ is a set of edges $M \subseteq E$ such that
   - For any two $e_i, e_j \in M$, $e_i$ has no end-vertices in common with $e_j$. This is called the *vertex disjointness property*.
   - The set $M$ is maximal with respect to this property. In other words, the graph $G'$ spanned by $E \setminus M$ consists of isolated vertices.

2. A maximal matching, $M$, is *perfect* if all vertices of $G$ occur in the subgraph induced by the edges in $M$.

3. If $G$ is a weighted graph then a *minimum weight maximal matching*, $M \subseteq E$ is a maximum matching such that the total weight of the edges in $M$ is minimal (among all possible maximum matchings).

Note that not every graph has a perfect matching — for instance the number of vertices must be *even*[1]. Figure VII.2 shows a graph with a perfect matching — the edges in the matching are darker than the other edges.

In the past, the term "maximal matching" has often been used to refer to a matching with the *largest possible number of edges*. Finding that form of maximal matching is

---

[1]Since the edges in $M$ are vertex-disjoint, and each edge has exactly two ends.

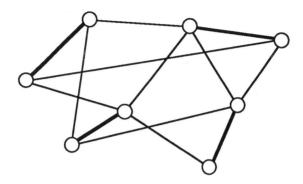

FIGURE VII.2. A graph with a perfect matching

much more difficult than finding one as *we* define it[2]. There is clearly a simple greedy sequential algorithm for finding a maximum matching in *our* sense. The two different definitions are connected via the concept of minimum-weight maximal matching — just give every edge in the graph a weight of $-1$.

Sequential algorithms exist for solving the "old" form of the maximum matching problem — the book of Lovász and Plummer, [104], and the paper of Edmonds, [47] give a kind of survey of the techniques.

The parallel algorithms for solving the maximal matching problem are generally based upon the theorem of Tutte proved in 1947 — see [152]. It proved that a graph has a perfect matching if and only if a certain matrix of *indeterminates*, called the *Tutte matrix*, is non-singular (has an inverse). A matrix is non-singular if and only if its determinant is nonzero — see the definition of a determinant in 1.5 on page 159. The first algorithm based on this result was due to Lovász in 1979 — see [102]. This algorithm determines whether the Tutte matrix is nonsingular by a Monte Carlo algorithm — namely 1.5 on page 445. Since it can be difficult to efficiently compute the determinant of a matrix of indeterminates, the algorithm of Lovász turns the Tutte matrix into a matrix of numbers by plugging random values into all of the indeterminates. There are several efficient parallel algorithms for deciding whether a matrix of numbers is nonsingular:

1. Use Csanky's algorithm for the determinant in § 1.5 of chapter V. This is actually not very efficient (but it is an **NC** algorithm).
2. Use the results in § 1.3 of the same chapter to decide whether the matrix has an inverse. This is the preferred method.

However we decide whether the modified Tutte matrix is nonsingular, we get a probabilistic algorithm for deciding whether the original graph had a perfect matching: the random values we plugged into the indeterminates *might* have been a "root" of the determinant (regarded as a polynomial in the indeterminates). In other words, we might get a "No" answer to the question of whether there exists a perfect matching,

---

[2]I.e., a matching such that adding any additional edge to destroys the vertex disjointness property.

even though the determinant of the original Tutte matrix is nonzero, as a polynomial (indicating that the graph does have a perfect matching). Lovász's algorithm is, consequently, a **Yes**-biased Monte Carlo algorithm, and can be regarded as an m-RNC algorithm.

The first m-RNC algorithm for maximal matchings was discovered by Karp, Upfal and Wigderson in 1985 — see [82] and [83]. It solves all forms of the maximal matching problem — including the weighted maximal matching problems[3].

We will present a simpler (and ingenious) algorithm for this problem developed by Ketan Mulmuley, Umesh Vazirani and Vijay Vazirani in 1987 — see [117]. This algorithm computes a maximal matching in a very interesting way:

1. It assigns weights to the edges of a graph in a random, but controlled fashion (this will be clarified later).
2. It computes the weight of a minimum-weight maximum matching in such a way we can determine the only edges that could possibly have participated in this matching. In other words, we assign weights that are not entirely random, but with the property that numerical values of the sum of any subset of the weights uniquely determines the weights that could have participated in the summation.

2.3.1. *A Partitioning Lemma.* We will begin with some technical results that allow us to "randomly" assign weights that have this property.

DEFINITION 2.7. A *set system* $(S, F)$ consists of a finite set $S$ of elements $\{x_1, \ldots, x_n\}$, and a family, $F$, of subsets of $S$, so $F = \{S_1, \ldots, S_k\}$ with $S_i \subseteq S$.

If we assign a weight $w_i$ to each element $x_i$ of $S$, we define the weight $w(S_i)$ of a set $S_i$ by

$$w(S_i) = \sum_{x_j \in S_i} w_j$$

LEMMA 2.8. Let $(S, F)$ be a set system whose elements are assigned integer weights chosen uniformly and independently from $[1 .. 2n]$. Then the probability that there exists a unique minimum-weight set in $F$ is $1/2$.

PROOF. Pick some value of $i$ and fix the weights of all of the elements except $x_i$. We define the *threshold* for element $x_i$ to be the number $\alpha_i$ such that:

1. if $w_i \leq \alpha_i$, then $x_i$ is contained in some minimum weight subset $S_j$, and
2. if $w_i > \alpha_i$, then $x_i$ is in no minimum-weight subset.

Clearly, if $w_i < \alpha_i$, then the element $x_i$ must be in *every* minimum-weight subset. Thus *ambiguity* about element $x_i$ occurs if and only if $w_i = \alpha_i$, since in this case there is

---

[3]Which implies that it also solves the "old" form of the maximal matching problem.

some minimum-weight subset that contains $x_i$ and another that does not. In this case, we will say that the element $x_i$ is *singular*.

Now we note that the threshold $\alpha_i$ was defined independently of the weight $w_i$. Since $w_i$ was selected randomly, uniformly, and independently of the other $w_j$, we get:

**Claim:** The probability that $x_i$ is singular is $\leq \dfrac{1}{2n}$.

Since $S$ contains $n$ elements:

**Claim:** The probability that there exists a singular element is $\leq n \cdot (1/2n) = 1/2$.

It follows that the probability that there is *no* singular element is $1/2$.

Now we observe that if no element is singular, then minimum-weight sets will be *unique*, since, in this case, every element will either be in *every* minimum-weight set or in *no* minimum-weight set.  □

2.3.2. *Perfect Matchings.* Now we will consider the special case of *perfect matchings* in *bipartite graphs*. Recall the definition of a bipartite graph in 8.8 on page 104. This turns out to be a particularly simple case. The methods used in the general case are essentially the same but somewhat more complex.

We will assume given a bipartite graph $G = G(V_1, V_2, E)$ with $2n$ vertices and $m$ edges. In addition, we will assume that $G$ *has* a perfect matching. As remarked above, this is a highly nontrivial assumption. We will give an **RNC** algorithm for finding a perfect matching.

We regard the edges in $E$ and the set of perfect matchings in $G$ as a set-system. We assign random integer weights to the edges of the graph, chosen uniformly and independently from the range $[1 .. 2m]$. Lemma 2.8 on page 453 implies that the minimum-weight perfect-matching is unique is $1/2$. In the remainder of this discussion, we will assume that we have assigned the random weights in such a way that the minimum-weight perfect-matching in $G$ is unique. We suppose that the weight of the edge connecting vertex $i$ and $j$ (if one exists), if $w(i, j)$.

DEFINITION 2.9. If $A$ is an $n \times n$ matrix, we will define $\bar{A}_{i,j} = \det(A'_{i,j})$, where $A'_{i,j}$ is the $n - 1 \times n - 1$ matrix that results from deleting the $i^{\text{th}}$ row and the $j^{\text{th}}$ column from $A$. We define the *adjoint* of $A$, denoted $\hat{A}$ by

$$\hat{A}_{i,j} = (-1)^{i+j} \bar{A}_{i,j}$$

for all $i$ and $j$ between 1 and $n$.

Cramer's Rule states that $A^{-1} = \hat{A}^{\text{tr}} / \det(A)$.

In order to describe this algorithm, we will need the concept of an *incidence matrix* of a bipartite graph.

DEFINITION 2.10. Let $G = G(U, V, E)$ be a bipartite graph with $U = \{u_1, \ldots, u_m\}$ and $V = \{v_1, \ldots, v_n\}$. The incidence matrix of $G$ is an $m \times n$ matrix $A$ defined by

$$A_{i,j} = \begin{cases} 1 & \text{if there exists an edge from } u_i \text{ to } v_j \\ 0 & \text{otherwise} \end{cases}$$

This is closely related to the *adjacency matrix* of $G$, defined in 2.4 on page 322: if $G$ is a bipartite graph with incidence matrix $D$ and adjacency matrix $A$, then

$$A = \begin{pmatrix} 0_{m,n} & D \\ D^{\mathrm{tr}} & 0_{n,m} \end{pmatrix}$$

where $D^{\mathrm{tr}}$ is the transpose of $D$ and $0_{a,b}$ denotes a matrix with $a$ rows and $b$ columns, all of whose entries are zero. Note that, if a bipartite graph has a perfect matching, both of its vertex-sets will be exactly the same size, and its incidence matrix will be a square matrix.

We will construct a matrix, $D$, associated with $G$ by the following sequence of operations:

1. Let $C$ be the incidence matrix of $G$ (defined above);
2. If $C_{i,j} = 1$ (so there is an edge connecting vertex $i$ and vertex $j$), set $D_{i,j} \leftarrow 2^{w(i,j)}$.
3. If $C_{i,j} = 0$ set $D_{i,j} \leftarrow 0$.

This matrix $D$ has the following interesting properties:

LEMMA 2.11. Suppose the minimum-weight perfect matching in $G(U, V, E)$ is unique. Suppose this matching is $M \subseteq E$, and suppose its total weight is $w$. Then

$$\frac{\det(D)}{2^w}$$

is an *odd number* so that:

1. $\det(D) \neq 0$
2. the highest power of 2 that divides $\det(D)$ is $2^w$.

PROOF. We analyze the terms that enter into the determinant of $D$. Recall the definition of the determinant in 1.5 on page 159:

(118) $$\det(D) = \sum_{\substack{i_1, \ldots, i_n \\ \text{all distinct}}} \wp(i_1, \ldots, i_n) D_{1,i_1} \cdots D_{n,i_n}$$

where $\wp(i_1, \ldots, i_n)$ is the parity of the permutation $\begin{pmatrix} 1 & \cdots & n \\ i_1 & \cdots & i_n \end{pmatrix}$ (defined in 1.4 on page 159).

Now we note that every perfect matching in $G$ corresponds to a permutation of the numbers $\{1, \ldots, n\}$. Each such perfect matching, consequently, corresponds to a term in equation (118). Suppose $\sigma$ is a permutation that corresponds to a perfect matching in $G$ — this means that there is an edge connecting vertex $i$ to vertex $\sigma(i)$ and we get

$$t(\sigma) = \wp(\sigma) D_{1,\sigma(1)} \cdots D_{n,\sigma(n)}$$
$$= \wp(i_1, \ldots, i_n) 2^{w(1,\sigma(1))} \ldots 2^{w(n,\sigma(n))}$$
$$= \wp(i_1, \ldots, i_n) 2^{\sum_{j=1}^n w(j,\sigma(j))}$$

If a permutation $\sigma$ doesn't correspond to a perfect matching, then $t(\sigma) = 0$, since it will have some factor of the form $D_{i,\sigma(i)}$, where there is no edge connecting vertex $i$ with vertex $\sigma(i)$.

Now we divide $\det(D)$ by $2^w$. The quotient will have one term of the form $\pm 1$, corresponding to the minimum-weight matching and other terms corresponding to higher-weight matchings. There will only be one term corresponding to the minimum-weight matching because it is *unique* (by assumption). The higher-weight matchings will give rise to terms in the quotient that are all *even* because the numbers in the $D$-matrix were powers of two that corresponded to the weights (which were higher, for these matchings). It follows that the number

$$\frac{\det(D)}{2^w}$$

is an *odd number*.    □

The second lemma allows us to determine which edges lie in this minimum-weight matching:

LEMMA 2.12. As before, suppose $M \subseteq E$ is the unique minimum-weight perfect matching in $G(U, V, E)$, suppose its total weight is $w$. Then an edge $e \in E$ connecting vertex $u_i \in U$ with vertex $v_j \in V$ is in $M$ if and only if

$$\frac{2^{w(i,j)} \det(\bar{D}_{i,j})}{2^w}$$

is an odd number.

PROOF. This follows by an argument like that used in 2.11 above and the fact that determinants can be computed by expanding using *minors*. It is not hard to give a direct proof that has a more graph-theoretic flavor. Recall that $\bar{D}_{i,j}$ is the matrix that result from deleting the $i^{\text{th}}$ row and $j^{\text{th}}$ column of $D$ — it is the form of the $D$-matrix that corresponds to the result of deleting $u_i$ from $U$ and $v_j$ from $V$. Call this new, smaller bipartite graph $G'$. It is not hard to see that $G'$ *also* has a unique minimum-weight matching — namely the one that result from deleting edge $e$ from $M$. Consequently, lemma 2.11 above, implies that

$$\frac{\det(\bar{D}_{i,j})}{2^{w-w(i,j)}}$$

is an odd number. Here $w - w(i,j)$ is the weight of this unique minimum-weight perfect matching in $G'$ that result by deleting $e$ from $M$. But this proves the result.   $\square$

All of the statements in this section depend upon the assumption that the minimum-weight perfect matching of $G$ is unique. Since lemma 2.8 on page 453 implies that the probability of this is $1/2$, so we have a probabilistic algorithm. In fact, we have a Las Vegas algorithm, because it is very easy to verify whether the output of the algorithm constitutes a perfect matching. This means we have an **RNC** algorithm, since we need only execute the original probabilistic algorithm over and over again until it gives us a valid perfect matching for $G$.

To implement this algorithm in parallel, we can use Csanky's **NC** algorithm for the determinant — this requires $O(n \cdot n^{2.376})$ processors and executes in $O(\lg^2 n)$ time. Since we must compute one determinant for each of the $m$ edges in the graph, our total processor requirement is $O(n \cdot m \cdot n^{2.376})$. Our *expected* execution-time is $O(\lg^2 n)$.

2.3.3. *The General Case.* Now we will explore the question of finding perfect matchings in general undirected graphs. We need several theoretical tools.

DEFINITION 2.13. Let $G = (V, E)$ be an undirected graph, such that $|V| = n$. The *Tutte matrix* of $G$, denoted $t(G)$, is defined via

$$t(G) = \begin{cases} x_{i,j} & \text{if there exists an edge connecting } v_i \text{ with } v_j \text{ and } i < j \\ -x_{i,j} & \text{if there exists an edge connecting } v_i \text{ with } v_j \text{ and } i > j \\ 0 & \text{if } i = j \end{cases}$$

Here the quantities $x_{i,j}$ with $1 \leq i < j \leq n$ are indeterminates — i.e., variables.

For instance, if $G$ is the graph in figure VII.3, then the Tutte matrix of $G$ is

$$\begin{pmatrix} 0 & x_{1,2} & x_{1,3} & 0 & x_{1,5} & x_{1,6} \\ -x_{1,2} & 0 & 0 & x_{2,4} & x_{2,5} & x_{2,6} \\ -x_{1,3} & 0 & 0 & x_{3,4} & x_{3,5} & 0 \\ 0 & -x_{2,4} & -x_{3,4} & 0 & x_{4,5} & 0 \\ -x_{1,5} & -x_{2,5} & -x_{3,5} & -x_{4,5} & 0 & 0 \\ -x_{1,6} & -x_{2,6} & 0 & 0 & 0 & 0 \end{pmatrix}$$

In [152], Tutte proved that these matrices have a remarkable property:

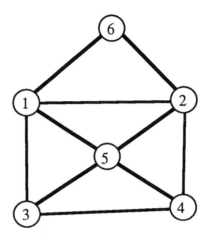

FIGURE VII.3

THEOREM 2.14. **Tutte's Theorem.** Let $G$ be a graph with Tutte matrix $t(G)$. If $G$ does not have a perfect matching then $\det(t(G) = 0$. If $G$ has perfect matchings, then

$$\det(t(g)) = (t_1 + \cdots + t_k)^2$$

where the $t_i$ are expressions of the form $\pm x_{\mu_1,\nu_1} \cdots x_{\mu_n,\nu_n}$, and the sets of edges $\{(\mu_1,\nu_1),\ldots,(\mu_n,\nu_n)\}$ **are** the perfect matchings of $G$.

Recall the definition of the determinant of a matrix in 1.5 on page 159. We will not prove this. It turns out that the proof that it vanishes is very similar to part of the proof of 2.11 on page 455. If we compute this determinant of the Tutte matrix of the graph in figure VII.3, we get:

$$\left(x_{1,6}x_{2,4}x_{3,5} + x_{1,5}x_{3,4}x_{2,6} + x_{1,3}x_{2,6}x_{4,5} - x_{1,6}x_{2,5}x_{3,4}\right)^2$$

so that this graph has precisely *three distinct* perfect matchings:

- $\{(1,6),(2,4),(3,5)\}$,
- $\{(1,3),(2,6),(4,5)\}$, and
- $\{(1,6),(2,5),(3,4)\}$

Given this theorem, we can generalize the results of the previous section to general graphs fairly easily. As before we assume that we have a graph $G$ that *has* a perfect matching, and we assign weights randomly to the edges of $G$. Now we define the $D$-matrix by the assignments

$$x_{i,j} \leftarrow 2^{w(i,j)}$$

in the Tutte matrix, for all $1 \le i < j \le n$. So our $D$-matrix is just the result of evaluating the Tutte matrix.

LEMMA 2.15. Suppose $G$ is an weighted undirected graph with $n$ vertices, that has a unique minimum-weight perfect matching. If $w$ is the total weight of this perfect matching then $w$. Then

$$\frac{\det(D)}{2^{2w}}$$

is an *odd number* so that:

1. $\det(D) \neq 0$
2. the highest power of 2 that divides $\det(D)$ is $2^{2w}$.

Note that we have to use $2^{2w}$ rather than $2^{w}$ because the sum of terms in the Tutte matrix is squared. The determinant of the incidence matrix of a bipartite graph had the same sum of terms, but the result wasn't squared.

PROOF. This proof is almost exactly the same as that of 2.11 on page 455 — Tutte's theorem has done most of the work for us. It shows that $\det(D)$ in the present case has exactly the same form as $\det(D)^2$ in 2.11. □

Similarly, we get:

LEMMA 2.16. As before, suppose $M \subseteq E$ is the unique minimum-weight perfect matching in $G$, and suppose its total weight is $w$. Then an edge $e \in E$ is in $M$ if and only if

$$\frac{2^{2w(i,j)} \det(\bar{D}_{i,j})}{2^{2w}}$$

is an odd number.

As before, we get a Las Vegas algorithm that executes in $O(\lg^2 n)$ time, using $O(n \cdot m \cdot n^{2.376})$ processors. This gives us an RNC algorithm with expected execution time of $O(\lg^2 n)$. All of this assumes that we know that the graph $G$ has a perfect matching. If we don't know this, Tutte's theorem implies that $\det(D) = 0$. Unfortunately, if we get we don't necessarily know that the graph didn't have a perfect matching – we might have picked *unlucky values* for the weights of the edges to produce this result even though the determinant of the Tutte matrix is *nonzero*. If we don't know whether the original graph had a perfect matching, our algorithm becomes an m-RNC algorithm that is biased in the direction of saying that there *is* a perfect matching.

## 2.4. The Maximal Independent-Set Problem.

**2.4.1.** *Definition and statement of results.* We will conclude this chapter by studying a very famous problem, whose first parallel solution represented something of a breakthrough.

DEFINITION 2.17. Let $G = (V, E)$ denote an undirected graph with $n$ vertices. A set, $S \subset V$, of vertices will be said to be *independent* if no two vertices in $S$ are adjacent.

An independent set of vertices, $S$, is said to be *maximal*, if every vertex $i \in V$ is adjacent to some element of $S$.

The problem of finding a maximal independent set has a trivial sequential solution — simply test all of the vertices of $G$, one by one. If a vertex is not adjacent to any element of $S$, then add it to $S$. It is not clear that there is any reasonable parallel solution that is **NC** — it appears that the choices one makes in any step of the sequential algorithm influences all future choices. This problem was long believed to be inherently sequential. Indeed, the *lexicographically first* version of the problem is known to be P-complete — see [30].

This algorithm was first discovered by Karp and Wigderson in 1984 — see [81]. They presented a probabilistic algorithm and then showed that, with a suitable choice of probability distribution, the algorithm became deterministic. Recall the notation of § 2 and § 4.3 in this chapter.

ALGORITHM 2.18. This is an algorithm for finding a maximal independent set within a graph, whose expected execution time is $O(\lg^2 n)$ with $O(n^2)$

- **Input:** Let $G = (V, E)$ be an undirected graph with $|V| = n$.
- **Output:** A maximal independent set $I \subseteq V$.

1. **Initialization:**
$$I \leftarrow \emptyset$$
$$G' \leftarrow G$$

2. **Main loop:**
   while $G' \neq \emptyset$ do
   $$I' \leftarrow \text{select}(G')$$
   $$I \leftarrow I \cup I'$$
   $$Y \leftarrow I' \cup \Gamma(I')$$
   $G' = (V', E')$ is the induced subgraph in $V' \setminus Y$
   **endwhile**

We have omitted an important detail in this description — the subroutine select:

ALGORITHM 2.19. **The select subroutine.** First version.
   select$(G' = (V', E'))$
   $$I' \leftarrow V'$$
   define a random function $\pi$ of $V' \to V'$ by

```
    for all i' ∈ V' do in parallel
        set π(i') = a random element of V'
    endfor
    for all edges (i, j) ∈ E' do
        if π(i) ≥ π(j) then
            I' ← I' \ {i}
        else
            I' ← I' \ {j}
    endfor
end (of select)
```

It is not hard to see that select does select an independent set in $G'$, so that the original algorithm selects a *maximal* independent set of the original graph. The only point that is still in question is the *execution time* of the algorithm. It will turn out that the expected number of iterations of the main loop in 2.18 is $O(\lg n)$ so the expected execution-time of the whole algorithm is $O(\lg^2 n)$.

The random choices in the select step of this algorithm are made independently of each other.

The precise meaning of this statement is:

DEFINITION 2.20. Let $\{E_1, \ldots, E_k\}$ be a finite sequence of events. These events are said to be *independent* if every subsequence $\{E_{j_1}, \ldots, E_{j_i}\}$ satisfies

$$\text{Prob}\,[E_{j_1} \cap \cdots \cap E_{j_i}] = \prod_{\ell=1}^{i} \text{Prob}\,[E_{j_\ell}]$$

Karp and Wigderson made the very important observation that it is really only necessary to make *pairwise independent* selections for the algorithm to work properly. "Pairwise independence" just means that, of the sequence of events $\{E_1, \ldots, E_k\}$, every *pair* of events satisfies the conditions of the definition above. In order to describe this we define:

DEFINITION 2.21. For each vertex $i' \in V'$ define:
1. $\deg(i') = |\Gamma(i')|$;
2. $\text{coin}(i')$ to be a $\{0, 1\}$-valued random variable such that
    a. if $\deg(i') \geq 1$ then $\text{coin}(i') = 1$ with probability

$$\frac{1}{2\deg(i')}$$

   and
    b. if $\deg(i') = 0$, then $\text{coin}(i')$ is always 1.

With this definition in mind, our new version of the select-step is

ALGORITHM 2.22. **The select subroutine.** Second version.

$\text{select}(G' = (V', E'))$
   $I' \leftarrow V'$
   $X \leftarrow \emptyset$
   **for** all $i' \in V'$ **do in parallel**
      compute $\deg(i')$
   **endfor**
   **for** all $i' \in V'$ **do in parallel**
      randomly choose a value for $\text{coin}(i')$
      **if** $\text{coin}(i') = 1$ **then** $X \leftarrow X \cup \{i'\}$
   **endfor**
   **for** all edges $(v_1, v_2) \in E'$ **do in parallel**
      **if** $v_1 \in X$ and $v_2 \in X$ **then**
         **if** $\deg(v_1) \leq \deg(v_2)$ **then**
            $I' \leftarrow I' \setminus \{v_1\}$
            **else** $I' \leftarrow I' \setminus \{v_2\}$
   **endfor**
 **end** (of select)

Note that this second algorithm is somewhat more complicated than the first. The proof that the algorithms work is based upon the fact that, on average, at least $1/16$ of the edges are eliminated in each iteration. This implies that all of the edges are eliminated in $O(\lg n)$ expected iterations.

The second, more complicated algorithm, has the interesting property that the algorithm continues to be valid if we restrict the set of random variables $\{\text{coin}(i)\}$, for all vertices $i$ in the graph, to some set of size $q^2$, where $q$ is some number that $\geq n$ but bounded by a polynomial in $n$. In other words there are $q^2$ possible values for the *set* of $n$ variables $\{\text{coin}(i)\}$, such that there is a nonzero probability that at least $1/16$ of the edges will be eliminated in an iteration of the algorithms that uses values of $\{\text{coin}(i)\}$ drawn from this set. Since the set is of size $q^2$, we can simply test *each member* of this set — i.e. we create processes that carry out an iteration of the algorithm using *each* of the members of the set. Since the probability of eliminating $1/16$ of the edges by using *some* member of this set is nonzero, it follows that at least one of the processes we have created *must* succeed in eliminating $1/16$ of the edges. Our **RNC** algorithm becomes a *deterministic* algorithm.

We will describe how this sample space is constructed. We want the probability conditions in 2.21 to be satisfied.

We construct an $n \times q$ matrix, $M$, where $q$ is the smallest prime $> \sum_{i \in V} 2d(i)$. We will construct this matrix in such a way that row $i$ (representing values of $\text{coin}(i)$ in different "coin tosses") has the value 1 in precisely $q/2d(i)$ entries, if $d(i) \geq 1$, and in all entries if $d(i) = 0$. The upshot will be that if we "draw" $q$ entries from this row (i.e., all of them), the probability of a 1 occurring is $1/2d(i)$, as specified in 2.21 above. The entries in $M$ are otherwise random.

Now we define a total of $q^2$ values of the set $\{\text{coin}(1), \ldots, \text{coin}(n)\}$, where we have numbered the vertices of the graph. Let $x$ and $y$ be integers such that $0 \leq x, y \leq q - 1$. Then there are $q^2$ possible values of the *pairs* $(x, y)$. We will index our sets $\{\text{coin}(1), \ldots, \text{coin}(n)\}$ by these pairs. Define

$$\text{coin}_{(x,y)}(i) = M_{i,(x+i \cdot y \bmod q)}$$

LEMMA 2.23. The probability that $\text{coin}(i) = 1$ satisfies the conditions in 2.21.

PROOF. Let the probability that $\text{coin}(i) = 1$ required by 2.21, be $R_i$. We have put $q \cdot R_i$ entries equal to 1 into the matrix $M$. For any value $j$ there are precisely $q$ pairs $(x, y)$ such that

$$(x + i \cdot y) \bmod q = j$$

— this just follows from 5.3 on page 424, which implies that given any value of $x$ we can *solve* for a corresponding value of $y$ that satisfies the equation: $y = i^{-1}(j - x) \bmod q$ — here $i^{-1}$ is the multiplicative inverse of $i \bmod q$. It follows that there are $q \cdot R_i$ values of $j$ that make $M_{i,(x+i \cdot y \bmod q)} = 1$. $\square$

LEMMA 2.24. Given $i, i'$, the probability that $\text{coin}(i) = 1$ and $\text{coin}(i') = 1$ simultaneously, is the product of the probabilities that each of them are 1 individually.

This lemma implies that the random variables generated by the scheme described above, are *pairwise independent*.

PROOF. We use the notation of the proof of 2.23 above. We must show that the probability that $\text{coin}(i) = 1$ and $\text{coin}(i') = 1$ occur simultaneously, is equal to $R_i R_{i'}$. Given any pair of numbers $j$ and $j'$ such that $0 \leq j, j' \leq q - 1$, the simultaneous equations

$$x + i \cdot y \bmod q = j$$
$$x + i' \cdot y \bmod q = j'$$

have a *unique* solution (for $x$ and $y$). If there are $a = R_i q$ entries in $M$ that produce a value of 1 for $\text{coin}(i)$ and $b = R_{i'} q$ entries that produce a value of 1, then there are $ab$ *pairs* $(j, j')$ that simultaneously produce 1 in $\text{coin}(i)$ and $\text{coin}(i')$. Each such pair (of entries of $M$) corresponds to a *unique* value of $x$ and $y$ that causes that pair to be selected. Consequently, there are $ab$ cases in the $q^2$ sets of numbers $\{\text{coin}(1), \ldots, \text{coin}(n)\}$. The probability that that $\text{coin}(i) = 1$ and $\text{coin}(i') = 1$ occur simultaneously is, therefore, $ab/q^2$, as claimed. $\square$

With these results in mind, we get the following deterministic version of the Maximal Independent Set algorithm:

ALGORITHM 2.25. **The Deterministic Algorithm.** This is an algorithm for finding a maximal independent set within a graph, whose execution time is $O(\lg^2 n)$ with $O(mn^2)$ processors (where $m$ is the number of edges of the graph).

- **Input:** Let $G = (V, E)$ be an undirected graph with $|V| = n$.
- **Output:** A maximal independent set $I \subseteq V$.

1. **Initialization:**
   $I \leftarrow \emptyset$
   $n \leftarrow |V|$
   **Compute** a prime $q$ such that $n \le q \le 2n$
   $G' = (V', E') \leftarrow G = (V, E)$

2. **Main loop:**
   **while** $G' \neq \emptyset$ **do**
       **for** all $i \in V'$ **do in parallel**
           **Compute** $d(i)$
       **endfor**
       **for** all $i \in V'$ **do in parallel**
           **if** $d(i) = 0$ **then**
               $I \leftarrow I \cup \{i\}$
               $V \leftarrow V \setminus \{i\}$
       **endfor**
       **find** $i \in V'$ such that $d(i)$ is a maximum
       **if** $d(i) \ge n/16$ **then**
           $I \leftarrow I \cup \{i\}$
           $G' \leftarrow$ graph induced on the vertices
               $V' \setminus (\{i\} \cup \Gamma(\{i\}))$
       **else for** all $i \in V'$, $d(i) < n/16$
       randomly choose $x$ and $y$ such that $0 \le x, y \le q - 1$
           $X \leftarrow \emptyset$
           **for** all $i \in V'$ **do in parallel**
               **compute** $n(i) = \lfloor q/2d(i) \rfloor$
               **compute** $l(i) = (x + y \cdot i) \bmod q$
               **if** $l(i) \le n(i)$ **then**
                   $X \leftarrow X \cup \{i\}$
           **endfor**
           $I' \leftarrow X$
           **for** all $i \in X, j \in X$ **do in parallel**
               **if** $(i, j) \in E'$ **then**

> **if** $d(i) \leq d(j)$ **then**
>     $I' \leftarrow I' \setminus \{i\}$
> $I \leftarrow I \cup I'$
> $Y \leftarrow I' \cup \Gamma(I')$
> $G' = (V', E')$ is the induced subgraph on $V' \setminus Y$
>     **endfor**
>   **endfor**
> **endwhile**

2.4.2. *Proof of the main results.* The following results prove that the expected execution-time of 2.19 is poly-logarithmic. We begin by defining:

DEFINITION 2.26. For all $i \in V'$ such that $d(i) \geq 1$ we define

$$s(i) = \sum_{j \in \Gamma(i)} \frac{1}{d(j)}$$

We will also need the following technical result:

LEMMA 2.27. Let $p_1 \geq \cdots \geq p_n \geq 0$ be real-valued variables. For $1 \leq k \leq n$, let

$$\alpha_\ell = \sum_{j=1}^{k} p_j$$

$$\beta_\ell = \sum_{j=1}^{k} \sum_{\ell=j+1}^{k} p_j \cdot p_\ell$$

$$\gamma_i = \alpha_i - c \cdot \beta_i$$

where $c$ is some constant $c > 0$. Then

$$\max\{\gamma_k | 1 \leq k \leq n\} \geq \frac{1}{2} \cdot \min\left\{\alpha_n, \frac{1}{c}\right\}$$

PROOF. We can show that $\beta_k$ is maximized when $p_1 = \cdots = p_n = \alpha_k/k$. This is just a problem of constrained maximization, where the constraint is that the value of $\alpha_k = \sum_{j=1}^{k} p_j$ is fixed. We solve this problem by Lagrange's Method of Undetermined Multipliers — we maximize

$$Z = \beta_k + \lambda \cdot \alpha_k$$

where $\lambda$ is some quantity to be determined later. We take the partial derivatives of $Z$ with respect to the $p_j$ to get

$$\frac{\partial Z}{\partial p_t} = \left(\sum_{j=1}^{k} p_j\right) - p_t + \lambda$$

(simply note that $p_t$ is paired with every *other* $p_j$ for $0 \le j \le k$ in the formula for $\beta_k$). If we set all of these to zero, we get that the $p_j$ must all be equal.

Consequently $\beta_k \le \alpha_k^2 \cdot (k-1)/2k$. Thus

$$\gamma_k \ge \alpha_k \cdot \left(1 - c \cdot \alpha_k \cdot \frac{(k-1)}{2k}\right)$$

If $\alpha_n \le 1/c$ then $\gamma_n \ge \alpha_n/2$. If $\alpha_k \ge 1/c$ then $\gamma_1 \ge 1/c$. Otherwise there exists a value of $k$ such that $\alpha_{k-1} \le 1/c \le \alpha_k \le 1/c \cdot k/(k-1)$. The last inequality follows because $p_1 \ge \cdots \ge p_n$. Then $\gamma_k \ge 1/2c$. $\square$

PROPOSITION 2.28. **Principle of Inclusion and Exclusion.** Let $E_1$ and $E_2$ be two events. Then:

1. If $E_1$ and $E_2$ are mutually exclusive then

$$\mathrm{Prob}\,[E_1 \cup E_2] = \mathrm{Prob}\,[E_1] + \mathrm{Prob}\,[E_2]$$

2. In general

$$\mathrm{Prob}\,[E_1 \cup E_2] = \mathrm{Prob}\,[E_1] + \mathrm{Prob}\,[E_2] - \mathrm{Prob}\,[E_1 \cap E_2]$$

3. and in more generality

$$\mathrm{Prob}\left[\bigcup_{i=1}^{k} E_i\right] = \sum_i \mathrm{Prob}\,[E_i] - \sum_{i_1 < i_2} \mathrm{Prob}\,[E_{i_1} \cap E_{i_2}]$$
$$+ \sum_{i_1 < i_2 < i_3} \mathrm{Prob}\,[E_{i_1} \cap E_{i_2} \cap E_{i_3} \cdots]$$
$$\mathrm{Prob}\left[\bigcup_{i=1}^{k} E_i\right] \ge \sum_i \mathrm{Prob}\,[E_i] - \sum_{i_1 < i_2} \mathrm{Prob}\,[E_{i_1} \cap E_{i_2}]$$

Essentially, the probability that $\mathrm{Prob}\,[E_i]$ occurs satisfies the conditions that

$$\mathrm{Prob}\,[E_1] = \mathrm{Prob}\,[E_1 \cap \neg E_2] + \mathrm{Prob}\,[E_1 \cap E_2]$$
$$\mathrm{Prob}\,[E_2] = \mathrm{Prob}\,[E_2 \cap \neg E_1] + \mathrm{Prob}\,[E_1 \cap E_2]$$

where $\mathrm{Prob}\,[E_1 \cap \neg E_2]$ is the probability that $E_1$ occurs, but $E_2$ *doesn't* occur. If we add $\mathrm{Prob}\,[E_1]$ and $\mathrm{Prob}\,[E_2]$, the probability $\mathrm{Prob}\,[E_1 \cap E_2]$ is counted twice, so be must subtract it out.

The third statement can be proved from the second by induction.

DEFINITION 2.29. **Conditional Probabilities.** Let $\mathrm{Prob}\,[E_1|E_2]$ denote the *conditional probability* that $E_1$ occurs, *given* that $E_2$ occurs. It satisfies the formula

$$\mathrm{Prob}\,[E_1|E_2] = \frac{\mathrm{Prob}\,[E_1 \cap E_2]}{\mathrm{Prob}\,[E_2]}$$

THEOREM 2.30. Let $Y_k^1$ and $Y_k^2$ be the number of edges in the graph $E'$ before the $k^{\text{th}}$ execution of the **while**-loop of algorithms 2.19 and 2.22, respectively. If $E(*)$ denotes expected value, then

1. $E[Y_k^1 - Y_{k+1}^1] \geq \frac{1}{8} \cdot Y_k^1 - \frac{1}{16}$.
2. $E[Y_k^2 - Y_{k+1}^2] \geq \frac{1}{8} \cdot Y_k^2$.

In the case where the random variables $\{\text{coin}(*)\}$ in 2.22 are only pairwise independent, we have

$$E[Y_k^2 - Y_{k+1}^2] \geq \frac{1}{16} Y_k^2$$

This result shows that the number of edges eliminated in each iteration of the **while**-loop is a fraction of the number of edges that existed.

PROOF. Let $G' = (V', E')$ be the graph before the $k^{\text{th}}$ execution of the body of the **while**-loop. The edges eliminated due to the $k^{\text{th}}$ execution of the body of the **while**-loop are the edges with at least one endpoint in the set $I' \cup \Gamma(I')$, i.e., each edge $(i, j)$ is eliminated either because $i \in I' \cup \Gamma(I')$ or because $j \in I' \cup \Gamma(I')$, due to the line $V' \leftarrow V' \setminus (I' \cup \Gamma(I'))$ in 2.18. Thus

$$E[Y_k^2 \setminus Y_{k+1}^2] \geq \frac{1}{2} \cdot \sum_{i \in V'} d(i) \cdot \text{Prob}\,[i \in I' \cup \Gamma(I')]$$

$$\geq \frac{1}{2} \cdot \sum_{i \in V'} d(i) \cdot \text{Prob}\,[i \in \Gamma(I')]$$

Here $\text{Prob}\,[i \in I' \cup \Gamma(I')]$ is the probability that a vertex $i$ is in the set $I' \cup \Gamma(I')$. We will now try to compute these probabilities. In order to do this, we need two additional results: □

LEMMA 2.31. For algorithm 2.19, and all $i \in V'$ such that $d(i) \geq 1$

$$\text{Prob}\,[i \in \Gamma(I')] \geq \frac{1}{4} \cdot \min\{s(i), 1\} \cdot \left(1 - \frac{1}{2n^2}\right)$$

PROOF. We assume that $\pi$ is a random permutation of the vertices of $V'$ — this occurs with a probability of at least $1 - 1/2n^2$. For all $j \in V'$ define $E_j$ to be the event that

$$\pi(j) < \min\{\pi(k)|k \in \Gamma(j)\}$$

(in the notation of 2.18). Set

$$p_i = \text{Prob}\,[E_i] = \frac{1}{d(i) + 1}$$

and

$$\Gamma(i) = \{1, \ldots, d(i)\}$$

Then, by the principle of inclusion-exclusion (Proposition 2.28 on page 466), for $1 \leq k \leq d(i)$,

$$\text{Prob}\left[i \in \Gamma(I')\right] \geq \text{Prob}\left[\bigcup_{j=1}^{k} E_j\right] \geq \sum_{j=1}^{k} p_j - \sum_{j=1}^{k} \sum_{\ell=j+1}^{k} \text{Prob}\left[E_j \cap E_\ell\right]$$

For fixed $j, \ell$ such that $1 \leq j < \ell \leq k$, let $E_j'$ be the event that

(119) $$\pi(j) < \min\{\pi(u) | u \in \Gamma(j) \cup \Gamma(\ell)\}$$

and let $E_\ell'$ be the event that

$$\pi(\ell) < \min\{\pi(u) | u \in \Gamma(j) \cup \Gamma(\ell)\}$$

Let

$$d(j, \ell) = |\Gamma(j) \cup \Gamma(\ell)|$$

Then,

$$\text{Prob}\left[E_j \cap E_\ell\right] \leq \text{Prob}\left[E_j'\right] \cdot \text{Prob}\left[E_\ell | E_j'\right] + \text{Prob}\left[E_\ell'\right] \cdot \text{Prob}\left[E_j | E_\ell'\right]$$

$$\leq \frac{1}{d(j, \ell) + 1} \cdot \left(\frac{1}{d(k) + 1} + \frac{1}{d(j) + 1}\right) \leq 2 \cdot p_j \cdot p_k$$

let $\alpha = \sum_{j=1}^{d(i)} p_j$. Then, by 2.27,

$$\text{Prob}\left[i \in \Gamma(I')\right] \geq \frac{1}{2} \cdot \min(\alpha, 1/2) \geq \frac{1}{4} \cdot \min\{s(i), 1\}$$

which proves the conclusion. $\square$

LEMMA 2.32. In algorithm 2.22 (page 462), for all $i \in V'$ such that $d(i) \geq 1$

$$\text{Prob}\left[i \in \Gamma(I')\right] \geq \frac{1}{4} \min\left\{\frac{s(i)}{2}, 1\right\}$$

PROOF. For all $j \in V'$ let $E_j$ be the event that $\text{coin}(j) = 1$ and

$$p_j = \text{Prob}\left[E_j\right] = \frac{1}{2d(i)}$$

Without loss of generality, assume that

$$\Gamma(i) = \{1, \ldots, d(i)\}$$

and assume that

$$p_1 \geq \cdots \geq p_{d(i)}$$

Let $E_1' = E_1$ and for $2 \leq j \leq d(i)$ let

$$E_j' = \left( \bigcap_{k=1}^{j-1} \neg E_k \right) \cap E_j$$

Note that $E_j'$ is the event that $E_j$ occurs and $\{E_1, \ldots, E_{j-1}\}$ do *not* occur. Let

(120)
$$A_j = \bigcap_{\substack{\ell \in \Gamma(j) \\ d(\ell) \geq d(j)}} \neg E_\ell$$

This is the probability that *none* of the coin($*$) variables are 1, for neighboring vertices $\ell$ with $d(\ell) \geq d(j)$. Then

$$\text{Prob}\left[i \in \Gamma(I')\right] \geq \sum_{j=1}^{d(i)} \text{Prob}\left[E_j'\right] \cdot \text{Prob}\left[A_j | E_j'\right]$$

But

$$\text{Prob}\left[A_j | E_j'\right] \geq \text{Prob}\left[A_j\right] \geq 1 - \sum_{\substack{\ell \in \Gamma(j) \\ d(\ell) \geq d(j)}} p_\ell \geq \frac{1}{2}$$

and

$$\sum_{j=1}^{d(i)} \text{Prob}\left[E_j'\right] = \text{Prob}\left[\bigcup_{j=1}^{d(i)} E_j\right]$$

(since the $E_j'$ are mutually exclusive, by construction).

For $k \neq j$, $\text{Prob}\left[E_j \cap E_k\right] = p_j \cdot p_k$ (since the events are independent). Thus, by the principle of inclusion-exclusion, for $1 \leq \ell \leq d(i)$,

$$\text{Prob}\left[\bigcup_{j=1}^{d(i)} E_j\right] = \text{Prob}\left[\bigcup_{j=1}^{\ell} E_j\right] \geq \sum_{j=1}^{\ell} p_j - \sum_{j=1}^{\ell} \sum_{k=j+1}^{\ell} p_j \cdot p_k$$

Let $\alpha = \sum_{j=1}^{d(i)} p_j$. The technical lemma, 2.27 on page 465, implies that

$$\text{Prob}\left[i \in \Gamma(I')\right] \geq \frac{1}{2} \cdot \min\left\{\sum_{j=1}^{d(i)} p_j, 1\right\}$$

It follows that $\text{Prob}\left[i \in \Gamma(I')\right] \geq \frac{1}{4} \cdot \min\{s(i)/2, 1\}$. $\square$

Now we will consider how this result must be modified when the random variables $\{\text{coin}(i)\}$ are only pairwise independent.

LEMMA 2.33. $\text{Prob}\left[i \in \Gamma(I')\right] \geq \frac{1}{8} \cdot \min\{s(i), 1\}$

PROOF. Let $\alpha_0 = 0$ and for $1 \leq \ell \leq d(i)$, let $\alpha_\ell = \sum_{j=1}^{\ell} p_j$. As in the proof of 2.32 above, we show that

$$\text{Prob}\left[i \in \Gamma(I')\right] \geq \sum_{j=1}^{d(i)} \text{Prob}\left[E_j'\right] \text{Prob}\left[A_j | E_j'\right]$$

where the $\{A_j\}$ are defined in equation (120) on page 469 and $E_j'$ is defined in equation (119) on page 468. We begin by finding a lower bound on $\text{Prob}\left[A_j | E_j'\right]$: $\text{Prob}\left[A_j | E_j'\right] = 1 - \text{Prob}\left[\neg A_j | E_j'\right]$. However

$$\text{Prob}\left[\neg A_j | E_j'\right] \leq \sum_{\substack{v \in \Gamma(j) \\ d(v) \geq d(j)}} \text{Prob}\left[E_v | E_j'\right]$$

and

$$\text{Prob}\left[E_v | E_j'\right] = \frac{\text{Prob}\left[E_v \cap \neg E_1 \cap \cdots \cap \neg E_{j-1}\right]}{\text{Prob}\left[\neg E_1 \cap \cdots \cap \neg E_{j-1} | E_j\right]}$$

The numerator is $\leq \text{Prob}\left[E_v | E_j\right] = p_v$ and the denominator is

$$1 - \text{Prob}\left[\bigcup_{\ell=1}^{j-1} E_\ell | E_j\right] \geq 1 - \sum_{\ell=1}^{j-1} \text{Prob}\left[E_\ell | E_j\right] = 1 - \alpha_{j-1}$$

Thus $\text{Prob}\left[E_v | E_j'\right] \leq p_v/(1 - \alpha_{j-1}))$. Consequently,

$$\text{Prob}\left[\neg A_j | E_j'\right] \leq \sum_{\substack{v \in \Gamma(j) \\ d(v) \geq d(j)}} \frac{p_v}{1 - \alpha_{j-1}} \leq \frac{1}{2(1 - \alpha_{j-1})}$$

and

$$\text{Prob}\left[A_j | E_j'\right] \geq 1 - \frac{1}{1 - \alpha_{j-1}} = \frac{1 - 2\alpha_{j-1}}{2(1 - \alpha_{j-1})}$$

Now we derive a lower bound on $\text{Prob}\left[E_j'\right]$:

$$\text{Prob}\left[E_j'\right] = \text{Prob}\left[E_j\right] \text{Prob}\left[\neg E_1 \cap \cdots \cap E_{j-1} | E_j\right]$$

$$= p_j \left(1 - \text{Prob}\left[\bigcup_{\ell=1}^{j-1} E_\ell | E_j\right]\right) \geq p_j(1 - \alpha_{j-1})$$

Thus, for $1 \leq \ell \leq d(v)$ and $\alpha_\ell < \frac{1}{2}$,

$$\text{Prob}\left[i \in \Gamma(I')\right] \geq \sum_{j=1}^{\ell} \frac{p_j(1 - 2\alpha_{j-1})}{2} = \frac{1}{2} \cdot \left(\sum_{j=1}^{\ell} p_j - 2 \cdot \sum_{j=1}^{\ell} \sum_{k=j+1}^{\ell} p_j \cdot p_k\right)$$

At this point 2.27 on page 465 implies

$$\text{Prob}\,[i \in \Gamma(I')] \geq \frac{1}{4} \cdot \min\{\alpha_{d(i)}, 1\}$$

□

### 3. Further reading

As mentioned on page 368, Aggarwal and Anderson found an RNC algorithm for depth-first search of general graphs. The algorithm for finding a maximal matching in § 2.3.3 on page 457 is an important subroutine for this algorithm.

There are a number of probabilistic algorithms for solving problems in linear algebra, including:

- Computing the rank of a matrix — see [18] by Borodin, von zur Gathen, and Hopcroft. This paper also gives probabilistic algorithms for greatest common divisors of elements of an algebraic number field.
- Finding various normal forms of a matrix. See the discussion of normal forms of matrices on page 197. In [79], Kaltofen, Krishnamoorthy, and Saunders present RNC algorithms for computing these normal forms.

In [128], Reif and Sen give RNC algorithms for a number of problems that arise in connection with computational geometry.

# A
# Answers to selected exercises

Chapter I, Section 1, 1. (page 13). We use induction on $k$. The algorithm clearly works for $k = 1$. Now we assume that it works for some value of $k$ and we prove it works for $k+1$. In the $2^{k+1}$-element case, the first $k$ steps of the algorithm perform the $2^k$-element version of the algorithm on the lower and upper *halves* of the set of $2^{k+1}$ elements. The $k + 1^{\text{st}}$ step, then adds the rightmost element of the cumulative sum of the lower half to *all* of the elements of the upper half.

Chapter II, Section 3, 1. (page 26). It is possible for a sorting algorithm to not be equivalent to a sorting network. This is usually true for *enumeration sorts*. These sorting algorithms typically perform a series of computations to determine the final position that each input data-item will have in the output, and then *move* the data to its final position. On a parallel computer, this final move-operation can be a single step. Neither the computations of final position, nor the mass data-movement operation can be implemented by a sorting network.

Chapter II, Section 3, 2. (page 26). The proof is based upon the following set of facts (we are assuming that the numbers are being sorted in ascending sequence and that larger numbers go to the right):

1. The rightmost 1 starts to move to the right in the first or second time unit;
2. after the rightmost 1 starts to move to the right it continues to move in each time until until it reaches position $n$ — and each move results in a zero being put into its previous position;
3. 1 and 2 imply that after the second time unit the second rightmost 1 plays the same role that the rightmost originally played — consequently it starts to move within 1 more time unit;
4. a simple induction implies that by time unit $k$ the $k - 1^{\text{st}}$ rightmost 1 has started

to move to the right and will continue to move right in each time unit (until it reaches its final position);

5. the $k - 1^{\text{st}}$ rightmost 1 has a maximum distance of $n - k + 1$ units to travel — but this is also equal to the maximum number of program steps that the 1 will move to the right by statement d above. Consequently it will be sorted into its proper position by the algorithm;

Chapter II, Section 3, 3. (page 27).

1. The answer to the first part of the question is no. In order to see this, consider a problem in which all of the alternative approaches to solving the problem have roughly the same expected running time.

2. The conditions for super-unitary speedup were stated in a rough form in the discussion that preceded the example on page 16. If we have $n$ processors available, we get super-unitary speedup in an AI-type search problem whenever the expected *minimum* running time of $n$ distinct alternatives is $< 1/n$ of the average running time of all of the alternatives.

Chapter II, Section 3, 4. (page 27).

PROOF. We make use of the 0-1 Principle. Suppose the $A$-sequence is $\underbrace{\{1,\ldots,1,0,\ldots,0\}}_{r\ 1\text{'s}}$, and the $B$-sequence $\underbrace{\{1,\ldots,1,0,\ldots,0\}}_{s\ 1\text{'s}}$. Then

1. $\{A_1, A_3, \cdots, A_{2k-1}\}$ is $\underbrace{\{1,\ldots,1,0,\ldots,0\}}_{\lceil(r+1)/2\rceil\ 1\text{'s}}$;

2. $\{A_2, A_4, \cdots, A_{2k}\}$ is $\underbrace{\{1,\ldots,1,0,\ldots,0\}}_{\lceil r/2\rceil\ 1\text{'s}}$;

3. $\{B_1, B_3, \cdots, B_{2k-1}\}$ is $\underbrace{\{1,\ldots,1,0,\ldots,0\}}_{\lceil(s+1)/2\rceil\ 1\text{'s}}$;

4. $\{B_2, B_4, \cdots, B_{2k}\}$ is $\underbrace{\{1,\ldots,1,0,\ldots,0\}}_{\lceil s/2\rceil\ 1\text{'s}}$;

Now, if we correctly merge $\{A_1, A_3, \cdots, A_{2k-1}\}$ and $\{B_1, B_3, \cdots, B_{2k-1}\}$, we get $\underbrace{\{1,\ldots,1,0,\ldots,0\}}_{\lceil(r+1)/2\rceil+\lceil(s+1)/2\rceil\ 1\text{'s}}$. Similarly, the result of merging the two even-sequences together results in $\underbrace{\{1,\ldots,1,0,\ldots,0\}}_{\lceil r/2\rceil+\lceil s/2\rceil\ 1\text{'s}}$. There are three possibilities now:

1. $r$ and $s$ are both even. Suppose $r = 2u$, $s = 2v$. Then $\lceil(r+1)/2\rceil+\lceil(s+1)/2\rceil = u + v$ and $\lceil r/2\rceil + \lceil s/2\rceil = u + v$. These are the positions of the rightmost 1's

in the merged odd and even sequences, so the result of shuffling them together will be $\{1,\ldots,1,0,\ldots,0\}$, the correct merged result.

$$\underbrace{\qquad\qquad}_{r+s\ 1's}$$

2. One of the quantities $r$ and $s$ is odd. Suppose $r = 2u - 1$, $s = 2v$. Then $\lceil (r + 1)/2 \rceil + \lceil (s + 1)/2 \rceil = u + v$ and $\lceil r/2 \rceil + \lceil s/2 \rceil = u + v - 1$, so the rightmost 1 in the merged even sequence is one position to the *left* of the rightmost 1 of the merged odd sequence. In this case we will still get the correct result when we shuffle the two merged sequences together. This is due to the fact that the individual terms of the even sequence get shuffled to the right of the terms of the odd sequence.

3. Both $r$ and $s$ are odd. Suppose $r = 2u - 1$, $s = 2v - 1$. Then $\lceil (r+1)/2 \rceil + \lceil (s + 1)/2 \rceil = u + v$ and $\lceil r/2 \rceil + \lceil s/2 \rceil = u + v - 2$. In this case the rightmost 1 in the even sequence is *two* positions to the left of the rightmost 1 of the odd sequence. After shuffling the sequences together we get $\{1,\ldots,1,0,1,0,\ldots,0\}$. In this case we must interchange a pair of adjacent elements to put the result in correct sorted order. The two elements that are interchanged are in positions $2(u + v - 1) = r + s - 1$ and $r + s$.

$\square$

**Chapter II, Section 5, 1.  (page 33).** We have to modify the second part of the CRCW write operation. Instead of merely selecting the lowest-numbered processor in a *run* of processors, we compute the sum, using the parallel algorithm on page 9. The execution-time of the resulting algorithm is asymptotically the same as that of the original simulation.

**Chapter II, Section 7, 2.  (page 51).** We need only simulate a single comparator in terms of a computation network. This is trivial if we make the vertices of the computation network compute max and min of two numbers.

**Chapter III, Section 3, 1. (page 77).** The idea here is to transmit the information to one of the low-ranked processors, and then transmit it *down* to all of the other processors. First send the data to the processor at rank 0 in column 0.

**Chapter III, Section 4, 1. (page 83).** The idea is to prove this via induction on the *dimension* of the hypercube and the *degree* of the butterfly network. The statement is clearly true when these numbers are both 1. A degree $k + 1$ butterfly network decomposes into two degree-$k$ butterflies when the vertices of the $0^{\text{th}}$ rank are deleted (see statement 3.1 on page 72). A corresponding property exists for hypercubes:

A $k + 1$-dimensional hypercube is equal to two $k$-dimensional hypercubes with corresponding vertices connected with edges.

The only difference between this and the degree $k+1$ butterfly network is that, when we restore the $0^{\text{th}}$ rank to the butterfly network (it was deleted above) corresponding columns of the two degree-$k$ sub-butterflies are connected by *two* edges — one at a 45° angle, and the other at a 135° angle, in figure III.2 on page 72).

Chapter III, Section 8, 3. (page 107). First consider the trivial case of degree 1:

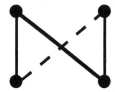

Here the lightly shaded edge represents the edge that has to be removed from the butterfly to give the CCC.

In degree 2 we can take two degree 1 CCC's with opposite edges removed and combine them together to get:

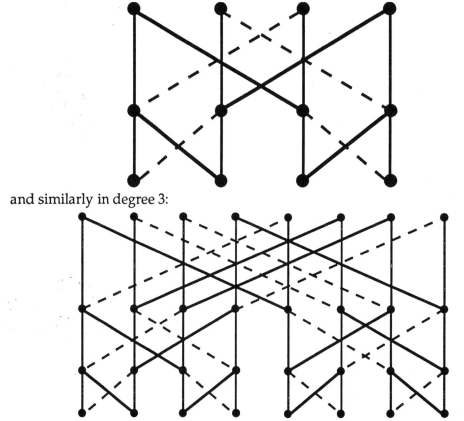

and similarly in degree 3:

Note that column 0 has the property that all ascending diagonal lines have been deleted. The fact that it must be possible to simulate the full butterfly by traversing two edges of the subgraph implies that every column whose number is of the form $2s$ must have all of its ascending diagonals intact. The fact that every node can have at most 3 incident edges implies that every column whose number is a sum of two distinct powers of 2 must have its ascending diagonals deleted. The general rule is as follows:

> The ascending diagonal communication lines of column $t$ are deleted if and only if the number of 1's in the binary representation of the number $t$ is even.

Chapter IV, Section 2, 1. (page 144). The CM-2 computer, in its simplest version, has four sequencers. These fit onto front-end computers and transmit instructions to the processors. Consequently, it is possible to subdivide a CM-2 computer into four partitions of 8192 processors each, and to have four distinct programs running on it simultaneously. (The system software allows the user to request any amount of processors he or she desires, up to the 64K limit.) Consequently, the number $K$ can be set to 4. The amount of pipelining that is performed depends upon the font-end computer, so $K'$ is essentially undefined (unless the characteristics of the front-end computer are known), but $\geq 1$. There is no pipelining of arithmetic operations[1]. The word-size parameters are hard to define, since all operations can use variable-size words, but if we program in C*, we can assume 32 bits. We get:

$$< 4 \times 1, 8912 \times 1, 32 \times 1 >$$

Chapter IV, Section 2, 3. (page 144). We must allocate a shape of processors that has one dimension for each level of nested indexing. If a variable is defined at a given level, and we must refer to it at deeper levels, we must use explicit left indices for all of the levels at which it *is* defined. In the code fragment above, we must define $A$ within a two-dimensional shape, and $t$ in a one-dimensional shape. When we refer to $t$ we must explicitly refer to its left-index, since we are using it (in the algorithm) at a deeper level of nested parallelism than that at which it was defined — we get:

```
with(twoshape)
  {
  [pcoord(0)]t=MAXINT;
  where (A==2)
    [pcoord(0)]t <?= pcoord(1);
  }
```

---

[1] This is even a complex issue — some operations are the pipelined result of several other operations

Chapter V, Section 1, 1. (page 166). This follows by straight computation — let $v = \alpha v_1 + \beta v_2$ be a linear combination of $v_1$ and $v_2$. Then

$$
\begin{aligned}
A(\alpha v_1 + \beta v_2) &= (\lambda_1 \alpha v_1 + \lambda_2 \beta v_2) \\
&= (\lambda_1 \alpha v_1 + \lambda_1 \beta v_2) \text{ (since } \lambda_1 = \lambda_2) \\
&= \lambda_1 (\alpha v_1 + \beta v_2) \\
&= \lambda_1 v
\end{aligned}
$$

So $v$ is an eigenvector of $\lambda_1$.

Chapter V, Section 1, 2. (page 166). Suppose that

(121)
$$
S = \sum_{i=1}^{\ell} \alpha_{i_j} v_{i_j} = 0
$$

is the *smallest possible* expression with *none* of the $\alpha_{i_j} = 0$ — the fact that the vectors are nonzero implies that any expression of size 1 cannot be zero (if it has a nonzero coefficient), so there is a lower bound to the possible size of such expressions. Now multiply this entire expression on the left by $A$ — the definition of an eigenvalue implies that the result is

(122)
$$
\sum_{i=1}^{\ell} \lambda_{i_j} \cdot \alpha_{i_j} v_{i_j} = 0
$$

Now we use the fact that the eigenvalues $\{\lambda_{i_j}\}$ are all *distinct*. We subtract $\lambda_{i_1} \times$ equation (121) from equation (122) to get

$$
\sum_{i=2}^{\ell} (\lambda_{i_j} - \lambda_{i_1}) \alpha_{i_j} v_{i_j} = 0
$$

— the first term has canceled out, and the remaining terms are nonzero because the eigenvalues are all distinct. This is a smaller expression that the one we started with that has nonzero coefficients. This contradicts our assumption that we had the smallest possible such expression and implies that no such expression exists.

Chapter V, Section 1, 6. (page 166). We compute the polynomial $\det(A - \lambda \cdot I) = -\lambda^3 + 14\lambda + 15$, and the roots are $\{-3, 3/2 + 1/2\sqrt{29}, 3/2 - 1/2\sqrt{29}\}$ — these are the eigenvalues. To compute the eigenvectors just solve the equations:

$$
Ax = \lambda x
$$

for each of these. The eigenvectors are well-defined up to a scalar factor. The eigenvector for $-3$ is the solution of the equation $Ax = -3x$, so $x$ is $[-2, 1, 2]$. The eigenvector for $3/2 + 1/2\sqrt{29}$ is $[1/2 + 1/6\sqrt{29}, -1/6 + 1/6\sqrt{29}, 1]$, and the eigenvector for $3/2 - 1/2\sqrt{29}$ is $[1/2 - 1/6\sqrt{29}, -1/6 - 1/6\sqrt{29}, 1]$.

The spectral radius of this matrix is $3/2 + 1/2\sqrt{29}$.

Chapter V, Section 1, 7. (page 166). We will use define the 2-norm to be $\|A\|_2 = \max_{\|v\|_2=1} \|Av\|_2$. We must maximize the expression

$$(x + 2y)^2 + (3x - y)^2 = 10\,x^2 - 2\,xy + 5\,y^2$$

subject to the condition that $x^2 + y^2 = 1$. We use the method of Lagrange Undetermined multipliers — we try to maximize the expression

$$10\,x^2 - 2\,xy + 5\,y^2 + \mu(x^2 + y^2 - 1)$$

where $x$ and $y$ are not subject to any condition. The derivative with respect to $x$ is

$$(20 + 2\,\mu)\,x - 2\,y$$

and the derivative with respect to $y$ is

$$(10 + 2\,\mu)\,y - 2\,x$$

If we set both of these to 0 we get the equations:

$$(20 + 2\mu)x = 2y$$
$$(10 + 2\mu)y = 2x$$

or

$$(10 + \mu)x = y$$
$$(5 + \mu)y = x$$

If we plug one equation into the other we get $(10 + \mu)(5 + \mu) = 50 + 15\,\mu + \mu^2 = 1$, and we get the following quadratic equation for $\mu$:

$$\mu^2 + 15\mu + 49 = 0$$

The solutions to this equation are

$$\{\frac{-15 \pm \sqrt{29}}{2}$$

Now we can solve for $x$ and $y$. The value $\mu = \frac{-15+\sqrt{29}}{2}$ gives

$$\frac{5 + \sqrt{29}}{2}x = y$$

and when we plug this into $x^2 + y^2 = 1$ we get

$$x^2 \left(1 + \left(\frac{5 + \sqrt{29}}{2}\right)^2\right) = 1$$

so

$$x = \pm \frac{\sqrt{2}}{\sqrt{29 + 5\sqrt{29}}}$$

$$y = \pm \frac{\sqrt{2}\left(27 + 5\sqrt{29}\right)}{2\sqrt{29 + 5\sqrt{29}}}$$

The value of the expression we want to maximize is:

$$10\,x^2 - 2\,xy + 5\,y^2$$
$$= \frac{145 + 23\sqrt{29}}{29 + 5\sqrt{29}}$$
$$= 4.807417597$$

Now we consider the second possibility for $\mu$: $\mu = \frac{-15-\sqrt{29}}{2}$. Here

$$x = \pm \frac{\sqrt{2}i}{\sqrt{5\sqrt{29} - 29}}$$

$$y = \pm -\frac{i\left(\sqrt{29} - 5\right)\sqrt{2}}{2\sqrt{5\sqrt{29} - 29}}$$

where $i = \sqrt{-1}$. We get

$$10\,x^2 - 2\,xy + 5\,y^2$$
$$= \frac{23\sqrt{29} - 145}{5\sqrt{29} - 29}$$
$$= 10.19258241$$

The norm of $A$ is the square root of this, or 3.192582405.

Chapter V, Section 1, 8. (page 167). This follows by direct computation:

$$(\alpha A + \beta I)\, v_i = \alpha A v_i + \beta I v_i$$
$$= \alpha \lambda_i v_i + \beta v_i$$
$$= (\alpha \lambda_i + \beta)\, v_i$$

Chapter V, Section 1, 1. (page 182). The eigenvalues of $Z(A)$ turn out to be .9543253815, $-.8369505436$, and $-.1173748380$, so the spectral radius is .9543253815 $< 1$, and all of the iterative methods converge. The optimal relaxation coefficient for the SOR method is $2/(1 + \sqrt{1 - .9543253815^2}) = 1.539919425$.

Chapter V, Section 1, 2. (page 182). If the matrix is consistently ordered, its associated graph has a *linear coloring* — 1.31 on page 177. In other words the graph associated with the matrix can be colored in such a way that the color-graph is a linear array of vertices. Now *re-color* this color-graph — simply *alternate* the first two colors. We can also re-color the graph of the matrix in a similar way. The result is a coloring with only two colors and the proof of the first statement of 1.31 on page 177 implies the conclusion.

Chapter V, Section 1, 1. (page 191). It *is* possible to determine whether the matrix $A$ was singular by examining the norm of $R(B)$.

- If $A$ was singular, then all norms of $R(B)$ must be $\geq 1$. Suppose $A$ annihilates form vector $v$. Then $R(B)$ will leave this same vector unchanged, so that for any norm $\| * \|$, $\|R(B)v\| \geq \|v\|$.
- We have proved that, if $A$ is nonsingular, then $\|R(B)\|_2 < 1$,

Chapter V, Section 1, 3. (page 191). In this case $\lg^2 n = 4$, so the constant must be approximately 3.25.

Chapter V, Section 1, 5. (page 191). The fact that the matrix is symmetric implies that $\|A\|_1 = \|A\|_\infty$. By 1.42 on page 188 $(n^{-1/2})\|A\|_1 = (n^{-1/2})\|A\|_\infty \leq \|A\|_2$, and $\|A\|_2^2 \leq \|A\|_1 \cdot \|A\|_\infty = \|A\|_\infty^2$, so we have the inequality

$$(n^{-1/2})\|A\|_\infty \leq \|A\|_2 \leq \|A\|_\infty$$

Since the matrix is symmetric, we can diagonalize it. In the symmetric case, the 2-norm of the matrix is equal to the maximum eigenvalue, $\lambda_0$, so $(n^{-1/2})\|A\|_\infty \leq \lambda_0 \leq \|A\|_\infty$. It follows that

$$\frac{n^{1/2}}{\|A\|_\infty} \geq \frac{1}{\lambda_0} \geq \frac{1}{\|A\|_\infty}$$

Now suppose that $U^{-1}AU = D$, where $D$ is a diagonal matrix, and $U$ is unitary (so it preserves 2-norms)

$$D = \begin{pmatrix} \lambda_0 & 0 & \cdots & 0 \\ 0 & \lambda_1 & \cdots & 0 \\ \vdots & \ddots & \ddots & \vdots \\ 0 & \cdots\cdots & & \lambda_n \end{pmatrix}$$

where $\lambda_0 > \lambda_1 > \cdots > \lambda_n > 0$. Then $U^{-1}(I - BA)U = I - BD$, and $I - BD$ is

$$D = \begin{pmatrix} 1 - \frac{\lambda_0}{\|A\|_\infty} & 0 & \cdots & 0 \\ 0 & \lambda_1 & \cdots & 0 \\ \vdots & & \ddots & \vdots \\ 0 & & \cdots\cdots & 1 - \frac{\lambda_n}{\|A\|_\infty} \end{pmatrix}$$

The inequalities above imply that

$$1 \geq \frac{\lambda_0}{\|A\|_\infty} \geq \frac{1}{n^{1/2}}$$

so

$$0 \leq 1 - \frac{\lambda_0}{\|A\|_\infty} \leq 1 - \frac{1}{n^{1/2}}$$

Now the smallest eigenvalue is $\lambda_n$ so

$$\frac{\lambda_n}{\lambda_0} \geq \frac{\lambda_n}{\|A\|_\infty} \geq \frac{\lambda_n}{\lambda_0}\frac{1}{n^{1/2}}$$

and

$$1 - \frac{\lambda_n}{\lambda_0} \leq 1 - \frac{\lambda_0}{\|A\|_\infty} \leq 1 - \frac{\lambda_n}{\lambda_0}\frac{1}{n^{1/2}}$$

The maximum of the eigenvalues of $I - BD$ (which is also the 2-norm of $I - BD$ and of $I - BA$) is thus $1 - \frac{\lambda_n}{\lambda_0 n^{1/2}}$. It turns out that $\lambda_n/\lambda_0 = 1/\operatorname{cond}(A)$.

Chapter V, Section 2, 3. (page 217). Define the sequence $\{B_0, \ldots, B_{n-1}\}$ to be $\{f(0), f(2\pi/n), \ldots, f(2(n-1)\pi/n)\}$. Perform the Discrete Fourier Transform on this sequence, using $e^{2\pi i/n}$ as the primitive $n^{\text{th}}$ root of 1. The result will be the first $n$ coefficients of the Fourier series for $f(x)$.

Chapter V, Section 2, 4. (page 221). The determinant of a matrix is equal to the product of its eigenvalues (for instance look at the definition of the characteristic polynomial of a matrix in 1.10 on page 160 and set $\lambda = 0$). We get

$$\det(A) = \prod_{i=0}^{n-1} \mathcal{F}_\omega(f)(i)$$

in the notation of 2.10 on page 219.

**Chapter V, Section 2, 5. (page 221).** The spectral radius is equal to the absolute value of the eigenvalue with the largest absolute value — for the $\mathcal{Z}(n)$ this is 2 for all values of $n$.

**Chapter V, Section 3, 1. (page 229).** Here is a Maple program for computing these functions:

```
c0 := 1/4+1/4*3^(1/2);
c1 := 3/4+1/4*3^(1/2);
c2 := 3/4-1/4*3^(1/2);
c3 := 1/4-1/4*3^(1/2);
p1 := 1/2+1/2*3^(1/2);
p2 := 1/2-1/2*3^(1/2);
rphi := proc (x) option remember;
    if x <= 0 then RETURN(0) fi;
    if 3 <= x then RETURN(0) fi;
    if x = 1 then RETURN(p1) fi;
    if x = 2 then RETURN(p2) fi;
      simplify(expand(c0*rphi(2*x)+c1*rphi(2*x-1)+
              c2*rphi(2*x-2)+c3*rphi(2*x-3)))
    end;
w2 := proc (x)
        simplify(expand(c3*rphi(2*x+2)-c2*rphi(2*x+1)+
              c1*rphi(2*x)-c0*rphi(2*x-1)))
    end;
```

Although Maple is rather slow, it has the advantage that it performs *exact* calculations, so there is no roundoff error.

**Chapter V, Section 3, 2. (page 229).** For a degree of precision equal to $1/P$, a parallel algorithm would require $O(\lg P)$-time, using $O(P)$ processors.

**Chapter V, Section 3, 2. (page 244).** We could assume that all numbers are being divided by a large power of 2 — say $2^{30}$. We only work with the numerators of these fractions, which we can declare as integers. In addition, we can compute with numbers like $\sqrt{3}$ by working in the extension of the rational numbers $\mathbb{Q}[\sqrt{3}]$ by defining an element of this extension to be a pair of integers: $(x, y) = x + y\sqrt{3}$. We define addition and multiplication via:

1. $(x_1, y_1) + (x_2, y_2) = (x_1 + x_2, y_1 + y_2)$;

2. $(x_1, y_1) * (x_2, y_2) = (x_1 + y_1\sqrt{3})(x_2 + y_2\sqrt{3}) = (x_1x_2 + 3y_1y_2 + x_1y_2\sqrt{3} + x_2y_1\sqrt{3}) = (x_1x_2 + 3y_1y_2, x_1y_2 + x_2y_1)$.

We could even define division via[2]:

$$(x_1, y_1)/(x_2, y_2) = \frac{x_1 + y_1\sqrt{3}}{x_2 + y_2\sqrt{3}}$$

$$= \frac{x_1 + y_1\sqrt{3}}{x_2 + y_2\sqrt{3}} \cdot \frac{x_1 - y_1\sqrt{3}}{x_1 - y_1\sqrt{3}}$$

$$= \frac{x_1x_2 - 3y_1y_2 + x_1y_2\sqrt{3} - x_2y_1\sqrt{3}}{x_2^2 - 3y_2^2}$$

The denominator is never 0 because $\sqrt{3}$ is irrational.

$$= \left((x_1x_2 - 3y_1y_2)/(x_2^2 - 3y_2^2), (x_1y_2 - x_2y_1)/(x_2^2 - 3y_2^2)\right)$$

Chapter V, Section 5, 3. (page 274). No, it is not EREW. It is not hard to make it into a calibrated algorithm, however. All of the data-items are used in each phase of the execution, so at program-step $i$ each processor expects its input data-items to have been written in program-step $i - 1$. It would, consequently, be *fairly straightforward* to write a MIMD algorithm that carries out the same computations as this. This is essentially true of all of the algorithms for solving differential equations presented here.

Chapter V, Section 5, 4. (page 274). The answer to the first part is no. In order to put it into a self-adjoint form, we re-write it in terms of a new unknown function $u(x, y)$, where $\psi(x, y) = u(x, y)/x$. We get:

$$\frac{\partial \psi}{\partial x} = -\frac{u}{x^2} + \frac{1}{x}\frac{\partial u}{\partial x}$$

$$\frac{\partial^2 \psi}{\partial x^2} = +\frac{2u}{x^3} - \frac{2}{x^2}\frac{\partial u}{\partial x} + \frac{\partial^2 u}{\partial x^2}$$

$$\frac{\partial^2 u}{\partial y^2} = \frac{1}{x}\frac{\partial^2 u}{\partial y^2}$$

so our equation becomes:

$$\frac{2u}{x^3} - \frac{2}{x^2}\frac{\partial u}{\partial x} + \frac{\partial^2 u}{\partial x^2} + \frac{2}{x}\left(-\frac{u}{x^2} + \frac{1}{x}\frac{\partial u}{\partial x}\right) + \frac{1}{x}\frac{\partial^2 u}{\partial y^2} = \frac{1}{x}\frac{\partial^2 u}{\partial x^2} + \frac{1}{x}\frac{\partial^2 u}{\partial y^2}$$

---

[2]We assume that $(x_2, y_2) \neq (0, 0)$.

Since the original equation was set to 0, we can simply clear out the factor of $1/x$, to get the ordinary Laplace equation for $u(x, y)$.

Chapter V, Section 5, 5. (page 274). This equation *is* essentially self-adjoint. It is not hard to find an integrating factor: We get

$$
\log \Phi = \int \left\{ \frac{2}{x+y} \right\} \, dx + C(y)
$$
$$
= 2 \log(x+y) + C(y)
$$

so

$$
\Phi(x, y) = c'(y)(x+y)^2
$$

Substituting this into equation (67) on page 273, we get that $c'(y)$ is a *constant* so that we have already completely solved for $\Phi(x, y)$. The the self-adjoint form of the equation it:

$$
\frac{\partial}{\partial x} \left( (x+y)^2 \frac{\partial \psi}{\partial x} \right) + \frac{\partial}{\partial y} \left( (x+y)^2 \frac{\partial \psi}{\partial y} \right) = 0
$$

Chapter V, Section 5, 1. (page 289). Recall the Schrödinger Wave equation

$$
-\frac{\hbar^2}{2m} \nabla^2 \psi + V(x, y)\psi = i\hbar \frac{\partial \psi}{\partial t}
$$

and suppose that

$$
\left| \frac{\partial \psi}{\partial t} - \frac{\psi(t+\delta t) - \psi(t)}{\delta t} \right| \le E_1
$$
$$
\left| \frac{2n(\psi - \psi_{\text{average}})}{\delta^2} - \nabla^2 \psi \right| \le E_2
$$
$$
|V| \le W
$$

Our numeric form of this partial differential equation is

$$
-\frac{\hbar^2}{2m} \frac{4(\psi_{\text{average}} - \psi)}{\delta^2} + V(x, y)\psi = i\hbar \frac{\psi(t+\delta t) - \psi}{\delta t}
$$

We re-write these equations in the form

$$-\frac{\hbar^2}{2m}\nabla^2\psi + V(x,y)\psi - i\hbar\frac{\partial\psi}{\partial t} = 0$$

$$-\frac{\hbar^2}{2m}\frac{4(\psi_{\text{average}} - \psi)}{\delta^2} + V(x,y)\psi - i\hbar\frac{\psi(t+\delta t) - \psi}{\delta t} = Z$$

where $Z$ is a measure of the error produced by replacing the original equation by the numeric approximation. We will compute $Z$ in terms of the error-estimates $E_1$, $E_2$, and $W$. If we form the difference of these equations, we get:

$$Z = \left(-\frac{\hbar^2}{2m}\frac{2n(\psi_{\text{average}} - \psi)}{\delta^2} + V(x,y)\psi - i\hbar\frac{\psi(t+\delta t) - \psi}{\delta t}\right)$$
$$- \left(-\frac{\hbar^2}{2m}\nabla^2\psi + V(x,y)\psi - i\hbar\frac{\partial\psi}{\partial t}\right)$$

So we get

$$Z = \frac{\hbar^2}{2m}\left(\nabla^2\psi - \frac{4(\psi_{\text{average}} - \psi)}{\delta^2}\right) + i\hbar\left(\frac{\partial\psi}{\partial t} - \frac{\psi(t+\delta t) - \psi}{\delta t}\right)$$

so

$$|Z| \leq \left|\frac{\hbar^2}{2m}\right|E_1 + |\hbar|E_2$$

Now we compute $\psi(t + \delta t)$ in terms of other quantities:

$$\psi(t+\delta t) = \left(1 - \frac{2i\hbar\delta t}{m\delta^2} - \frac{i\delta t V(x,y)}{\hbar}\right)\psi + \frac{3i\hbar\delta t}{m\delta^2}\psi_{\text{average}} + \frac{i\delta t Z}{\hbar}$$

so the error in a single iteration of the numerical algorithm is

$$\leq \frac{\delta t}{\hbar}\left|\frac{\hbar^2}{2m}\right|E_1 + \frac{\delta t}{\hbar}|\hbar|E_2 = \frac{\delta t\hbar}{2m}E_1 + \delta t E_2$$

This implies that

(123)    $$\psi(t+\delta t) - \psi = -\left(\frac{2i\hbar\delta t}{m\delta^2} + \frac{i\delta t V(x,y)}{\hbar}\right)\psi + \frac{2i\hbar\delta t}{m\delta^2}\psi_{\text{average}} + \frac{i\delta t Z}{\hbar}$$

Now we estimate the long-term behavior of this error. The error is multiplied by the matrix in equation (123). We can estimate the 2-norm of this matrix to be

$$\left|\frac{2i\hbar\delta t}{m\delta^2} + \frac{i\delta t V(x,y)}{\hbar}\right| \leq \frac{4\hbar\delta t}{m\delta^2} + \frac{2\delta t W}{\hbar}$$

Our conclusion is that the errors dominate the solution quickly unless

$$\frac{4\hbar\delta t}{m\delta^2} + \frac{2\delta tW}{\hbar} < 1$$

In general, this means that the iterative algorithm will diverge unless

$$\delta t < \frac{1}{\frac{4\hbar}{m\delta^2} + \frac{2W}{\hbar}}$$

Chapter VI, Section 1, 1. (page 310). One way to handle this situation is to note that the recurrence is *homogeneous* — which means that it has no *constant terms*. This implies that we can divide right side of the recurrence by a number $\beta$ to get

$$S_k' = -\frac{a_1}{\beta}S_{k-1}' - \frac{a_2}{\beta}S_{k-2}' - \cdots - \frac{a_n}{\beta}S_{k-n}'$$

and this recurrence will have a solution $\{S_i'\}$ that is asymptotic to $\alpha_1^i/\beta^i$. Consequently, if we can find a value of $\beta$ that makes the limit of the second recurrence equal to 1, we will have found the value of $\alpha_1$. We can:

- carry out computations with the original recurrence until the size of the numbers involved becomes too large;
- use the ratio, $\beta$ computed above, as an estimate for $\alpha_1$ and divide the original recurrence by $\beta$. The resulting recurrence will have values $\{S_i'\}$ that approach 1 as $i \to \infty$, so that it will be much better-behaved numerically.

Chapter VI, Section 1, 2. (page 312). Just carry out the original algorithm $\lg n$ times.

Chapter VI, Section 1, 3. (page 313). Here we use the Brent Scheduling Algorithm on page 307. We subdivide the $n$ characters in the string into $n/\lg n$ substrings of length $\lg n$ each. Now one processor to each of these substrings and process them *sequentially* using the algorithm 1.10 of page 312. This requires $O(\lg n)$ time. This reduces the amount of data to be processed to $n/\lg n$ items. This sequence of items can be processed in $O(\lg n)$ time via the parallel version of algorithm 1.10.

Chapter VI, Section 2, 3. (page 320). Assign 1 to each element of the Euler Tour that is on the end of a *downward* directed edge (i.e., one coming from a parent vertex) and assign $-1$ to the end of each *upward edge*.

Chapter VI, Section 2, 1. (page 325). If some edge-weights are negative, it is possible for a *cycle* to have negative total weight. In this case the entire *concept* of shortest-path becomes meaningless — we can make any path shorter by running around a negative-weight cycle sufficiently many times. The algorithm for shortest paths given in this section would never terminate. The proof proof that the algorithm terminates within $n$ ("exotic") matrix-multiplications makes use of the fact that a shortest path has at most $n$ vertices in it — when negative-weight cycles exist, this is no longer true.

Chapter VI, Section 2, 2. (page 325). It is only necessary to ensure that no negative-weight cycles exist. It suffices to show that none of the cycles in a *cycle-basis* (see § 2.7) have negative total weight. Algorithm 2.24 on page 366 is a parallel algorithm for computing a cycle-basis. After this has been computed, it is straightforward to compute the total weight of each cycle in the cycle-basis and to verify this condition.

Chapter VI, Section 2, 3. (page 367). If we cut the graph out of the plane, we get a bunch of small polygons, and the entire plane with a polygon removed from it. This last face represents the *boundary* of the embedded image of the graph in the plane. It is clearly equal to the sum of all of the other face-cycles. Given any simple cycle, $c$, in the graph, we can draw that cycle in the embedded image of the plane. In this drawing, the image of $c$ encloses a number of face of the graph. It turns out that $c$ is equal to the sum of the cycles represented by these faces.

Chapter VI, Section 2, 7. (page 369). The problem with this procedure is that the undirected spanning trees constructed in step 1 might not satisfy the requirement that each vertex has *at least two* children. In fact the tree might simply be one long path. In this case the algorithm for directing the tree no longer runs in $O(\lg n)$-time — it might run in linear time.

Chapter VI, Section 3, 1. (page 382). Not quite, although it is tantalizing to consider the connection between the two problems. The circuit described on page 44 *almost* defines a parse tree of a boolean expression. Unfortunately, it really only defines an *acyclic* directed graph. This is due to the fact that the expressions computed in sequence occurring in the Monotone Circuit Problem can be used in *more than one* succeeding expression. Consider the Monotone Circuit:

$$\{t_0 = T, t_1 = F, t_2 = t_0 \vee t_1, t_3 = t_0 \wedge t_2, t_4 = T, t_5 = t_0 \vee t_3 \wedge t_4\}$$

This defines the acyclic graph depicted in figure A.1.

It is interesting to consider how one could convert the acyclic graph occurring in the Circuit Value Problem into a tree — one could duplicate repeated expressions. Unfortunately, the resulting parse tree may contain (in the worst case) an *exponential number* of leaves — so the algorithm described in § 3 of chapter VI doesn't necessarily help. As before, the question of the parallelizability of the Circuit Value Problem

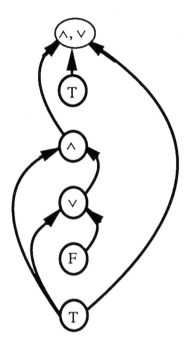

FIGURE A.1. Acyclic graph derived from a monotone circuit

remains open — see § 6.2 and page 44.

Chapter VI, Section 5, 1. (page 439). Fermat's little theorem says that $a^{p-1} \equiv 1$ (mod $p$), if $a \not\equiv 0$ (mod $p$). This implies that $a^{p-2} \equiv a^{-1}$ (mod $p$). It is possible to calculate $a^{p-2}$ in $O(\lg(p-2))$-time with one processor, by repeated squaring.

Chapter VI, Section 5, 3. (page 440). Here is a Maple function that does this:

```
np := proc(p)
    local i,tp,diff,prim,pr;
        tp := 2^p;
        for i from 2 to 2*p do
            diff := nextprime(i*tp)−i*tp;
            if diff = 1 then
              prim:=i*tp+1;
              pr:=primroot(1,prim)&^i mod prim;
              print(i,tp,prim,pr, pr&^(−1) mod prim,
                  tp&^(−1) mod prim) fi
        od
    end;
```

This function prints out most of the entries that appeared in table VI.1 for *several* different primes. It is called by typing np(i) in Maple, where i exponent of 2 we want to use in the computations (for instance, it represents the leftmost column of table VI.1).

Chapter VII, Section 2, 1. (page 447). No. lv-RNC=RNC, because a Las Vegas algorithm produces a correct answer in a fixed expected number of executions — see 1.7 on page 445.

Chapter VII, Section 2, 1. (page 450). Infinite. If we are extraordinarily unlucky there is no reason for *any* list-elements to *ever* be deleted.

# B
# Index of notation

1. $\Gamma(S)$ — if $S$ is a set of vertices in a graph this denotes the set of *neighbors* of vertices in $S$. First used on page 403
2. $\ll$ — $A \ll B$ means that $A$ is sufficiently smaller than $B$. First used on page 405

3. $P_1 \propto P_2$ — this notation means that $P_1$ is *reducible* to $P_2$. See page 43. First used on page 43
4. $P_1 \propto_{\text{logspace}} P_2$ — this notation means that $P_1$ is *logspace reducible* to $P_2$. See page 43. First used on page 43
5. $\mathcal{F}_\omega(A)$ — the Discrete Fourier Transform of a sequence of numbers $A = \{a_0, \ldots, a_{n-1}\}$, formed with respect to a primitive $n^{\text{th}}$ root of 1, $\omega$. $\mathcal{F}_\omega(A) = \{b_0, \ldots, b_{n-1}\}$, where

$$b_i = \sum_{j=0}^{n-1} a_j \omega^{ij}$$

First used on page 202
6. $\leftarrow$ — this is the assignment operator in pseudocode. First used on page 31
7. $\lfloor a \rfloor$ — the greatest integer $< a$. First used on page 206
8. $\text{Prob}\,[E]$ — denotes the probability that event $E$ occurs. First used on page 441
9. $\setminus$ — If $A$ and $B$ are sets, then $A \setminus B$ represents the *set-difference*. First used on page 34
10. $\star$ — a *generic* associative operation used in the statement and proof of 1.4 First used on page 306
11. $\subseteq$ — If $A$ and $B$ are sets, then $A \subseteq B$ if $A$ is a subset, or equal to $B$. First used on page 11
12. $A^{\text{H}}$ — If $A$ is a matrix, this is the *Hermitian transpose* of $A$. $A^{\text{H}}_{i,j} = \bar{A}_{j,i}$, where $\bar{A}$ denotes the complex conjugate. First used on page 161
13. $\approx$ — "approximately equal to". First used on page 17

14. $\lceil a \rceil$ is the smallest integer $> a$. First used on page 474
15. $\vee$ — the boolean **OR** operation. First used on page 382
16. $\wedge$ — the boolean **AND** operation. First used on page 382
17. $\| * \|$ — notation for the *norm*, where $*$ is a vector or a matrix. First used on page 160
18. $\|v\|_1 = \sum_i |v_i|$. If $v$ is a vector, this is the 1-norm. First used on page 160
19. $\|v\|_2 = \sqrt{\sum_i |v_i|^2}$. If $v$ is a vector, this is the 2-norm. First used on page 160
20. $\|v\|_\infty = \max_i |v_i|$. If $v$ is a vector, this is the $\infty$-norm. First used on page 160
21. $a \equiv b \pmod{c}$ – Here $a$, $b$, and $c$ are integers. This means that $a - b$ is an integer-multiple of $c$. This is equivalent to saying that $a \bmod c = b \bmod c$, where we are using mod in the sense of the Pascal **mod** function. First used on page 84
22. $n! = n(n-1)(n-1) \cdots 2 \cdot 1$. First used on page 106
23. $\binom{m}{q} = \frac{m!}{q!(m-q)!}$. First used on page 106
24. $\nabla^2 \psi = \frac{\partial^2 \psi}{\partial x_1^2} + \frac{\partial^2 \psi}{\partial x_2^2} + \cdots + \frac{\partial^2 \psi}{\partial x_n^2}$ — the Laplacian operator. First used on page 257
25. $\partial/\partial x$ — partial derivative with respect to $x$. First used on page 256

26. $G(A, B, E)$ — this is the notation for a bipartite graph with two vertex-sets $A$ and $B$, and edge-set $E$. First used on page 104

27. $\operatorname{cond} A = \|A\|_2 \cdot \|A^{-1}\|_2 \geq \|I\| = 1$ — This is the *condition number* of the matrix $A$. First used on page 161

28. $\det(A)$ denotes the determinant of the square matrix $A$. First used on page 159

29. $(V, E)$ — notation for a graph with vertex-set $V$ and edge-set $E$. First used on page 99

30. $\lg(x) = \log_2(x)$. First used on page 6
31. $\ln(x) = \log_e(x)$, where $e = 2.7182818\ldots$. First used on page 404

32. $a \bmod b$ — Here $a$ and $b$ are integers. This is the *remainder* that results when $b$ is divided by $a$. We assume that $0 \geq a \bmod b < b$ — if necessary, we add or subtract sufficiently many copies of $b$ to put the remainder into this range. This is roughly analogous to the Pascal **mod** function function. First used on page 84

33. **NC** — "Nick's class". This is the class of problems that are solvable on a parallel computer in time that is $O(\lg^k n)$, for a suitable value of $k$ (where $n$ is proportional to the size of the input) using $O(n^\ell)$ processors, for a suitable value of $\ell$. See § 6.1 in chapter II (page 33). First used on page 11

34. $O(*)$ — If $g(x)$ is some function then $g(x) = O(f(x))$ if there exists constants $a > 0$, $\bar{x}$ such that $g(x) \le af(x)$ for all $x > \bar{x}$. First used on page 15
35. $\Omega(*)$ — If $g(x)$ is a function, then $g(x) = \Omega(f(x))$ if there exist constants $a > 0$, $b > 0$, and $\bar{x}$ such that, $af(x) < g(x) < bf(x)$ for all values of $x > \bar{x}$. First used on page 6

36. **P** — the class of problems solvable in polynomial time. See § 6.1 in chapter II (page 33). First used on page 11
37. $\phi(*)$ — Euler *phi*-function. If $n$ is an integer, $\phi(n)$ is the number of values $k$ such that $0 < k < n$ and $k$ is relatively prime to $n$. First used on page 425
38. $\pi = 3.1415926\ldots$ First used on page 441

39. $\rho(*)$ — If $A$ is a matrix $\rho(A)$ is the *spectral radius* of $A$. First used on page 161

40. $t(g)$ — Tutte matrix of the graph $G$. First used on page 457
41. $\operatorname{tr}(A) = \sum_{i=1}^{n} A_{ii}$, where $A$ is a square matrix. This is called the *trace* of $A$. First used on page 194

# Bibliography

[1] Milton Abramowitz and Irene Stegun, *Handbook of mathematical functions*, Dover Publications, New York, NY, 1965.

[2] Alfred Aho, Ravi Sethi, and Jeffrey Ullman, *Compilers: Principles, Techniques, and Tools*, Addison-Wesley Publishing Company, 1986.

[3] M. Ajtai, J. Komlos, and E. Szemeredi, *An $O(n \log n)$-sorting network*, Symposium on the Theory of Computing, vol. 15, ACM, 1983, pp. 1–9.

[4] _____, *Sorting in $c \log n$ parallel steps*, Combinatorica **3** (1983), no. 1, 1–19.

[5] D. A. Alton and D. M. Eckstein, *Parallel graph-processing using depth-first search*, Conference on Theoretical Computer Science, University of Waterloo, 1977, pp. 21–29.

[6] W. F. Ames, *Numerical methods for partial differential equations*, Academic Press, New York, 1977.

[7] Richard J. Anderson and Gary L. Miller, *Deterministic parallel list ranking*, (John Reif, ed.), 1988 Aegean Workshop on Computing, Lecture Notes in Computer Science, vol. 319, Springer-Verlag, 1988, pp. 81–90.

[8] _____, *A simple randomized parallel algorithm for list-ranking*, unpublished, 1988.

[9] G. Arens, I. Fourgeau, D. Girard, and J. Morlet, *Wave propagation and sampling theory*, Geophysics **47** (1982), 203–236.

[10] François Baccelli and Thierry Fleury, *On parsing arithmetic expressions in a multiprocessor environment*, Acta Informatica **17** (1982), 287–310.

[11] François Baccelli and Philippe Mussi, *An asynchronous parallel interpreter for arithmetic expressions and its evaluation*, IEEE Trans. Comput. **c-35** (1986), no. 3, 245–256.

[12] Ilan Bar-On and Uzi Vishkin, *Optimal parallel generation of a computation tree form*, Transactions on Programming Languages and Systems **7** (1985), 348–357.

[13] Michael J. Beckerle, James E. Hicks, Gregory M. Papadopoulos, and Kenneth R. Traub, *Overview of the Monsoon project*, Proceedings of the International Conference on Computer Design (Cambridge, MA), IEEE, 1991.

[14] L. Bers, *Mathematical aspects of subsonic and transonic gas dynamics*, John Wiley, 1958.

[15] Dimitri P. Bertsekas and John N. Tsitsiklis, *Parallel and distributed computation*, Prentice-Hall, Englewood Cliffs, NJ, 1989.

[16] Sandeep N. Bhatt, Fan R. K. Chung, Jia-Wei Hong, F. Thomson Leighton, and Arnold L. Rosenberg, *Optimal simulations by butterfly networks*, Symposium on the Theory of Computing, ACM, 1988, pp. 192–204.

[17] A. Bijaoui, G. Mars, and E. Slezak, *Identification of structures from Galaxy counts — use the wavelet transform*, Astronomy and Astrophysics **227** (1990), no. 2, 301–316.

[18] Allan Borodin, Joachim von zur Gathen, and John Hopcroft, *Fast parallel matrix and GCD computations*, Inform. and Control **52** (1982), 241–256.

[19] O. Borůvka, *O jistém problému minimálním*, Práca Moravské Přírodovědecké Společnosti **3** (1926), 37–58.

[20] P. R. Brent, *The parallel evaluation of general arithmetic expressions*, Journal of the ACM **22** (1974), 201–206.

[21] R. Brent, D. J. Kuck, and K. M. Maruyama, *The parallel evaluation of arithmetic expressions without divisions*, IEEE Trans. Comput. **c-22** (1973), 532–534.

[22] Georges Louis Leclerc Buffon, *Résolution des problèms qui regardent le jeu du franc carreau*, 1733.

[23] A. Chandra, *Maximum parallelism in matrix-multiplication*, Rc 6193, IBM, 1975.

[24] Jik H. Chang, Oscar Ibarra, and Michael A. Palis, *Parallel parsing on a one-way array of finite-state machines*, IEEE Trans. Comm. **c-36** (1987), no. 1, 64–75.

[25] N. S. Chang and K. S. Fu, A Study on Parallel Parsing of Tree Languages and its Application to Syntactic Pattern Recognition, 107–129, Plenum Press, New York, NY, 1981, pp. 107–129.

[26] F. Y. Chin, J. Lam, and I-Ngo Chen, *Optimal parallel algorithms for the connected components problem*, Proceedings International Conference on Parallel Processing, IEEE, 1981, pp. 170–175.

[27] _____, *Efficient parallel algorithms for some graph problems*, Comm. ACM **25** (1982), 659–665.

[28] Richard Cole, *Parallel merge sort*, SIAM Journal on Computing **17** (1988), 770–785.

[29] Stephen A. Cook, *An overview of computational complexity*, Comm. ACM **26** (1983), no. 6, 400–409.

[30] _____, *A taxonomy of problems with fast parallel algorithms*, Inform. and Control **64** (1985), 2–22.

[31] Stephen A. Cook and P. McKenzie, *Problems complete for deterministic logarithmic space*, J. Algorithms **8** (1987), no. 3, 385–394.

[32] J. W. Cooley and J. W. Tuckey, *An algorithm for the machine calculation of complex Fourier series*, Math. Comp. **19** (1965), no. 90, 297–301.

[33] Don Coppersmith and Schmuel Winograd, *Matrix multiplication via arithmetic progressions*, Symposium on the Theory of Computing (New York City), vol. 19, ACM, 1987, pp. 1–6.

[34] Thomas H. Corman, Charles E. Leiserson, and Ronald L. Rivest, *Introduction to algorithms*, MIT Press - McGraw-Hill, 1991.

[35] D. G. Corneil and E. Reghbati, *Parallel computations in graph theory*, SIAM J. Comput. **7** (1978), 230–237.

[36] R. Courant and K. O. Friedrichs, *Supersonic flow and shock waves*, Interscience, New York, 1948.

[37] R. Courant and D. Hilbert, *Methods of mathematical physics*, Interscience, New York, 1962.

[38] L. Csanky, *Fast parallel matrix inversion algorithm*, SIAM J. Comput. **5** (1976), 618–623.

[39] N. G. Danielson and C. Lanczos, *Some improvements in practical Fourier analysis and their application to X-ray scattering from liquids*, Journal of the Franklin Institute **233** (1942), 365–380, 435–452.

[40] Karabi Datta, *Parallel complexities and computations of Cholesky's decomposition and QR-factorization*, Internat. J. Comput. Math. **18** (1985), 67–82.

[41] Ingrid Daubechies, *Orthonormal bases of compactly supported wavelets*, Comm. Pure Appl. Math. **41** (1988), 909–996.

[42] P. J. Davis and P. Rabinowitz, *Numerical integration*, Blaisdell, 1967.

[43] Eliezer Dekel and Sartaj Sahni, *Parallel generation of the postfix form*, Proceedings of the 1982 International Symposium on Parallel Processing, 1982, pp. 171–177.

[44] Denenberg, *Computational complexity of logical problems*, Ph.D. thesis, Harvard University, 1984.

[45] Z. Deng, *An optimal parallel algorithm for linear programming in the plane*, Inform. Process. Lett. **35** (1990), 213–217.

[46] E. W. Dijkstra, *Hierarchical ordering of sequential processes*, Acta Inform. **1** (1971), no. 2, 115–138.

[47] J. Edmonds, *Paths, trees and flowers*, Canad. J. Math. **17** (1965), 449–467.

[48] V. Faber, O. M. Lubeck, and A. B. White Jr., *Superlinear speedup of an efficient sequential algorithm is not possible*, Parallel Comput. **3** (1986), 259–260.

[49] D. K. Faddeev and V. N. Faddeeva, *Computational methods of linear algebra*, W. H. Freeman, San Francisco, CA, 1963.

[50] M. J. Flynn, *Very high-speed computing systems*, Proceedings of the IEEE **54** (1966), no. 12, 1901–1909.

[51] S. Fortune and J. Wyllie, *Parallelism in random access machines*, ACM Symposium on the Theory of Computing, vol. 10, 1978, pp. 114–118.

[52] Zvi Galil and Wolfgang Paul, *An efficient general-purpose parallel computer*, Journal of the ACM **30** (1983), 360–387.

[53] K. Gallivan, W. Jalby, S. Turner, A. Veidenbaum, and H. Wijshoff, *Preliminary performance analysis of the cedar multiprocessor memory system*, Proceedings of the 1991 International Conference on Parallel Processing (Boca Raton, FL), vol. I, Architecture, CRC Press, 1991, pp. I–71–I–75.

[54] P. R. Garabedian, *Estimation of the relaxation factor for small mesh size*, Math. Tables Aids Comput. **10** (1956), 183–185.

[55] Michael Garey and David Johnson, *Computers and intractability*, W. H. Freeman and Company, 1979.

[56] Alan Gibbons and Wojciech Rytter, *Optimal parallel algorithms for dynamic expression evaluation and context-free recognition*, Information and Computation, vol. 81, 1989, pp. 32–45.

[57] L. Goldschlager, *The monotone and planar circuit value problems are log space complete for P*, SIGACT News **9** (1977), no. 2, 25–29.

[58] L. Goldschlager, L. Shaw, and J. Staples, *The maximum flow problem is logspace-complete for P*, Theoret. Comput. Sci. **21** (1982), 105–115.

[59] Raymond Greenlaw, H. James Hoover, and Walter L. Ruzzo, *A compendium of problems complete for P*, (file name tr91-11.dvi.Z), aval. via anonymous ftp thorhild.cs.ualberta.ca, 1991.

[60] P. Gropillaud, A. Grossman, and J. Morlet, *Cycle-octave and related transforms in seismic signal analysis*, Geoexploration **23** (1984), 85–102.

[61] Torben Hagerup, *Planar depth-first search in $O(\log n)$ parallel time*, SIAM J. Comput. **19** (1990), no. 4, 678–704.

[62] W. Händler, *The impact of classification schemes on computer architecture*, Proceedings of the 1977 International Conference on Parallel Processing (New York), IEEE, 1977, pp. 7–15.

[63] Frank Harary, *Graph theory*, Addison-Wesely Publishing Company, 1972.

[64] Phil Hatcher, Walter F. Tichy, and Michael Philippsen, *A critique of the programming language C\**, Interner bericht nr. 17/91, Universität Karlsruhe Fakultät für Informatik, Postfach 6980, D-7500 Karlsruhe 1 Germany, 1991.

[65] D. Heller, *On the efficient computation of recurrence relations*, Department of computer science report, Carnegie-Mellon University, Pittsburgh, PA, 1974.

[66] ———, *A survey of parallel algorithms in numerical linear algebra*, SIAM Rev. **20** (1978), 740–777.

[67] Christian G. Herter, Michael Philippsen, and Walter F. Tichy, *Modula-2\* and its compilation*, No. 1421, programme 2 (calcul symbolique, programmation et génie logiciel), INRIA, Domaine de Voluceau Rocquencourt B. P. 105 78153 LeChesnay Cedex FRANCE, 1991.

[68] Christian G. Herter and Walter F. Tichy, *Modula-2\*: An extension of modula-2 for highly parallel, portable programs*, Interner bericht nr. 4/90, Universität Karlsruhe Fakultät für Informatik, Postfach 6980, D-7500 Karlsruhe 1 Germany, 1990.

[69] Christian G. Herter, Walter F. Tichy, Michael Philippsen, and Thomas Warschko, *Project Triton: Towards improved programmability of parallel machines*, Interner bericht nr. 1/92, Universität Karlsruhe

Fakultät für Informatik, Postfach 6980, D-7500 Karlsruhe 1 Germany, 1992.

[70] D. Hirschberg, A. Chandra, and D. Sarawate, *Computing connected components on parallel computers*, Comm. ACM **22** (1979), 461–464.

[71] John Hopcroft and Jeffrey Ullman, *Introduction to automata theory, languages and computation*, Addison-Wesley, 1979.

[72] T. L. Huntsberger and B. A. Huntsberger, *Hypercube algorithm for image decomposition and analysis in the wavelet representation*, Proceedings of the Fifth Distributed Memory Conference (South Carolina), 1990, pp. 171–175.

[73] O. H. Ibarra, S. M. Sohn, and T.-C. Pong, *Parallel parsing on the hypercube*, Tr 88-23, University of Minnesota, Computer Science Department, 1988.

[74] Jr. J. E. Dennis and R. B. Schnabel, *Numerical methods for unconstrained optimization and nonlinear equations*, Prentice-Hall, Englewood Cliffs, NJ, 1983.

[75] Joseph Ja' Ja' and Janos Simon, *Parallel algorithms in graph theory: planarity testing*, SIAM J. Comput. **11** (1982), 314–328.

[76] J. Johnson, *Some issues in designing algebraic algorithms for CRAY X-MP*, Academic Press, New York, 1989.

[77] R. W. Johnson, C. Lu, and R. Tolimieri, *Fast Fourier transform algorithms for the size N, a product of distinct primes*, unpublished notes.

[78] W. Kahan, *Gauss-Seidel methods of solving large systems of linear equations*, Ph.D. thesis, University of Toronto, Toronto, Canada, 1958.

[79] Erich Kaltofen, M. S. Krishnamoorthy, and B. David Saunders, *Parallel algorithms for matrix normal forms*, Linear Algebra Appl. **136** (1990), 189–208.

[80] M. Karchmer and J. Naor, *A fast parallel algorithm to $\omega$ color a graph with $\Delta$ colors*, J. Algorithms **9** (1988), 83–91.

[81] R. M. Karp and A. Wigderson, *A fast parallel algorithm for the maximal independent set problem*, Proc. 16th Symposium on the Theory of Computing, ACM, 1984, pp. 266–272.

[82] R. W. Karp, E. Upfal, and A. Wigderson, *Constructing a perfect matching is in random NC*, Proc. 17th Symposium on the Theory of Computing, ACM, 1985, pp. 22–32.

[83] _____, *Constructing a perfect matching is in random NC*, Combinatorica **6** (1986), no. 1, 35–48.

[84] Richard Karp and Vijaya Ramachandran, *A survey of parallel algorithms for shared-memory machines*, Technical report ucb/csd 88/408, Publications Office, Computer Science Division Publications, Univ. of California, Berkeley, CA 94720, 1988.

[85] S. Khuller, *Extending planar graph algorithms to $K_{3,3}$-free graphs*, Int. Comput. **84** (1990), 13–25.

[86] S. K. Kim and A. T. Chronopoulos, *An efficient parallel algorithm for eigenvalues of sparse nonsymmetric matrices*, Supercomputer Applications **6** (1992), no. 1, 98–111.

[87] P. Klein and J. Reif, *An efficient parallel algorithm for planarity*, 27th IEEE Symposium on the Foundations of Computer Science, vol. 27, 1986, pp. 465–477.

[88] G. Knowles, *A VLSI architecture for the discrete wavelet transform*, Electronics Letters **26** (1990), no. 15, 1184–1185.

[89] Donald Knuth, *The art of computer programming, vol. III: Sorting and searching*, Addison-Wesley, 1973.

[90] H. König and C. Runge, *Die grundlehren der Mathematischen Wissenschaften*, vol. 11, Springer-Verlag, Berlin, 1924.

[91] S. Rao Kosaraju, *Speed of recognition of context-free languages by array automata*, SIAM J. Comput. **4** (1975), no. 3, 331–340.

[92] C. Kruskal, *Searching, merging, and sorting in parallel computation*, IEEE Trans. Comput. **C-32** (1983), 942–946.

[93] D. J. Kuck, *Evaluating arithmetic expressions of n atoms and k divisions in* $\alpha(\log_2 n + 2\log_2 k) + c$ *steps*, unpublished manuscript, 1973.

[94] H. T. Kung, *Why systolic architectures?*, Computer **15** (1982), 37–45.

[95] R. E. Ladner, *The circuit problem is log space complete for P*, SIGACT News **7** (1975), no. 1, 18–20.

[96] T. Lai and S. Sahni, *Anomalies in parallel branch-and-bound algorithms*, Communications of the Association for Computing Machinery **27** (1984), no. 6, 594–602.

[97] Thomas J. LeBlanc, Michael L. Scott, and Christopher M. Brown, *Large-scale parallel programming: Experience with the BBN butterfly parallel processor*, SIGPLAN Notes **23** (1988), no. 9, 161–172.

[98] S. Levialdi, *On shrinking binary picture patterns*, Comm. ACM **15** (1972), 7–10.

[99] Ming Li and Yaacov Yesha, *New lower bounds for parallel computation*, Symposium on the Theory of Computing, vol. 18, ACM, 1986, pp. 177–187.

[100] Michael Littman and Chris Metcalf, *An exploration of asynchronous data-parallelism*, Technical report 684, Yale University, 1988.

[101] F. A. Lootsma and K. M. Ragsdell, *State of the art in parallel nonlinear optimization*, Parallel Comput. **6** (1988), 131–155.

[102] L. Lovász, *On Determinants, Matchings and Random Algorithms*, Akademia-Verlag, Berlin, 1979.

[103] _____, *Computing ears and branchings in parallel*, Symposium on the Theory of Computing, vol. 17, ACM, 1985, pp. 464–467.

[104] L. Lovász and M. Plummer, *Matching theory*, Academic Press, Budapest, Hungary, 1988.

[105] A. Lubotzky, R. Phillips, and P. Sarnak, *Ramanujan conjecture and explicit constructions of expanders and super-concentrators*, Symposium on the Theory of Computing, vol. 18, ACM, 1986, pp. 240–246.

[106] Michael Luby, *Removing randomness in parallel computation without a processor penalty*, 29th Annual Symposium on the Foundations of Computer Science, IEEE, 1988, pp. 162–173.

[107] S. G. Mallat, *A theory for multiresolution signal decomposition: the wavelet representation*, IEEE Trans. Pattern Anal. and Machine Intel. **11** (1989), 674–693.

[108] D. Marr and E. Hildreth, *Theory of edge-detection*, Proc. Royal Soc. **207** (1980), 187–217.

[109] Y. Matsumoto, *Parsing gapping grammars in parallel*, Tr-318, Institute for New Generation Computer Technology, JAPAN, 1987.

[110] R. Mehotra and E. F. Gehringer, *Superlinear speedup through randomized algorithms*, Proceedings of the 1985 International Conference on Parallel Processing, 1985, pp. 291–300.

[111] Y. Meyer, *Ondelettes et opérateurs*, Hermann, Paris, FRANCE.

[112] G. Miller and J. Reif, *Parallel tree-contractions and its applications*, 17th ACM Symposium on the Theory of Computing, vol. 17, 1985, pp. 478–489.

[113] _____, *Parallel tree-contractions Part I: Fundamentals*, (Greenwich, CT) (S. Micali, ed.), Randomness in Computation, vol. 5, JAI Press, Greenwich, CT, 1989, pp. 47–72.

[114] _____, *Parallel tree contractions Part 2: Further applications*, SIAM J. Comput. **20** (1991), no. 6, 1128–1147.

[115] Gary L. Miller, Vijaya Ramachandran, and Erich Kaltofen, *Efficient parallel evaluation of straight-line code and arithmetic circuits*, SIAM J. Comput. **17** (1988), no. 4, 687–695.

[116] Philip M. Morse and Herman Feshbach, *Methods of theoretical physics*, vol. I and II, McGraw-Hill, New York, 1953.

[117] Ketan Mulmuley, Umesh V. Vazirani, and Vijay V. Vazirani, *Matching is as easy as matrix inversion*, 19th Annual Symposium on the Theory of Computing (New York City), ACM, 1987, pp. 345–354.

[118] Joseph Naor, Moni Naor, and Alejandro A. Schäffer, *Fast parallel algorithms for chordal graphs*, 19th Annual Symposium on the Theory of Computing (New York City), ACM, 1987, pp. 355–364.

[119] J. T. Oden and J. N. Reddy, *An introduction to the mathematical theory of finite elements*, John Wiley and Sons, New York, 1976.

[120] S. E. Orcutt, *Computer organization and algorithms for very high-speed computations*, Ph.D. thesis, Stanford University, Stanford, CA, 1974.

[121] V. Pan and J. Reif, *Efficient parallel solutions of linear systems*, Symposium on the Theory of Computing, vol. 17, ACM, 1985, pp. 143–152.

[122] D. Parkinson, *Parallel efficiency can be greater than unity*, Parallel Comput. **3** (1986), 261–262.

[123] Franco Preparata, *New parallel-sorting schemes*, IEEE Trans. Comput. **c-27** (1978), no. 7, 669–673.

[124] Franco Preparata and J. Vuillemin, *The cube-connected cycles: A versatile network for parallel computation*, Foundations of Computer Science Symposium (Puerto Rico), vol. 20, IEEE, 1979, pp. 140–147.

[125] _____, *The cube-connected cycles: A versatile network for parallel computation*, Comm. ACM **24** (1981), no. 5, 300–309.

[126] V. Ramachandran and J. Reif, *An optimal parallel algorithm for graph planarity*, 30th Annual Conference on the Foundations of Computer Sceince, IEEE, 1989, pp. 282–287.

[127] John H. Reif, *Depth-first search is inherently sequential*, Inform. Process. Lett. **20** (1985), 229–234.

[128] John H. Reif and Sandeep Sen, *Optimal randomized parallel algorithms for computational geometry*, Algorithmica **7** (1992), 91–117.

[129] Wojciech Rytter, The complixity of two-way pushdown automata and recursive programs, 341–356, Springer-Verlag, Berlin, 1985, pp. 341–356.

[130] _____, *On the recognition of context-free languages*, (Berlin) (A. Skowron, ed.), Computation Theory, Lecture Notes in Computer Science, vol. 208, Springer-Verlag, Berlin, 1985, pp. 318–325.

[131] _____, *Parallel time log n recognition of unambiguous cfl's*, , Fundamentals of Computation Theory, Lecture Notes in Computer Science, Springer-Verlag, Berlin, 1985, pp. 380–389.

[132] Wojciech Rytter and Raffaele Giancarlo, *Optimal parallel parsing of bracket languages*, Theoret. Comput. Sci. **53** (1987), 295–306.

[133] Carla Savage, *Parallel algorithms for graph theoretic problems*, Ph.D. thesis, Department of Mathematics, University of Illinois, Urbana, IL, 1977.

[134] T. Schwartz, *Ultracomputers*, Trans. Prog. Lang. Sys. **2** (1980), no. 4, 484–521.

[135] Satoshi Sekiguchi, Kei Hiraki, and Toshio Shimada, *A design of pratical dataflow language DFCII and its data structures*, Tr-90-16, Electrotechnical Laboratory, Tsukuba Science City, Ibaraki 305, Japan, 1990.

[136] Satoshi Sekiguchi, Tsutomu Hoshino, and Toshitsugu Yuba, *Parallel scientific computer research in Japan*, Tr-88-7, Computer Systems Division, Electrotechnical Laboratory, Tsukuba Science City, Ibaraki 305, Japan, 1988.

[137] Satoshi Sekiguchi, Masaaki Sugihara, Kei Hiraki, and Toshio Shimada, *A new parallel algorithm for the eigenvalue problem on the SIGMA-1*, Tr-88-3, Electrotechnical Laboratory, Tsukuba Science City, Ibaraki 305, Japan, 1988.

[138] Atle Selberg, *An elementary proof of the prime number theorem for arithmetic progressions*, Canad. J. Math. **2** (1950), 66–78.

[139] Jean-Pierre Serre, *A course in arithmetic*, Springer-Verlag, New York, Heidelberg, 1973.

[140] Gregory Shannon, *A linear-processor algorithm for depth-first search in planar graphs*, Inform. Process. Lett. **29** (1988), no. 3, 119–124.

[141] Yossi Shiloach and Uzi Vishkin, *An $O(\log n)$ parallel connectivity algorithm*, J. Algorithms **3** (1982), 57–67.

[142] George Simmons, *Differential equations with applications and historical notes*, McGraw-Hill, 1972.

[143] Justin Smith, *Parallel algorithms for depth-first searches I. The planar case*, SIAM J. Comput. **15** (1986), no. 3, 814–830.

[144] I. M. Sobel', *The Monte Carlo Method*, University of Chicago Press, Chicago, IL, 1974.

[145] Gilbert Strang, *Wavelets and dilation equations: A brief introduction*, SIAM J. Comput. **31** (1989), no. 4, 614–627.

[146] Wim Sweldens and Robert Piessens, *An introduction to wavelets and efficient quadrature formulae for the calculation of the wavelet decomposition*, Tw 159, Department of Computer Science Katholieke Universiteit Leuven, Celestijnenlaan 200A — B-3001 Leuven Belgium, 1991.

[147] Robert Tarjan, *Depth-first search and linear graph algorithms*, SIAM J. Comput. **1** (1972), 146–160.

[148] _____, *Finding optimum branchings*, Networks **7** (1977), 25–35.

[149] _____, *Data structures and network algorithms*, SIAM, CBMS Series, Philadelphia, PA, 1983.

[150] Robert Tarjan and Uzi Vishkin, *Finding biconnected components and computing tree functions in logarithmic parallel time*, 25$^{th}$ IEEE Symposium on the Foundations of Computer Science, vol. 25, 1984, pp. 12–20.

[151] Yung H. Tsin and Francis Y. Chin, *Efficient parallel algorithms for a class of graph theoretic problems*, SIAM J. Comput. **13** (1984), no. 3, 580–599.

[152] W. T. Tutte, *The factorization of linear graphs*, J. London Math. Soc. (2) **22** (1947), 107–111.

[153] Jeffrey Ullman, *Computational aspects of VLSI*, Computer Science Press, 1984.

[154] Eli Upfal, *A probabilistic relation between desirable and feasible models of parallel computation*, Symposium on the Theory of Computing, vol. 16, ACM, 1984, pp. 258–265.

[155] Eli Upfal and Avi Wigderson, *How to share memory in a distributed system*, 25$^{th}$ IEEE Symposium on the Foundations of Computer Science, vol. 25, 1984, pp. 171–180.

[156] L. G. Valiant, *Parallelism in comparison problems*, SIAM J. Comput. (1975), 348–355.

[157] L. G. Valiant and G. J. Brebner, *Universal schemes for parallel communication*, SIAM J. Comput. **13** (1981), 263–277.

[158] Stephen Vavasis, *Gaussian elimination with pivoting is P-complete*, SIAM J. Discr. Math. **2** (1989), 413–423.

[159] U. J. J. Le Verrier, *Sur les variations seculaires des elementes elliptiques des sept planets principales*, J. Math. Pures Appl. **5** (1840), 220–254.

[160] Uzi Vishkin, *Implementation of simultaneous memory access in models that forbid it*, J. Algorithms **4** (1983), 45–50.

[161] _____, *Randomized speed-ups in parallel computation*, Symposium on the Theory of Computing, vol. 16, ACM, 1984, pp. 230–238.

[162] Christopher Wilson, *On the decomposability of $\mathcal{NC}$ and $\mathcal{AC}$*, SIAM J. Comput. **19** (1990), no. 2, 384–396.

[163] Jet Wimp, *Computation with recurrence relations*, Pitman Advanced Publishing Program, Boston, London, Melbourne, 1984.

[164] Markus G. Wloka, *Parallel VLSI synthesis*, Ph.D. thesis, Brown University, 1991.

[165] David M. Young, *Iterative solution of large linear systems*, Academic Press, New York and London, 1971.

[166] Yixin Zhang, *Parallel algorithms for minimal spanning trees of directed graphs*, Internat. J. Para. Proc. **18** (1985), no. 3.

# Index